Ductile Design of Steel Structures

Ductile Design of Steel Structures

Michel Bruneau
Chia-Ming Uang
Andrew Whittaker

McGraw-Hill

New York San Francisco Washington, D.C. Auckland Bogotá
Caracas Lisbon London Madrid Mexico City Milan
Montreal NewDelhi San Juan Singapore
Sydney Tokyo

McGraw-Hill

A Division of The McGraw·Hill Companies

Copyright © 1998 by The McGraw-Hill Companies, Inc. All rights reserved. Printed in the United States of America. Except as permitted under the United States Copyright Act of 1976, no part of this publication may be reproduced or distributed in any form or by any means, or stored in a data base or retrieval system, without the prior written permission of the publisher.

1 2 3 4 5 6 7 8 9 0   FGR/FGR   9 0 2 1 0 9 8 7

ISBN 0-07-008580-3

The sponsoring editor was Larry Hagar, the editing supervisor was Wade Strickland, the editing liaison was Patricia Amoroso, and the production supervisor was tina Cameron. It was set in Century Schoolbook by Benchmark Productions.

Printed and bound by Quebecor/Fairfield.

> Information contained in this work has been obtained by The McGraw-Hill Companies, Inc. (McGraw-Hill) from sources believed to be reliable. However, neither McGraw-Hill nor its authors guarantee the accuracy or completeness of any information published herein and neither McGraw-Hill nor its authors shall be responsible for any errors, omissions, or damages arising out of use of this information. This work is published with the understanding that McGraw-Hill and its authors are supplying information but are not attempting to render engineering or other professional services. If such services are required, the assistance of an appropriate professional should be sought.

 This book is printed on recycled, acid-free paper containing a minimum of 50 percent recycled de-inked fiber.

McGraw-Hill books are available at special quantity discounts to use as premiums and sales promotions, or for use in corporate training programs. For more information, please write to the Director of Special Sales, McGraw-Hill, 11 West 19th Street, New York, NY 10011. Or contact your local bookstore.

Contents

Preface   xi

Chapter 1. Introduction
References   5

Chapter 2. Structural Steel
2.1 Introduction   7
2.2 Common properties of steel materials   8
   2.2.1 Engineering stress-strain curve   8
   2.2.2 Effect of temperature on stress-strain curve   8
   2.2.3 Effect of temperature on ductility and notch-toughness   12
   2.2.4 Strain rate effect on tensile and yield strengths   17
   2.2.5 Probable yield strength   21
2.3 Plasticity, hysteresis, Bauschinger effects   22
2.4 Metallurgical process of yielding, slip planes   24
2.5 Brittleness in welded sections   27
   2.5.1 Metallurgical transformations during welding, heat-affected zone, preheating   27
   2.5.2 Hydrogen embrittlement   29
   2.5.3 Carbon equivalent   29
   2.5.4 Flame cutting   31
   2.5.5 Weld restraints   32
   2.5.6 Lamellar tearing   34
   2.5.7 Thick steel sections   35
   2.5.8 Fracture mechanics   39
   2.5.9 Partial penetration welds   40
2.6 Low-cycle versus high-cycle fatigue   42
   2.6.1 High-cycle fatigue   42
   2.6.2 Low-cycle fatigue   42
2.7 Material models   43
   2.7.1 Rigid plastic model   43
   2.7.2 Elasto-plastic models   45
   2.7.3 Power, Ramberg-Osgood, and Menegotto-Pinto functions   46
2.8 Advantages of plastic material behavior   55
2.9 References   60

Chapter 3. Plastic Behavior at the Cross-Section Level

- 3.1 Pure flexural yielding — 63
  - 3.1.1 Doubly symmetric sections — 64
  - 3.1.2 Sections having a single axis of symmetry — 69
  - 3.1.3 Impact of some factors on inelastic flexural behavior — 71
  - 3.1.4 Behavior during cyclic loading — 78
- 3.2 Combined flexural and axial loading — 80
  - 3.2.1 Rectangular cross section — 82
  - 3.2.2 Wide-flange sections: strong axis bending — 84
  - 3.2.3 Wide-flange sections: weak axis bending — 87
  - 3.2.4 Moment-curvature relationships — 88
- 3.3 Combined flexural and shear loading — 88
- 3.4 Combined flexural, axial, and shear loading — 92
- 3.5 Pure plastic torsion: sand-heap analogy — 96
  - 3.5.1 Review of important elastic analysis results — 96
  - 3.5.2 Sand-heap analogy — 98
- 3.6 Combined flexure and torsion — 99
- References — 101

Chapter 4. Concepts of Plastic Analysis

- 4.1 Introduction to simple plastic analysis — 103
- 4.2 Simple plastic analysis methods — 106
  - 4.2.1 Event-to-event calculation (step-by-step method) — 106
  - 4.2.2 Equilibrium method (statical method) — 110
  - 4.2.3 Kinematic method (virtual-work method) — 114
- 4.3 Theorems of simple plastic analysis — 119
  - 4.3.1 Upper bound theorem — 120
  - 4.3.2 Lower bound theorem — 121
  - 4.3.3 Uniqueness theorem — 121
- 4.4 Application of the kinematic method — 121
  - 4.4.1 Basic mechanism types — 122
  - 4.4.2 Combined mechanism — 124
  - 4.4.3 Mechanism analysis by center of rotation — 132
  - 4.4.4 Distributed loads — 136
- 4.5 Shakedown theorem (deflection stability) — 145
- References — 152

Chapter 5. Systematic Methods of Plastic Analysis

- 5.1 Number of basic mechanisms — 153
- 5.2 Direct combination of mechanisms — 154
  - Example 5.2.1: One-bay, one-story frame — 159
  - Example 5.2.2: Two-story frame with overhanging bay — 162
- 5.3 Method of inequalities — 165
- References — 173

Chapter 6. Applications of Plastic Analysis

- 6.1 Moment redistribution design methods — 176
  - 6.1.1 Statical method of design — 176
  - 6.1.2 Autostress design method — 177

6.2	Capacity design	181
	6.2.1 Concepts	181
	6.2.2 Shear failure protection	182
	6.2.3 Protection against column hinging	187
6.3	Push-over analysis	190
	6.3.1 Monotonic push-over analysis	191
	6.3.2 Cyclic push-over analysis	197
6.4	Seismic design using plastic analysis	198
References		199

Chapter 7. Design of Ductile Braced Frames

7.1	Introduction	201
7.2	Concentrically braced frames	202
	7.2.1 General	202
	7.2.2 Development of CBFs	203
	7.2.3 Cyclic axial load response	205
	7.2.4 Nonlinear response of concentric braces	207
	7.2.5 CBF design philosophy	220
	7.2.6 CBF design example	236
7.3	Eccentrically braced frames	246
	7.3.1 General	246
	7.3.2 Development of EBFs	248
	7.3.3 EBF design philosophy	250
	7.3.4 EBF frame geometry	252
	7.3.5 Kinematics of the EBF	252
	7.3.6 Link behavior and length	255
	7.3.7 Link strength and deformation calculations	257
	7.3.8 Link details	259
	7.3.9 Frame design outside links	262
	7.3.10 EBF design example	265
References		270

Chapter 8. Design of Ductile Moment-Resisting Frames

8.1	Introduction	273
8.2	Basic response of ductile moment-resisting frames to lateral loads	275
	8.2.1 Internal forces during seismic response	275
	8.2.2 Plastic rotation demands	277
	8.2.3 Lateral bracing and local buckling	278
8.3	Ductile moment-frame column design	279
	8.3.1 Axial forces in columns	279
	8.3.2 Considerations for column splices	279
	8.3.3 Strong-column/weak-beam philosophy	280
	8.3.4 Effect of axial forces on column ductility	284
8.4	Panel zone	285
	8.4.1 Flange distortion and column web yielding/crippling prevention	285
	8.4.2 Forces on panel zones	288
	8.4.3 Behavior of panel zones	290
	8.4.4 Modeling of panel zone behavior	295
	8.4.5 Design of panel zone	300

viii Contents

    8.5 Beam-to-Column Connections   302
        8.5.1 Knowledge and practice prior to the 1994 Northridge earthquake   302
        8.5.2 Damage during the Northridge earthquake   312
        8.5.3 Causes for failures   326
        8.5.4 Reexamination of pre-Northridge practice   335
        8.5.5 Post-Northridge beam-to-column connections design strategies for new buildings   337
        8.5.6 International significance of the Northridge moment-connection failures   362
    8.6 Design of a ductile moment frame   364
        8.6.1 Generic design procedure   364
        8.6.2 RBS design example   370
    References   376

Chapter 9. Limit State Philosophy in Seismic Design Provisions

    9.1 Introduction   381
    9.2 Seismic limit state philosophy   381
    9.3 Seismic design procedures in modern codes   382
        9.3.1 General   382
        9.3.2 Japanese Building Standard Law (BSL)   382
        9.3.3 NEHRP Recommended Provisions   384
        9.3.4 Uniform Building Code (UBC)   385
        9.3.5 National Building Code of Canada   388
    9.4 Seismic force reduction and displacement amplification factors   389
        9.4.1 General   389
        9.4.2 Structural response and design seismic forces   389
        9.4.3 Definition of relevant terms   390
        9.4.4 Formulation of R (or Rw) and Cd factors   392
        9.4.5 Seismic force reduction factor—a comparison of seismic provisions   392
        9.4.6 Displacement amplification factor—a comparison of seismic provisions   393
    9.5 Comparison of service limit state requirements   395
        9.5.1 Introduction   395
    9.6 Future directions   399
    9.7 Historical perspective on force reduction factors   403
    Reference   407

Chapter 10. Stability and Rotation Capacity of Steel Beams

    10.1 Introduction   411
    10.2 Plate elastic and post-elasticbuckling behavior   414
    10.3 General description of inelastic beam behavior   419
        10.3.1 Beams with uniform bending moment   419
        10.3.2 Beams with moment gradient   421
        10.3.3 Comparison of beam behavior under uniform moment and moment gradient   423
    10.4 Inelastic flange local buckling   424
        10.4.1 Modelling Assumptions   424
        10.4.2 Buckling of an Orthotropic Plate   425
        10.4.3 Torsional buckling of a restrained rectangular plate   427

10.5	Web local buckling	433
10.6	Inelastic lateral-torsional buckling	436
	10.6.1 General	436
	10.6.2 Beam under uniform moment	437
	10.6.3 Beam under moment gradient	442
10.7	Code comparisons	447
10.8	Interaction of beam buckling modes	448
10.9	Cyclic beam buckling behavior	455
10.10	References	460

Chapter 11. Passive Energy Dissipation Systems

11.1	Introduction	463
11.2	Response modification due to added damping	464
11.3	Types of damper hardware	466
	11.3.1 Displacement-dependent dampers	466
	11.3.2 Velocity-dependent dampers	467
11.4	Steel-yielding dampers	469
11.5	Implementation of dampers in building frames	477
References		478

Index 481

Preface

The design practice of structural engineering is currently undergoing a considerable amount of fundamental change. For example, in recent years, load and resistance factor design has been introduced in the United States for the design of steel structures, and second-order analysis has become mandatory in Canada. However, other more dramatic changes in the practice of structural steel design are attributable to a shift in design philosophy as evidenced by the introduction of numerous new codes and standards that now require design for ductile response.

Many practicing engineers have wrongly believed for years that the ductile nature of the structural steel material directly translates into inherently ductile structures. Although steel remains the most ductile of the modern structural engineering materials, research in the last 20 years has clearly demonstrated that special care is necessary to ensure ductile structural behavior, even when structural stability is not a concern. Most of this recent knowledge is now finding its way into structural steel design codes and standards worldwide and, to a large extent, into the earthquake-resistant design clauses of these documents.

In North America, prior to 1988, there were no specific code regulations for detailing of earthquake-resistant steel structures. Since then, many new code detailing requirements have been introduced to ensure seismic resistance. For example, new steel provisions have been added to the *Recommended Lateral Force Requirements* of the Structural Engineers Association of California (traditionally the basis of seismic code regulations in the United States), the *Uniform Building Code* (widely used in the western United States), and the *National Earthquake Hazard Reduction Program Recommended Provisions for the Development of Seismic Regulations for New Buildings* issued by the Federal Emergency Management Administration (this document is rapidly developing into a national standard for seismic resistant construction in the United States). Moreover, for the first time, in 1990, the American Institute for Steel

Construction (AISC) developed new seismic provisions for structural steel buildings (updated in 1992), and Canada adopted mandatory detailing requirements for seismic resistant steel structures. Finally, regulations for seismic resistant construction are being adopted in the central and eastern United States, where seismic resistant structures have traditionally not been designed.

Thus, North American code requirements for the seismic design of steel structures have changed tremendously in the last few years. However, much of the information necessary to understand the principles of structural steel design for ductile response is scattered in research reports, journal articles, and conference proceedings. This textbook seeks to summarize the relevant existing information on this topic into tightly woven chapters on material, cross-section, component, and system response, while providing useful guidance. Students who are enrolled in graduate structural engineering programs and must learn how to design ductile steel structures need such a reference document. Similarly, many practicing engineers will welcome this educational aid that conceptually and practically explains the nature of this design approach and provides practical examples and illustrations of the resulting design and consequences.

The emphasis of this book is on earthquake-resistant design because the provision of ductile structures is crucial to ensure seismic survival. However, there exist many other important applications of the principles and design approaches outlined in this textbook. The growing market of structural rehabilitation of existing buildings in North America is one such major example. The revitalization activities taking place in many city centers in answer to actual commuters' frustrations, the aging North American infrastructure, the projected North American population growth patterns, the goals of historical or heritage building preservation, and other societal trends, have led to a considerable increase in renovation and rehabilitation activities in both seismic and nonseismic regions. For existing construction, plastic analysis can provide a much better estimate of a structure's actual strength than can procedures based on elastic analysis, which in turn can be used advantageously to minimize the extent of the needed rehabilitation.

Other possible applications of ductile steel design include offshore structures subjected to extreme wave and ice loads and bridges that can now be designed to carry normal traffic using a new alternative bridge design procedure (the Autostress method) that relies heavily on ductile response and requires a good understanding of the shakedown theory. Thus, although the focus of this text is earthquake engineering, the information presented herein will be broadly applicable to the ductile design of steel structures.

The primary audience for this book is design professionals and senior undergraduate and graduate students working in the field of structural and earthquake engineering. This book is written with a focus on concepts and references a number of the world's most widely used codes. It is assumed that the reader has a background knowledge of steel design.

We gratefully acknowledge Q.S. Yu for his assistance in identifying some of the relevant literature referred to in Chapter 10, as well as for his constructive comments for that chapter. Technical proofreading of an early draft of this book by Jocelyn Paquette and Majid Sarraf is sincerely appreciated. We thank Professor Andre Filiatrault of Ecole Polytechnique in Montreal for his comments on Chapters 2 and 7, and Professor Michael Engelhardt of the University of Texas in Austin for providing complementary technical material used in Chapter 8. We sincerely thank our spouses and children for their patience, care, support, and love throughout this project that stole valuable hours from our family life, and to them, we dedicate this book.

Dr. Michel Bruneau, P. Eng.

Ottawa-Carleton Earthquake Engineering Research Centre,
Department of Civil Engineering, University of Ottawa, Ottawa,
Ontario, Canada.

Dr. Chia-Ming Uang

Charles Lee Powell Structural Systems Laboratories,
Department of Civil Engineering, University of California,
San Diego, California, U.S.A.

Dr. Andrew Whittaker, S. E.

Earthquake Engineering Research Center, University of
California, Berkeley, California, USA.

Ductile Design of Steel Structures

Chapter 1

Introduction

In the context of structural engineering, a "ductile" material is one that is capable of undergoing large inelastic deformations without losing its strength. More formally, the *Metal Handbook of the American Society for Metals* (ASM 1964) defines "ductility" as "the ability of a material to deform plastically without fracture." "Brittleness," on the other hand, is the "quality of a material that leads to crack propagation without plastic deformations." In that perspective, structural steel is the most ductile of the widely used engineering materials. This advantageous property of steel is often implicitly used by design professionals. For example, many simple modern connections must plastify to perform as intended by the designer, and large ductile deformations are sometimes relied upon to provide early warnings of unexpected overloads. Many such beneficial incidental effects of plasticity are already integrated into steel design requirements and design practice.

However, there are many situations in which an explicit approach to the design of ductile steel structures is necessary because the inherent material ductility alone is not sufficient to provide the desired ultimate performance. For example, in nearly all buildings designed today, survival in large earthquakes depends directly on the ability of their framing system to dissipate energy hysteretically while undergoing large inelastic (i.e., plastic) deformations. To achieve this ductile response, one must recognize and avoid conditions that may lead to brittle failures and adopt appropriate design strategies to allow for stable and reliable hysteretic energy-dissipation mechanisms.

This sort of thinking is relatively new in structural engineering. Indeed, many practicing engineers have believed for years, albeit incorrectly, that steel structures were immune to earthquake-induced damage as a consequence of the material's inherent ductile properties.

However, earthquake engineering research in the last 20 years has clearly demonstrated that special care must be taken to ensure ductile structural behavior. Moreover, numerous steel structures suffered damage during the Richter magnitude 6.8 Northridge (Los Angeles) earthquake of January 17, 1994, and the Richter magnitude 7.2 Kobe (Japan) earthquake that occurred, coincidentally, exactly one year later (January 17, 1995). Both earthquakes occurred in highly developed urban areas in nations known for their leadership in earthquake engineering, and both confirmed previous research findings that material ductility alone is not a guarantee of ductile structural behavior when steel components and connections can fail in a brittle manner (e.g., AIJ 1995; Bruneau et al. 1996; EERC 1995; EERI 1995, 1996; Tremblay et al. 1995, 1996). In particular, these earthquakes raised important concerns regarding the ductile behavior of some standard beam-to-column moment connections (as described in Chapter 8).

Fortunately, most of the body of knowledge on the design of ductile steel structures to resist earthquakes is being implemented in structural steel design codes and standards worldwide. Prior to 1988, there were no specific code regulations for detailing seismic-resistant steel structures in North America, but in the years since, many new codes and standards detailing regulations have been introduced to ensure ductile response during earthquakes. And, as a side benefit, many other applications (e.g., offshore oil platforms) have benefited from the new knowledge on the ultimate cyclic behavior of steel structures generated by earthquake engineers, and from the recent emphasis on ductile design requirements for earthquake resistance.

The road to design and detailing of ductile structures has been a long one. A comprehensive historical description of this evolution is beyond the scope of this book. Nonetheless, an overview is worthwhile to help appreciate how the past has shaped today's structural engineering practice.

From a materials perspective (Timoshenko 1983), work on the plastic behavior of ductile metals can be traced back at least to the 1860s, with observations by Lüder that visible lines appear on the surface of specimens stretched beyond their elastic limit, as well as some mathematical papers by Tresca in 1868 and Saint-Venant in the 1870s. Bauschinger reported the first results from inelastic cyclic tests of mild steel coupons in 1886 and subsequently tested (with Tetmajer) short columns buckling in the inelastic range, which led to the development of an empirical column-strength equation that became widely used throughout Europe at that time.

In the first half of the 20th century, researchers investigated the plastic properties of steel and of steel cross-sections subjected to various stress conditions, with experimental work reported as early as

1914 by some sources (ASCE 1971). Much of that activity took place in Germany. In particular, the law of plastic behavior most commonly used in structural steel, the Huber-Hencky-Mises yield condition (also known as Von Mises yield criterion), results from work conducted by these researchers in 1904, 1913, and 1925, respectively (Popov 1968). The 1950s and 1960s witnessed a phenomenal growth of research activity in this area (ASCE 1971). Active research groups in England and the United States, complemented by valuable contributions by researchers in other countries, spearheaded the development and perfected the tools of plastic analysis and design in structural engineering, with the intent of providing a possible replacement for the allowable stress method (also known as working stress method), which was perceived as fraught with shortcomings and limitations. Extensive analytical and experimental research was conducted on the ultimate strength of members as well as on entire structural frameworks, henceforth establishing plastic design as a viable alternative design procedure. During those decades, a few buildings were designed using plastic design principles, and special provisions for plastic design were introduced in many structural steel design specifications (such as those of the American Institute of Steel Construction and Canadian Institute of Steel Construction, among many) and remain to this day.

However, until the end of the 1960s, research on plastic design did not address earthquake-induced loads. Consequently, when earthquake-resistant design became a major concern in some parts of the world, the field started to grow in two seemingly opposite directions as a result of diverging research interests. On one hand, for researchers not concerned with design for earthquakes, research focused on developing better models of member behavior, with a considerable effort being placed on understanding the requirements for member and structural stability. Plastic strength was perceived as the special case to which a member stability problem converges when proper bracing is provided. Out of that research endeavor evolved ultimate design methods (i.e., the Load and Resistance Factor Design, and Limit States Design), in use at the time of this writing, that rely on elastic structural analysis but consider plasticity at the member level when determining ultimate member capacities. On the other hand, in the second research direction, emphasis was placed on developing design and detailing requirements that could ensure stable plastic behavior under the extreme cyclic inelastic deformation demands resulting from severe earthquake excitations. From that perspective, the development of ductile details for steel construction was paramount, and stringent requirements were enforced to avoid (or make as ductile as possible) stability-related failures. From that research effort evolved ductile detailing requirements for earthquake engineering applications.

Interestingly, the two research paths seem destined to meet in the near future. For example, capacity design and push-over analyses (which are expressions of full-structure plastic design, as shown in Chapter 6) are being frequently conducted now as part of earthquake engineering projects, while engineers involved in nonseismic applications are advocating advanced analysis and design methods that recognize the ultimate capacity of structures, as well as of members. In some areas, the convergence of the two paths has already begun (Fukumoto and Lee 1992).

The design of steel structures for ductile response requires (1) material ductility, (2) cross-section and member ductility, and (3) structural ductility (it is generally preferable to protect connections against yielding, although, in some instances such as with semirigid connections, connection ductility may be unavoidable and substitute for member ductility). Further, a hierarchy of yielding must be imposed on a structure to ensure a desirable failure mode. The material presented in this book follows this order.

Chapter 2 focuses on the structural steel material, with a particular emphasis on its desirable ductile properties as well as factors that can have a negative impact on that behavior, such as temperature, notch toughness, steel strength, strain rate, material thickness, and welding-related problems such as hydrogen embrittlement, weld restraints, and lamellar tearing. The cyclic elastic-plastic behavior of common steels is described through various material models, and some of the advantages of plastic analysis over elastic analysis are illustrated by an example.

In Chapter 3, element ductility is reviewed through descriptions of the plastic strength of cross-sections subjected to axial force, flexure, shear, torsion, and combinations thereof. The influence of residual stresses, strain hardening, and some other factors on cross-sectional strength is discussed.

Chapters 4 and 5 present plastic analysis at the structural level: the concepts of simple plastic analysis are enunciated and demonstrated in the former, and systematic methods of plastic analysis for more complex structures are formulated in the latter. The fundamental upper bound, lower bound, and uniqueness theorems of plastic analysis are presented in Chapter 4, together with the classical step-by-step method, equilibrium method, and kinematic method whose applications are illustrated by examples. Chapter 4 concludes with a presentation of the shake-down theorem. Methodical determination of the plastic collapse load of structures using direct combination of mechanisms and the method of inequalities are presented in Chapter 5.

A few nonseismic applications of plastic design are described in Chapter 6. These include the plastic moment redistribution method, the Autostress design method, the capacity design strategy, and the push-over (nonlinear static) analysis method.

Emphasis on earthquake engineering applications begins with Chapters 7 and 8, in which the requirements for earthquake-resistant ductile braced frames and ductile moment-resisting frames, respectively, are presented. In both cases, the plastic cyclic behavior of these structural systems is extensively described to ensure a good understanding of the philosophy supporting the code requirements, and the advantage of a well-defined hierarchy of yielding will become evident.

A review of different international philosophies regarding the permissible severity of cyclic inelastic deformations during earthquakes is presented in Chapter 9, together with a comparison of the limit state philosophy supporting the seismic design procedures adopted in various design codes worldwide. Similarities and differences are emphasized. Additional detailing requirements that are needed to ensure satisfactory plastic behavior are presented in Chapter 10, along with recent research findings on this topic.

An overview of special energy dissipation systems is presented in the last chapter, to provide the reader with an introduction to a field that will be the subject of much future research and that promises new applications for ductile steel structures and steel energy dissipation devices.

Finally, although many types of ductile steel structures, such as steel plate shear walls, are not covered in this book, it is believed that the information presented herein will provide the reader with sufficient background to successfully tackle analysis or design problems related to many types of ductile steel structures.

References

1. AIJ. 1995. *Preliminary Reconnaissance Report of the 1995 Hyogoken-Nanbu Earthquake*. Architectural Institute of Japan. Nakashima, M., and Bruneau, M., Editors of English Edition. Tokyo, Japan.
2. ASCE. 1971. *Plastic Design in Steel — Manual and Reports on Engineering Practice, No. 41*. American Society of Civil Engineers. New York, NY.
3. ASM. 1964. *The Metal Handbook*, Vol. 2, 8th Edition. Materials Park, Ohio: American Society for Metals.
4. Bruneau, M.; Wilson, J.C.; and Tremblay, R. 1996. "Performance of Steel Bridges during the 1995 Hyogoken-Nanbu (Kobe, Japan) Earthquake." *Canadian Journal of Civil Engineering*, Vol. 23, No. 3, 678–713.
5. EERC. 1995. *Seismological and Engineering Aspects of the 1995 Hyogoken-Nanbu (Kobe) Earthquake*. Earthquake Engineering Research Center, Report UCB/EERC-95/10. California: University of California at Berkeley.
6. EERI. 1995. *The Hyogo-ken Nanbu Earthquake, January 17, 1995: Preliminary Reconnaissance Report*. Oakland, California: Earthquake Engineering Research Institute.
7. EERI. 1996. *Northridge Earthquake Reconnaissance Report*, Vol. 2. Earthquake Spectra, Supplement C to Vol. 11.

8. Fukumoto, Y. and Lee, G. 1992. *Stability and Ductility of Steel Structures under Cyclic Loading*. Boca Raton, Florida: CRC Press.
9. Popov, E.P. 1968. *Introduction to Mechanics of Solids*. Englewood Cliffs, New Jersey: Prentice-Hall.
10. Timoshenko, S.P. 1983. *History of Strength of Materials*. New York: Dover Publications.
11. Tremblay, R.; Timler, P.; Bruneau, M.; and Filiatrault, A. 1995. "Performance of Steel Structures during the January 17, 1994, Northridge Earthquake." *Canadian Journal of Civil Engineering*, Vol. 22, No. 2: 338–360.
12. Tremblay, R.; Bruneau, M.; Nakashima, M.; Prion, H.G.L.; Filiatrault, A.; and DeVall, R. 1996. "Seismic Design of Steel Buildings: Lessons from the 1995 Hyogo-ken Nanbu Earthquake." *Canadian Journal of Civil Engineering*, Vol. 23, No. 3: 727–756.

Chapter 2

Structural Steel

2.1 Introduction

For the first half of the 20th century, there was essentially only one type of structural steel widely available in North America: Grade A-7 steel. (Although the names A-7 and A-9 were used for bridges and buildings before 1939, the two steels were virtually identical.) By the early 1960s, engineers needed to be familiar with five different types of structural steel, but they were likely to use only one or two for all practical purposes. However, with today's exponential growth of technology (and opening to the world's global market), the engineer is offered a "smorgasbord" of steel grades, and many engineers in some parts of North America already do not hesitate to commonly use two or more different steel grades for the main structural members in a given project.

Despite this growth in available products, the minimum requirements specified for structural steel grades remain relatively simple. They generally consist of a few limits on chemical composition, limits on some mechanical properties such as minimum yield and tensile strengths, and minimum percentage elongation prior to failure (definitely not very challenging for metallurgical engineers accustomed to dealing with the comprehensive lists of performance requirements specified for the steels needed in some specialty applications). As a result, in the last few decades, a wide variety of steel grades that meet the simple set of minimum requirements for structural steel has been produced. These steels will be adequate in many applications, but from the perspective of ductile response, the structural engineer is cautioned against hastily using unfamiliar steel grades without being fully aware of how these steels perform under a range of extreme conditions.

To increase this awareness, as well as to provide some fundamental material-related information for the design of ductile structures, this chapter reviews some of the lesser known properties of steel, along with information on the modeling of plastic material behavior.

2.2 Common properties of steel materials

2.2.1 Engineering stress-strain curve

Most engineers will remember testing a standard steel coupon in tension as part of their undergraduate studies and obtaining results similar to those presented in Figure 2.1, wherein the stress-strain curves for various structural steel grades tested at ambient temperature are plotted. For reference, engineering stress, σ, is calculated as the ratio of the applied force, P, to the cross-sectional area, A, and engineering strain, ε, is equal to $\Delta L/L$, where ΔL is the elongation measured over a specified gauge length, L. As shown in Figure 2.1, the operations necessary to achieve higher yield strengths (such as alloying or quenching and tempering [Van Vlack 1980]) generally reduce the maximum elongation at failure and the length of the plastic plateau. For the structural steels used in rolled shapes today, these side effects are negligible as sizable plastic deformation capacity remains beyond yield. The stress-strain curve for a uniaxially loaded steel specimen can therefore be schematically described as shown in Figure 2.2, with an elastic range up to a strain of ε_y, followed by a plastic plateau between strains ε_y and ε_{sh}, and a strain hardening range between ε_{sh} and ε_{ult}, where ε_y, ε_{sh}, and ε_{ult} are the strains at the onset of yielding, strain-hardening, and necking, respectively. Depending on the steel used, ε_{sh} generally varies between 5 to $15\varepsilon_y$, with an average value of $10\varepsilon_y$ typically used in many applications. For all structural steels, the modulus of elasticity can be taken as 200,000 MPa (29,000 ksi), and the tangent modulus at the onset of strain-hardening is roughly 1/30th of that value, or approximately 6700 MPa (970 ksi).

2.2.2 Effect of temperature on stress-strain curve

The shape of the stress-strain curve varies considerably at very high and very low temperatures. The yield and ultimate strengths of steel, as well as its modulus of elasticity, drop (while maximum elongation at failure marginally increases) as temperature increases, as shown in Figures 2.3 and 2.4 This drop is relatively slow and not too significant up to 500°F (260°C) and is almost linear until these properties reach approximately 80 percent of their initial value, at a temperature of approximately 800°F (425°C). Beyond that point, weakening

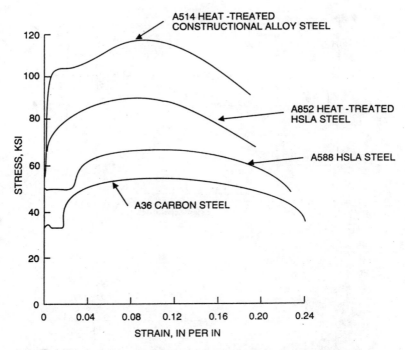

Figure 2.1 Stress-strain curves for some commercially available structural steel grades at ambient temperature. *(From R.L. Brockenbrough and F.S. Merritt,* Structural Steel Designer's Handbook, *2nd Edition, 1994, with permission of the McGraw-Hill Companies.)*

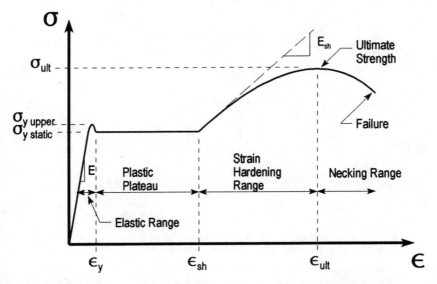

Figure 2.2 Schematic representation of stress-strain curve of structural steel.

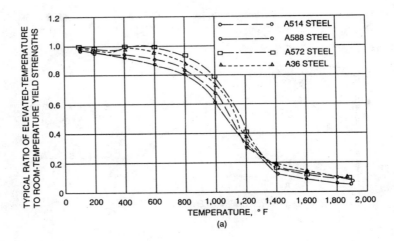

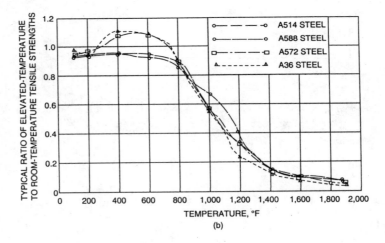

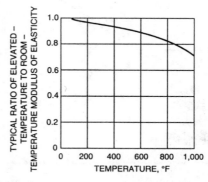

Figure 2.3 Effect of temperature on (a) yield strength, (b) tensile strength, and (c) modulus of elasticity of structural steels *(From R.L. Brockenbrough and B.G. Johnston,* USS Steel Design Manual, *with permission of U.S. Steel.)*

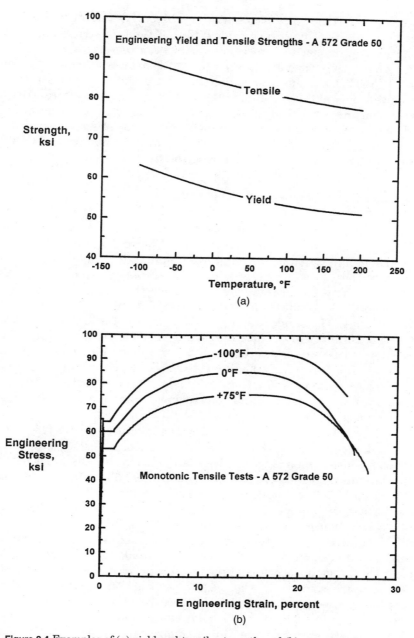

Figure 2.4 Examples of (a) yield and tensile strength and (b) stress-strain curves versus temperature. Composite curves constructed from data taken from several heats of ASTM A572 Grade 50 plate on the order of 1 in thick. *(Courtesy of H.S. Reemsnyder, Homer Research Laboratory, Bethlehem Steel Corporation, with permission. Note: The data in Figure 2.4 should not be considered typical, maxima, or minima for all plates and rolled sections of this grade. Properties will vary with chemistry, thermomechanical processing, thickness, and product form.)*

and softening of the steel accelerates significantly. A number of equations that capture this behavior for different types of structural steel are presented in documents concerned with fire resistance (ASCE 1992). Given that special high-temperature heat treatments (up to 1000°F (540°C) are often specified to reduce the residual stresses introduced by welding in industrial steel pressure vessels (particularly when a new steel plate is welded in place of an existing damaged or corroded one), a good knowledge of the high-temperature properties of structural steel is crucial to prevent collapse of these vessels under their own weight during the heat treatment.

2.2.3 Effect of temperature on ductility and notch-toughness

Temperatures below room temperature do not have an adverse impact on the yield strength of steel, as shown in Figure 2.4, but lower temperatures can have a substantial impact on ductility. Indeed, the ultimate behavior of steel will progressively transform from ductile to brittle when temperatures fall below a certain threshold and enter the appropriately labeled "Ductile-to-Brittle-Transition-Temperature" (DBTT) range. This undesirable property of structural steel led to a few notable failures in the late 1800s and early 1900s, but began to be fully appreciated only in the 1940s and 1950s when large cracks were discovered in more than 1000 all-welded U.S. steel ships built in that period. These included more than 200 cases of severe fractures and 16 ships lost at sea when their hulls unexpectedly "snapped" in two while they were cruising the Arctic sea.

The Charpy V-notch test (also known as a notch-toughness test) was developed to determine the DBTT range. In this test, a standard notched steel specimen is broken by a falling pendulum hammer fitted with a standard striking edge (Figures 2.5a and 2.5b). In principle, the energy absorbed by the specimen during its failure will translate into a loss of potential energy of the pendulum. Thus, a rough measure of this absorbed energy can be calculated from the difference between the initial height (h_1) of the pendulum when released and the maximum height (h_2) it reaches on the far side after breaking the specimen. A typical plot of the absorbed energy of a Charpy specimen as a function of temperature is shown in Figures 2.5c and 2.5d for a standard structural steel. Corresponding changes in fracture appearance and lateral expansion at failure are shown in Figures 2.5e and 2.5f.

Generally, structural engineering standards and codes will allow the use of only steels that exhibit a minimum energy absorption capability at a predetermined temperature—for example, 15 ft-lbs at 40°F (20 N-m at 4.5°C). Although it may first appear to the reader, from the Charpy data of Figure 2.5, that a level of energy dissipation of 15 ft-lbs corresponds to a nearly brittle material behavior and is insufficient to

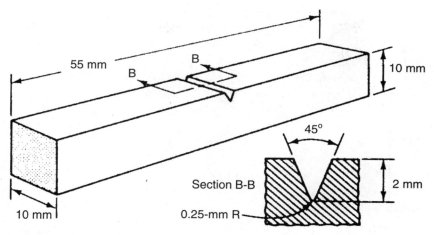

Figure 2.5 Charpy V-notch impact tests: (a) typical specimen.

ensure ductile response, it must be recognized that the Charpy V-notch test produces failures at very high strain rates. As shown in Figure 2.5c for A572 Grade 50 steel, there exists a shift of 125°F between the DBTT range obtained from the Charpy V-notch tests and that range obtained from failure tests conducted at an intermediate strain rate of 0.001/second, with more ductile behavior obtained at lower strain rates (see ASTM 1985a for details on tests at slower strain rates, and Rolfe and Barsom 1987 for techniques to compare different measures of fracture toughness obtained from different testing procedures). Hence, compliance with the specified Charpy V-notch 15 ft-lbs energy

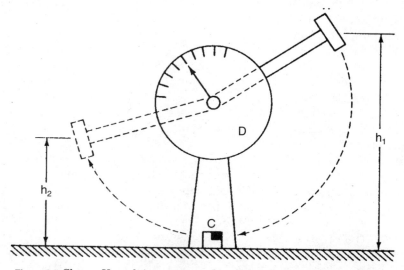

Figure 2.5 Charpy V-notch impact tests: (b) schematic of testing procedure.

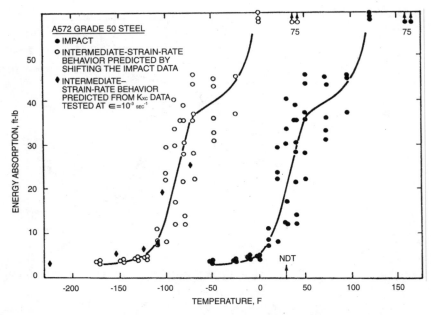

Figure 2.5 Charpy V-notch impact tests: (c) energy absorption behavior for impact loading and intermediate strain-rate loading for standard specimens for A572 Grade 50 steel.

absorption at 40°F should ensure ductile behavior over the practical range of service temperatures for structural elements subjected to such a strain rate. Tables 2.1 and 2.2, as well as Figure 2.6, summarize the Charpy V-notch test results for a number of commonly used steel grades and show that most comply with the above example specified Charpy limit. However, it should be noted from those tables that structural shapes having large flange and web thicknesses (such as those in AISC Groups 4 and 5 [AISC 1994]) generally have lower energy absorption capabilities. For specialty applications that require better performance than that which is commonly available, the structural engineer should request steel specially alloyed to provide the needed material properties at low temperatures, high strain rates, or both.

Beyond the energy absorption measure, a large number of descriptive indices exist in the literature to capture and quantify this ductile-to-brittle behavior. The Nil-Ductility-Transition (NDT) temperature, defined as the highest temperature at which a specimen fails in a purely brittle manner (or alternatively as the temperature at which a small crack will propagate to failure in a specimen loaded exactly to the yield stress), can also be obtained from the ASTM E208 test (ASTM 1985b). The Fracture Appearance Transition Temperature (FATT) is the temperature at which 50 percent of the fracture surface corresponds to a cleavage failure (i.e., a flat crystalline surface characteristic of brittle

Structural Steel 15

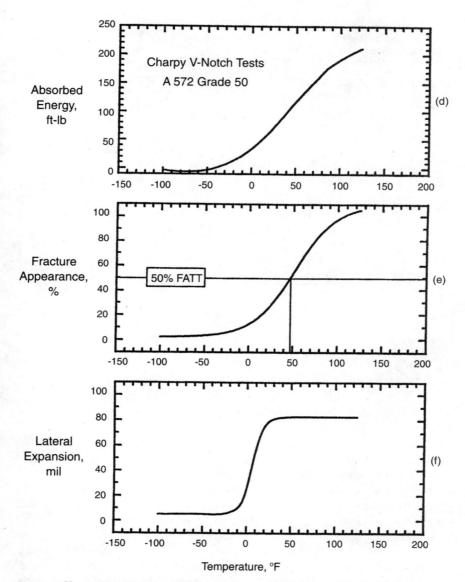

Figure 2.5 Charpy V-notch impact tests: Composite plots of (d) energy absorption, (e) fracture appearance, and (f) lateral expansion constructed from data taken from several heats of ASTM A572 Grade 50 plate on the order of 1 in thick.

(Figures a, b, and c from J.M. Barsom and S.T. Rolfe, Fracture and Fatigue Control in Structures—Applications of Fracture Mechanics, *with permission from J.M. Barsom and S.T. Rolfe; Figures d, e, and f courtesy of H.S. Reemsnyder, Homer Research Laboratory, Bethlehem Steel Corporation, with permission. Note: The data in Figure 2.5 should not be considered typical, maxima, or minima for all plates and rolled sections of this grade. Properties will vary with chemistry, thermomechanical processing, thickness, and product form.)*

TABLE 2.1 Compilation of Charpy V-notch test results for various structural steel grades, from data provided by North American producers (*From American Institute of Steel Construction, Statistical analysis of Charpy V-notch toughness for steel wide flange structural shapes, with permission*)

Steel Grade	Test Temp (°F)	ASTM Shape Group	Sample Size	Mode (ft lbs)	Minimum (ft lbs)	Mean (ft lbs)	Maximum (ft lbs)	First Quartile (ft lbs)	Median (ft lbs)	Third Quartile (ft lbs)
A36 OST	32	ALL	21	162	91	151	204	137	160	168
A36 HR	32	ALL	73	130	91	150	241	124	145	169
A36 HR	40	ALL	2011	66	16	112	286	65	98	147
		1	421	100	20	116	272	79	113	146
		2	1057	239	16	130	286	77	117	176
		3	315	36	19	86	240	63	77	98
		4	218	54	16	54	193	41	51	62
A36 HR	70	ALL	426	239	22	95	253	43	70	124
		1	262	43	25	59	177	36	50	75
		2	59	69	59	122	253	83	97	138
		3	24	239	25	200	240	235	239	239
		4	81	240	22	158	240	99	177	221
A572 Gr 50 OST	32	ALL	24	N/A	101	136	182	122	138	149
A572 Gr 50 HR	32	ALL	15	N/A	106	140	170	135	142	147
A572 Gr 50 HR	40	ALL	3930	79	16	91	288	64	80	97
		1	400	58	16	84	259	58	73	95
		2	2181	80	18	93	288	71	85	102
		3	813	86	16	83	280	65	82	96
		4	453	60	17	66	155	53	63	77
		5	83	49	29	62	155	47	59	71
A572 Gr 50 HR	70	ALL	598	47	15	61	241	31	51	74
		2	37	124	31	135	237	89	134	194
		3	90	74	31	76	202	56	68	91
		4	364	50	17	57	241	34	51	68
		5	104	26	15	33	116	23	27	32
A588 HR	40	ALL	223	21	16	140	290	71	129	204
		2	182	54	18	148	290	82	145	215
		3	41	N/A	16	103	249	58	75	155
A913 Gr65 OST	32	ALL	87	156	92	141	212	122	142	158
A913 Gr65 OST	70	ALL	34	48	33	61	108	45	57	79
Dual Certified	40	ALL	202	43	17	53	121	37	51	67
		1	142	38	17	51	121	35	48	63
		2	55	43	24	59	116	43	58	74
Dual Certified	70	ALL	368	65	15	59	131	41	55	74
		1	322	53	15	55	131	37	53	69
		2	46	65	60	86	123	67	91	99

NOTE: HR = Hot Rolled; QST = Quenched Self-Tempered

failure), with the remaining 50 percent showing shear-lips of fibrous appearance typical of ductile failures. Figure 2.7 illustrates that this transition in texture of the failure surface is a function of both temperature and strain rate. Further, it shows that alignment of the surfaces exhibiting similar texture, but tested at different strain rates, provides a striking visual expression of the aforementioned temperature shift of the DBTT range.

Although notch toughness and ductility for a given steel are closely related, it is important to realize that notch toughness is not related to yield strength. In fact, in very thick steel sections, the notch toughness of some steels is known to vary quite significantly throughout the cross-section of members, even when a variation in yield strength is not observed. This is particularly true at the core of the web-flange intersection of very thick sections in which a larger steel grain structure exists as a consequence of a lesser amount of cold forming achieved there during the rolling process. For this reason, the AISC (1994) specifies that samples for Charpy V-notch tests be taken from the core area, not the location specified in the ASTM standard.

TABLE 2.2 Probability of exceedance of two commonly specified Charpy V-notch energy absorption material requirements for various structural steel grades, from data provided by North American producers (*From American Institute of Steel Construction, Statistical analysis of Charpy V-notch toughness for steel wide flange structural shapes, with permission.*)

Steel Grade	Test Temp (°F)	ASTM Shape Group	Observed Probability of Exceedance 15 ft lbs (percent)	Observed Probability of Exceedance 20 ft lbs (percent)
A36 QST	32	ALL	100	100
A36 HR	32	ALL	100	100
A36 HR	40	ALL	99	97
		1	100	97
		2	100	99
		3	100	99
		4	96	95
A36 HR	70	ALL	100	95
		1	100	95
		2	100	100
		3	100	94
		4	100	97
A572 Gr 50 QST	32	ALL	100	100
A572 Gr 50 HR	32	ALL	100	100
A572 Gr 50 HR	40	ALL	100	99
		1	99	99
		2	100	99
		3	98	98
		4	100	99
		5	100	99
A572 Gr 50 HR	70	ALL	95	87
		2	100	100
		3	100	100
		4	96	89
		5	85	60
A588 HR	40	ALL	98	97
		2	99	98
		3	97	95
A913 Gr65 QST	32	ALL	100	100
A913 Gr65 QST	70	ALL	100	100
Dual Certified	40	ALL	99	94
		1	98	94
		2	100	99
Dual Certified	70	ALL	98	94
		1	98	93
		2	100	100

NOTE: HR = Hot Rolled; QST = Quenched Self-Tempered

2.2.4 Strain rate effect on tensile and yield strengths

Strain rate is another factor that affects the shape of the stress-strain curve. Typically, the tensile and yield strengths will increase at higher strain rates, as shown in Figure 2.8, except at high temperatures for which the reverse is true. Consideration of this phenomenon is crucial for blast-resistant design in which very high strain-rates are expected, but of little practical significance in earthquake-engineering applications. Past studies have repeatedly demonstrated that, for the steel grades commonly used, the expected increases of roughly 5 percent to 10 percent in yield strength at typical earthquake-induced strain rates

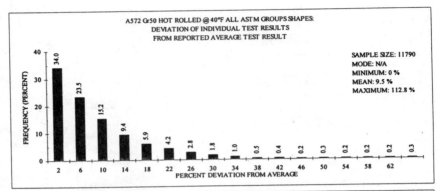
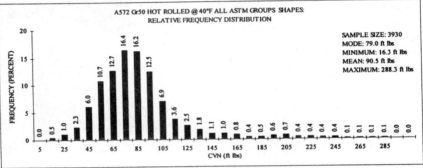
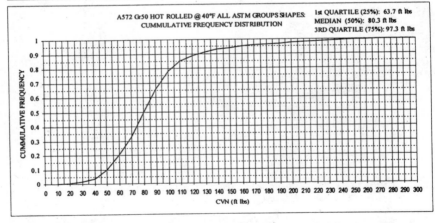

Figure 2.6 Compilation of Charpy V-notch test results for A572 Grade 50 structural steel, from data provided by North American producers. *(From American Institute of Steel Construction, Statistical Analysis of Charpy V-Notch Toughness for Steel Wide Flange Structural Shapes, with permission.)*

is negligible compared with the much greater uncertainties associated with the earthquake input. The effect of strain rate on notch toughness, as described above, is also considerably more significant.

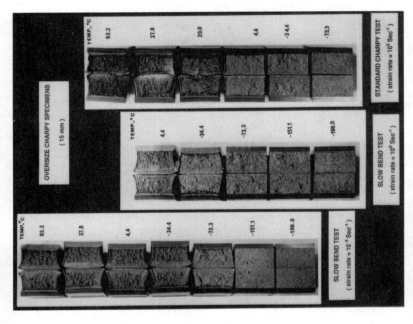
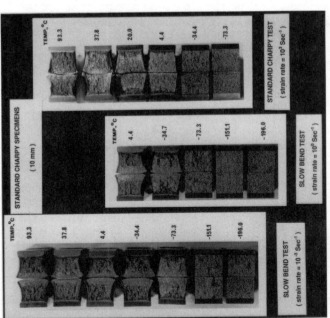

Figure 2.7 Failure surface texture of Charpy V-notch specimens at different test temperatures and strain rates. *(Courtesy of the Canadian Institute of Steel Construction, with permission)*

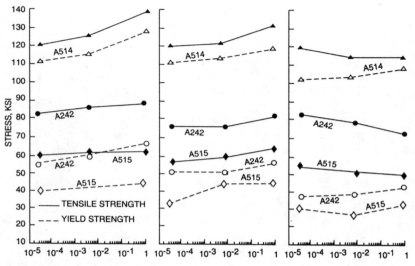

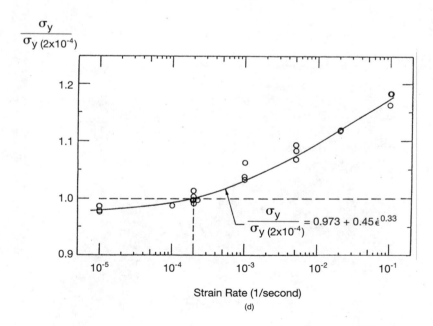

Figure 2.8 Yield stress as a function of strain rate and temperature for some structural steels. *(Figures a, b, and c, from R.L. Brockenbrough and B.G. Johnston,* USS Steel Design Manual, *with permission from U.S. Steel; Figure d from Moncarz, P.D. and Krawinkler H.,* Theory and Application of Experimental Model Analysis in Earthquake Engineering, *Report No. 50, John A. Blume Earthquake Engineering Research Center, Department of Civil Engineering, Stanford University, with permission.)*

2.2.5 Probable yield strength

In seismic design, as will be demonstrated later, knowledge of the maximum probable yield strength is equally important as knowledge of the minimum reliable yield strength. Recent studies have reported that the margin between the actual average yield strength and specified yield strength has progressively increased over the years for some structural steels, even though the steel specification itself remained unchanged. For example, a few decades ago, yield strengths of 255 to 270 MPa (37 to 39 ksi) were typically reported for ASTM-A36 steel by researchers studying the behavior of structural members and connections (e.g., Popov 1971, Galambos and Ravindra 1978, NBS 1980), whereas similar tests conducted 20 years later (e.g., Englehardt and Husain 1993) using the very same steel grade revealed a substantial increase of the yield strength, with values ranging from 325 to 360 MPa (47 ksi to 52 ksi). Yield stress values in excess of 420 MPa (60 ksi) have also occasionally been reported for that steel grade (SAC 1995). In fact, some steel mills have apparently adopted a dual-certification procedure for steel conforming to both the ASTM-A36 and A572 specifications. Although this higher strength translates into safer structures for nonseismic design, an unexpectedly higher yield strength can be disadvantageous for seismic design. For example, a specific structural component can be designed to yield, absorb energy, and prevent adjacent elements from being loaded above a predetermined level during an earthquake, thus acting much like a "structural fuse." A yield strength much higher than expected could prevent that structural fuse from yielding and overload the adjacent structural components (such as the welded joints in moment-resisting frames), with drastic consequences on the ultimate behavior of the structure (see Chapter 8).

As this book is written, North American structural shape producers are in the process of preparing a new ASTM specification for a grade 50 steel having a maximum yield strength in addition to the minimum value traditionally specified (SAC 1995). Such a steel having a specified yield strength of 345 MPa (50 ksi), an upper limit on yield strength of 448 MPa (65 ksi), and a specified maximum yield-to-tensile strength ratio of 0.85 is apparently already produced by some steel mills (AISC 1997a, 1997b). Japanese steel mills have already produced a special ductile steel having 40 percent to 60 percent elongation at failure and well-controlled maximum yield values, albeit with a low-yield strength of 120 MPa (17.5 ksi), and seismic design applications have already been found for that material (Nakashima et al. 1994).

Note, however, that greater-than-expected yield strength can also simply be the result of an inadvertent but well-intentioned substitution, by a supplier, from the originally requested steel grade to one of higher yield strength, to compensate, for example, for a temporary stock shortage.

Therefore, in those circumstances that warrant it, the engineer should indicate on all construction documents the key members for which substitution to a higher strength material is not acceptable.

Incidentally, high-performance steels are also being developed in North America (Wright 1996) to meet the following specifications: (a) high strength, with yield strengths of 480 MPa (70 ksi) and 690 MPa (100 ksi); (b) excellent weldability without any need for preheating (see Section 2.5); (c) extremely high toughness, with Charpy V-notch values of roughly 200 ft-lbs (270 N-m) at -10°F (-23°C), compared with the current bridge design requirements of 15 ft-lbs (20 N-m) at −10°F (-23°C); and (d) corrosion resistance comparable to that of weathering steel. Although these new steels are being developed for bridge applications, their outstanding properties might also make them equally appealing for general ductile design applications.

2.3 Plasticity, hysteresis, Bauschinger effects

After steel has been stressed beyond its elastic limit and into the plastic range, a number of phenomena can be observed during repeated unloading, reloading, and stress reversal. First, unloading to $\sigma=0$ and reloading to the previously attained maximum stress level will be elastic with a stiffness equal to the original stiffness, E, as shown in Figure 2.9. Then, as also shown, upon stress reversal (to $\sigma = -\sigma_y$), a sharp "corner" in the stress-strain curve is not found at the onset of yielding; instead, stiffness softening occurs gradually with yielding initiating earlier than otherwise predicted. This behavior, known as the Bauschinger effect, is a natural property of steel. If the stress reversal is initiated prior to attainment of the strain-hardening range when the steel is loaded in one direction, a yield plateau will eventually be found in the reversed loading direction as shown in Figure 2.9a. However, once the strain-hardening range has been entered in one loading direction, the yield plateau effectively disappears in both loading directions (Figure 2.9b).

A most important property of steels subjected to large cyclic inelastic loading is their ability to dissipate hysteretic energy. The energy needed to plastically elongate or shorten a steel specimen can be calculated as the product of the plastic force times the plastic displacement (i.e., the work done in the plastic range) and is called the hysteretic energy. Unlike kinetic and strain energy, hysteretic energy is a nonrecoverable dissipated energy. As shown in Figure 2.10a, under a progressively increasing loading, followed by subsequent unloading, the hysteretic energy, E_H, can be expressed as:

$$E_H = P_Y(\delta_{MAX} - \delta_Y) \qquad (2.1)$$

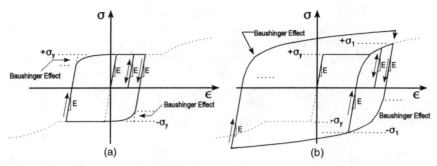

Figure 2.9 Cyclic stress-strain relationship of structural steel.

that is, the shaded area in this figure. For a full cycle of load reversal, the hysteretic energy will simply be the area enclosed by the loop of the force-displacement curve, as shown in Figure 2.10b, and approximately expressed as:

$$E_H \cong P_Y[(\delta_{MAX} - \delta_Y) + (\delta_{MAX} - \delta_{MIN} - 2\delta_Y)] \quad (2.2)$$

A more accurate calculation of hysteretic energy in this case would recognize the small loss of hysteretic energy at the rounded corners of the force-displacement curve due to the Bauschinger effect.

Under repeated cycles of loading, the energy dissipated in each cycle is simply summed to calculate the total energy dissipated. This cumulative energy dissipation capacity is a most important property that makes possible the survival of steel structures to rare but rather severe loading conditions, such as blast loading or earthquake loading.

Within the framework of this book, the above description of inelastic cyclic behavior is certainly adequate. However, it is noteworthy that a few additional minor phenomena also develop as steel undergoes

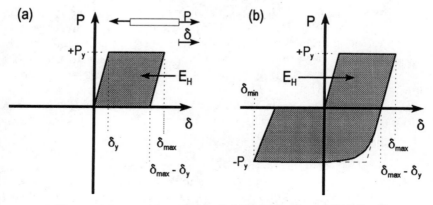

Figure 2.10 Hysteretic energy of structural steel: (a) half cycle and (b) full cycle.

numerous cycles of severe hysteretic behavior. For example, the threshold beyond which strain-hardening starts to develop, as well as the extent of the elastic range prior to onset of the Bauschinger effect, is a function of the prior plastic loading history. Mizuno et al. (1992), Dafalias (1992), and Lee et al. (1992) provide a good overview of these phenomena and progress on the development of constitutive relationships that can capture the complex behavior of structural steels subjected to arbitrary cyclic loading histories.

2.4 Metallurgical process of yielding, slip planes

As mentioned earlier, steel coupons tested axially exhibit a well-defined plastic plateau. In principle, the tangent modulus of elasticity of steel (i.e., the slope of the stress-strain curve) is effectively zero along this plateau, as shown in Figure 2.2. As a result, one may wonder how a steel member can ever reach the strain-hardening range in compression, knowing that the maximum stress that can be resisted by a steel plate prior to buckling is given by (Popov 1968):

$$\sigma_{cr} = \frac{k}{(b/t)^2} \left[\frac{\pi^2 E}{12(1-v^2)} \right] \quad (2.3)$$

Indeed, according to Equation 2.3, buckling should occur as soon as the strain exceeds ε_y and the tangent modulus drops to zero (i.e., $E=0$). To resolve this paradox, an understanding of the metallurgical process of yielding is needed.

Steel is a polycrystalline material that, when loaded beyond its elastic limit, develops slip planes at 45°. These visible yield lines, also known as Lüder lines, are a consequence of the development of slip planes within the material as yielding develops. A schematic representation of a slip plane is shown in Figure 2.11. At the precise location of the slip plane, the strains can be thought of as having "slipped" from ε_y to ε_{sh} in one single jump. Following this first slip, other planes will subsequently slip, one after the other in a random sequence as a function of the random distribution of various weaknesses and dislocations in the steel's crystalline structure, as their respective slip resistances are reached. Thus, under no perceptible variation in the applied stress, the number of sections that have jumped from ε_y to ε_{sh} will progressively increase until the entire length of member subjected to the yield stress has strain hardened to a strain equal to ε_{sh}. Assuming for convenience that at any given time during this process, all slip planes can be grouped together over a length ϕL, where L is the length of the specimen subjected to the yield stress as shown in Figure 2.11b, and, knowing that all slipped segments have reached ε_{sh} while the others are still at ε_y, the average strain over the specimen length can be expressed as:

$$\varepsilon_{av} = \frac{\Delta L}{L} \frac{\varepsilon_y(L-\phi L)+\phi\varepsilon_{sh}L}{L}$$

$$= (1-\phi)\varepsilon_y + \phi\varepsilon_{sh} \qquad (2.4)$$

$$= (1-\phi)\varepsilon_y + \phi s\varepsilon_y$$

where s (>1.0) is the ratio of ε_{sh} to ε_y. This clearly illustrates how the average strain can increase progressively without any apparent increase in the applied stress, while the actual stiffness of the steel material during this yield process varies from E (when $\phi=0.0$) to E_{sh} (when $\phi=1.0$), but never zero. Defining the effective stiffness as $\sigma_y/\varepsilon_{av}$ in the yield plateau, Figure 2.11c shows the variation of this effective stiffness over the range $0.0<\phi<1.0$.

Therefore, the plastic plateau on the stress-strain curve is simply a consequence of standard testing methods for which yielding must spread over a specified gage length (100 or 200 mm, for example) before higher loads can be applied to a given specimen. This also explains why strain hardening is usually reached before local buckling and other instabilities develop in a structural member (see Chapter 10).

An understanding of the slip plane phenomenon is also helpful in explaining the Bauschinger effect illustrated in Figure 2.9, recognizing that a piece of steel is actually an amalgam of randomly oriented steel crystals (Timoshenko 1983). Thus, a slip plane contains a large number of steel grains in which sliding has occurred along definite crystallographic planes of weakness (and hence contributed to the yield plateau) and others that have not. If a specimen is unloaded after yielding in tension but prior to attainment of ε_{sh}, the crystals that have slid are locked in their elongated position while others try to elastically return to their undeformed position. Because of continu-

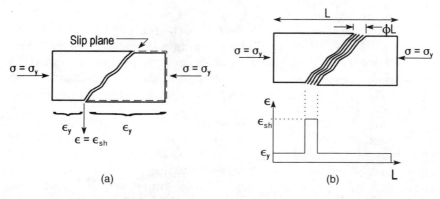

Figure 2.11 Schematic representation of slip planes during yielding: (a) single slip plane, (b) cumulative effect of multiple slip planes.

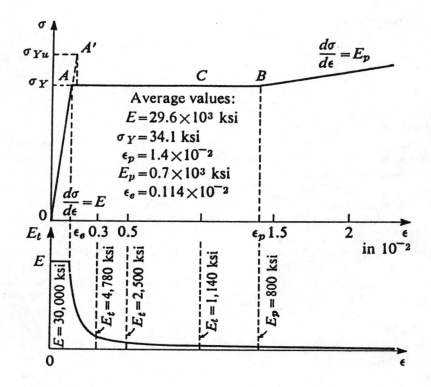

Figure 2.11 Schematic representation of slip planes during yielding for Grade A-7 steel: (c) effective modulus throughout development of slip planes.
(Figure c from Massonnet, C.E. and Save, M.A., Plastic Analysis and Design, Vol. 1: Beams and Frames.)

ity of the metal, this action generates internal residual stresses, with the slipped crystals compressed as they prevent the others from fully recovering their elastic deformations. If, after unloading, the specimen is subjected to a reversed loading (i.e., compression in this example), slipping of the most detrimentally oriented crystals triggers at a load much smaller than would have otherwise been necessary on a virgin specimen. This is logical because, prior to the application of the external compression load, these crystals are already in compression because of the residual stresses created upon unloading from the earlier tension yielding excursion.

This also partly explains why, on an experimentally obtained stress-strain curve, the yield stress corresponding to the observed yield plateau (denoted $\sigma_{y\text{-}static}$ in Figure 2.2) is sometimes preceded by a slightly higher yield value (denoted $\sigma_{y\text{-}upper}$ in Figure 2.2) in specimens machined to a very high tolerance to have a uniform cross-section over the gauge length where yielding is expected and tested on high precision equipment (see Lay 1965 for more details). A simplistic analogy between the development of slip planes and the actual physical behavior of friction between solids makes clear that it takes more energy to initiate a slip plane (or to start sliding a body whose only resistance against motion is provided by friction) than to maintain it once it has initiated. In any case, $\sigma_{y\text{-}static}$ is the yield value of engineering significance.

2.5 Brittleness in welded sections

Brittleness can occur as a result of a variety of influences, as discussed in the following sections.

2.5.1 Metallurgical transformations during welding, heat-affected zone, preheating

The welding process can embrittle the steel material located in the vicinity of the weld. Simple solutions exist to circumvent most problems, although these become more difficult to implement when extremely thick heavy rolled sections are welded.

An important first step in avoiding brittle failures in welded members is to recognize that welding is a complex metallurgical process (and not a "gluing" operation that performs miracles). Not only is new material deposited during welding, but sound fusion with the base metal is necessary to provide the desired continuity between the welded components.

The topology of a welded area consists of three important zones: the fusion zone, the heat-affected zone (HAZ), and the base metal (see Figure 2.12). The fusion zone consists of all the metal that is effectively melted during welding. Good penetration of the fusion zone into the base metal will generally provide a better-quality weld. This is one reason flux-cored arc welding (FCAW) provides better-quality welds than shielded metal arc welding (SMAC) for a given filler metal.

Immediately adjacent to the fusion zone lies the heat-affected zone, which, as the name implies, consists of steel whose grain structure has been modified by the high heat imparted during the welding process. The crystalline constitution of the HAZ will depend on the metallurgical content of the base metal and on the speed of cooling of the metal (Van Vlack 1980). Generally, if cooling is too rapid in high-strength steel, the metal in the HAZ will become a hard and brittle martensite layer that is highly susceptible to cracking in the presence of stress raisers or concentrations. Rapid cooling can be a significant problem in thicker steel sections because the heat introduced by welding will be more rapidly dissipated into larger volumes of colder steel, resulting in rapid cooling rates. To avoid introducing brittle marten-

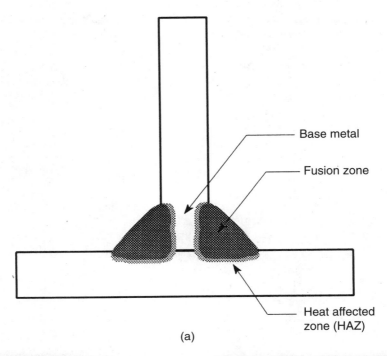

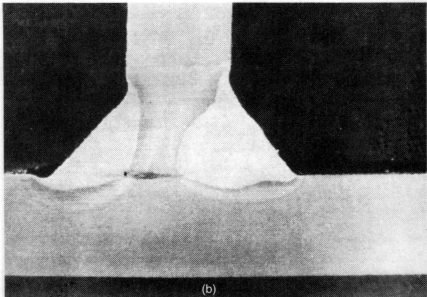

Figure 2.12 Topology of a welded area: (a) schematic illustration, (b) fillet welds by shielded metal arc welding (SMAW) on the left, and flux-cored arc welding (FCAW) on the right, in A36 steel (note the increased penetration of FCAW). *(Figure b from F.R. Preece and A.L. Collin*, Steel Tips: Structural Steel Construction in the '90s, *with permission from the Structural Steel Education Council.)*

site in the HAZ, it is generally recommended to preheat the base metal to a specified temperature prior to welding and to maintain that temperature (termed the interpass temperature) throughout the execution of the weld.

Table 2.3 provides information on heating requirements. Higher preheating and interpass temperatures are specified for thicker steels, in accordance with the above logic regarding heat dissipation. Additional and more extensive requirements are specified by the American Welding Society (AWS 1990) and American Institute of Steel Construction (AISC 1994) for standard conditions, and these documents should be consulted for more details on this matter. A welding engineer can help determine the preheating, interpass, and postwelding heating needs for specialty applications and unusually congested details.

2.5.2 Hydrogen embrittlement

An important distinction is made in Table 2.3 as to whether low hydrogen electrodes are used. Indeed, the introduction of hydrogen into the fusion or heat-affected zones increases the risk of embrittlement. Although the molten metal created during welding has a great propensity to absorb the surrounding hydrogen, much of this hydrogen is rejected during normal cooling. However, if cooling is too rapid, the hydrogen gas does not have sufficient time to escape and becomes entrapped at a high pressure within the steel, with the risk that microcracks will develop. Low hydrogen electrodes have been developed to lessen this problem. Combined with preheating, they can generally eliminate hydrogen embrittlement. Nonetheless, it must be recognized that hydrogen may originate from other sources, the most common being water. For that reason, low-hydrogen electrodes must be stored in a dry environment. Preheating is also useful to evaporate moisture at the surface of the base metal prior to welding.

2.5.3 Carbon equivalent

The "carbon equivalent" concept has been developed to convert into equivalent carbon content the effect of other alloys known to increase the hardness of steel. Although numerous compounds are added to increase the hardness of steel without causing much loss of ductility, increases in strength are always accompanied by a corresponding increase in hardness, some loss of ductility, and a reduced weldability. Most structural steels are generally alloyed to ensure their weldability, but some steels on the market still have rather high carbon equivalent content. The AWS (1990) mentions that for an equivalent carbon content above 0.40, there is a potential for cracking in the heat-affected

TABLE 2.3 Minimum pre-heat and interpass temperature requirements

Steel grade	Weld process					AWS D1.1-90/AISC Clause J2.7. Minimum temperature requirement in °C (°F) for various maximum thickness of parts at welding location			
	SMAW w/o LHE	SMAW w/LHE	SAW	GMAW	FCAW	0 to 18 mm	18 to 37 mm	37 to 63 mm	63 mm or more[a]
ASTM									
A36 A500[b] A501	X					None/10 (None/50)	65/65 (150/150)	110/120 (225/250)	150/150 (300/300)
A36 A500[b] A501 A572[c] A441		X	X	X	X	None/25 (None/75)	10/38 (50/100)	65/95 (150/200)	110/150 (225/300)
A572[d]		X	X	X	X	10/25 (50/75)	65/95 (150/200)	110/120 (225/250)	150/150 (300/300)

SMAW = Shielded Metal Arc Welding (commonly referred to as "stick" or manual welding)
SAW = Submerged Arc Welding
GMAW = Gas Metal Arc Welding
FCAW = Flux-Cored Arc Welding
LHE = Low Hydrogen Electrodes

[a] AISC Clause J2.7 also requires preheating/interpass temperature of 175°C (350°F) for steel plates thicker than 90 mm (3.5 inches)
[b] Grades A and B
[c] Grades 42 and 50
[d] Grades 60 and 65

zones near flame-cut edges and welds. Various formulas have been proposed to calculate the carbon equivalent content of certain steels. For structural steels, the American Welding Society (AWS 1990) recommends the following formula:

$$CE = C + \frac{(Mn + Si)}{6} + \frac{(Cr + Mo + V)}{5} + \frac{(Ni + Cu)}{15} \qquad (2.5)$$

where CE is the carbon equivalent measure, and C, Mn, Si, Cr, Mo, V, Ni, and Cu are the percentage of carbon, manganese, silicon, chromium, molybdenum, vanadium, nickel, and copper alloyed in the steel.

Interestingly, structural engineering standards do not specify carbon content equivalent limits for structural steels. Instead, control of their weldability is indirectly achieved through limiting the maximum percentage of certain alloys. This practice has endured largely because of the generally satisfactory performance of welded structures constructed with the weldable structural steels currently available on the market, even though some of those steels have a CE in excess of 50 and therefore a high cracking potential. However, as high-strength steels come into much greater use and reported instances of brittle failures continue to increase (e.g., Tuchman 1986, Fisher and Pense 1987), an awareness of the significance of the carbon content equivalent of steels is essential.

Finally, it is interesting that, although structural steel used to be produced directly from pig-iron in the 1960s and 1970s, steel is now often produced from scrap metal. As a result of this change in practice, a large number of "foreign" metals such as aluminum are now found in trace amounts in these steels. There is no evidence that this metallurgical variance could have undesirable consequences on the physical properties of steel, but some structural engineers have expressed concerns regarding the unknown consequences of this practice (SAC 1995).

2.5.4 Flame cutting

Some structural details can also introduce brittle conditions in steel structures. For example, weld-access holes are frequently flame-cut in the webs of beams near their flanges to facilitate welding. The weld-access hole may also relieve stresses by reducing the transverse restraint on the flange welds. However, flame-cutting generally creates an irregular surface along these holes and modifies the metallurgy of the steel into a brittle martensite up to a depth of 3 millimeters (0.12 inch) along the edge of the hole. This martensite transformation along the roughened surface promotes crack formation. Therefore, when these holes must be flame-cut, it has been recommended that the martensite region be ground smooth prior to welding (Bjorhovde 1987, Fisher and Pense 1987).

2.5.5 Weld restraints

Absorbed hydrogen, high carbon content, and flame cutting can all create an environment favorable for crack initiation and propagation, but an active external factor is usually needed to trigger fracture. The residual stresses induced by restrained weld shrinkage can sometimes provide this necessary additional factor (AISC 1973). The choice of welding sequence and weld configuration can severely restrain weld shrinkage.

To understand the nature and impact of these restraints, it is helpful to visualize welds as molten steel that solidifies when cooled. If unrestrained, the hot weld metal will shrink as it cools. It is well known that important distortions will occur as a result of this shrinkage in nonsymmetrical welds, as shown in Figure 2.13. Many examples of such distortions and information on their expected magnitude are available in the literature (e.g., Blodgett 1977a). However, if existing restraints prevent this distortion, internal stresses in self-equilibrium must develop. As shown in Figure 2.14 for single-pass welds, the weld metal will generally be in tension and the pieces being connected will generally be in compression where they are in contact. In multipass welds, some of the weld metal first deposited (usually at the root of the weld) will initially be in tension as it cools, but could eventually end up in compression after all the subsequently deposited weld material has cooled and compressed the previously deposited material. The weld material deposited last will generally be subjected to the largest residual tension stresses. Complex residual stress patterns will obviously exist in all but the simplest details. An example of how the residual stress condition in a detail can be qualitatively assessed is presented at the end of this section.

To minimize the tensile residual stresses in welds, at least for simple weld details such as those shown in Figure 2.14, it has been suggested (Blodgett 1977b) that a few soft wires could be inserted between the pieces to be welded. During welding, these wires will simply crush with little resistance, allowing the welds to shrink without restraints, thus preventing the development of residual stresses.

Assuming invariant material properties, it is generally preferable to choose a weld configuration that minimizes internal residual stresses. However, there are instances in which large residual stresses are bound to be present. For example, if a beam-to-column fully welded connection must be performed at both ends of a beam located between two braced frames, as shown in Figure 2.15, shrinkage of the welds is prevented by the rigid braced frames, and the likelihood of yielding (or cracking) these welds or the adjacent base metal is rather high. In such a case, preheating and postheating may be necessary, along with

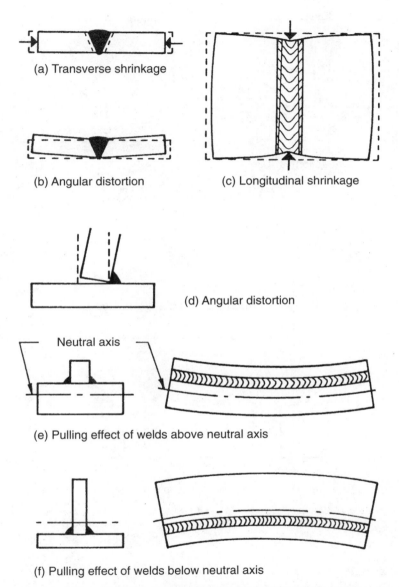

Figure 2.13 Examples of distortion and dimensional changes in assemblies due to unsymmetrical welds. *(From The Lincoln Electric Company, The Procedure Handbook or Arc Welding, 12th Edition, with permission from The Lincoln Electric Company.)*

a stringent inspection program to check the integrity of the resulting welds. Alternatively, changes to the construction sequence should be contemplated.

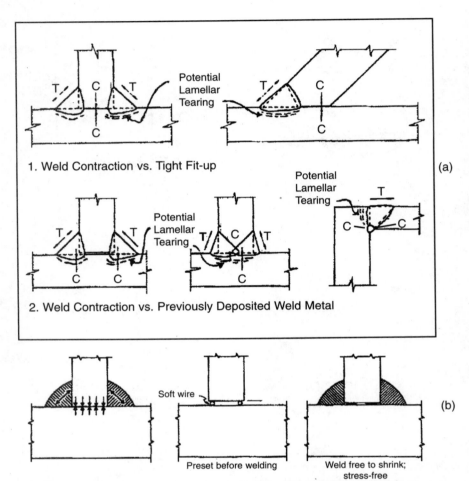

Figure 2.14 (a) Examples of internal stresses due to weld restraints. *(From F.R. Preece and A.L. Collin,* Steel Tips: Structural Steel Construction in the '90s, *with permission from the Structural Steel Education Council.)* (b) Example of measure to reduce the magnitude of such stresses. *(From O.W. Blodgett, Special Publication G230 of The Lincoln Electric Company:* Why Do Welds Crack, How Can Weld Cracks Be Prevented, *with permission from The Lincoln Electric Company.)*

2.5.6 Lamellar tearing

Steel is usually treated as an isotropic material. However, the available test data on steel plates generally demonstrates a significant anisotropy of strength and ductility (Figure 2.16a) because the properties of steel plates tested in the through-thickness direction vary significantly from those in any of the plane directions. The presence of small microscopic nonmetallic compounds (termed "inclusions") in the metal and flattened during the rolling process explains this difference. These flattened inclusions act as microcracks, of no consequence when

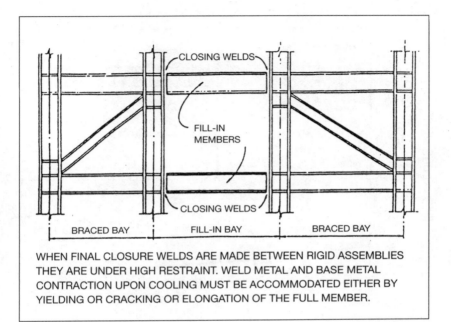

Figure 2.15 Closing welds for members between rigid assemblies as an example of poor construction sequence, leading to highly restrained welds. *(From F.R. Preece and A.L. Collin,* Steel Tips: Structural Steel Construction in the '90s, *with permission from the Structural Steel Education Council.)*

a steel plate is stressed in its plane directions, but that can grow and link when stresses are applied in the through-thickness direction (z-direction), as shown in Figure 2.16b, producing a brittle failure mechanism known as lamellar tearing. However, this phenomenon has been observed only in thick steel plates with highly restrained weld details (thicker steel sections generally have more nonmetallic inclusions). An effective solution to the problem of lamellar tearing is to detail the welded connection with bevels that penetrate deep into the cross-sections to be welded, thereby engaging the full thickness of the plates in the resistance mechanism instead of relying on through-thickness strength to resist the tension force at the surface of the plate. Examples of alternative weld details are presented in Figure 2.16c. The engineer can visualize, in each of these improved details, the weld pulling on all "layers" of the thick steel plate rather than on its surface.

2.5.7 Thick steel sections

Most of the aforementioned factors contributing to brittle failures will not lead to such failures in most cases. However, special care must be

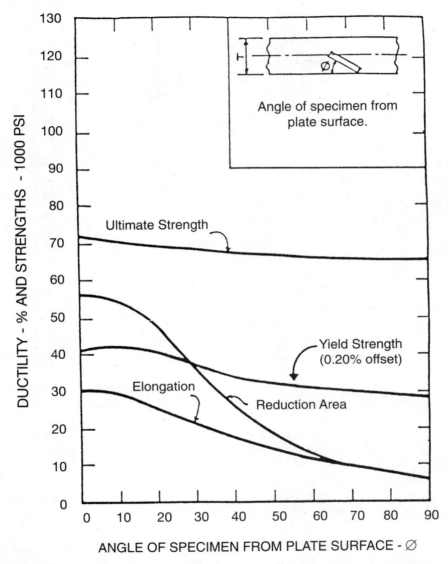

Figure 2.16 Lamellar tearing: (a) anisotropic properties of structural steel. (*Figure a, b, and c from F.R. Preece and A.L. Collin,* Steel Tips: Structural Steel Construction in the '90s, *with permission from the Structural Steel Education Council*).

taken with very thick high-strength steel sections, roughly defined as 60 mm (2.4 inches) and thicker, because these are known to be more prone to lamellar tearing and other fracture problems.

Likewise, the complex triaxial stresses induced by highly restrained weld details are of greater consequence for heavy thick steel sections. For example, brittle failures of fully welded heavy rolled wide-flange

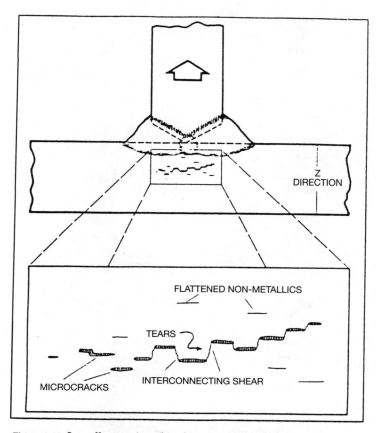

Figure 2.16 Lamellar tearing: (b) schematic of mechanism of lamellar tearing, starting with microfissures initiation at flattened nonmetallic particle inclusions, growth of cracks due to cross-thickness tension forces, and rupture by interconnecting tears between cracks on different planes, resulting in a stepped rupture surface.

beam splices have been reported (Fisher and Pense 1987, Fisher personal communication 1984) in which cracking initiated at the end of the flame-cut weld-access hole. Through use of existing qualitative information on weld shrinkage and distortion, orientation of the weld and the welding sequence can provide insight into the triaxial stress state at the end of a flame-cut hole. For the example shown in Figure 2.17a, the weld sequence is assumed to have been initiated with a root pass on each flange, followed by welding of the web, the outside of the flanges, and finally the inside of the flanges. Because of the presence of the root weld between the flanges, the web welds induce tension in the vertical plane (noted 1* on the triaxial stress diagram in Figure 2.17a), but the weld-access hole can effectively reduce these residual stresses. More critical is the welding of the outside of the flanges prior to the inside of the flanges. As shown on Figure 2.17b, this creates a

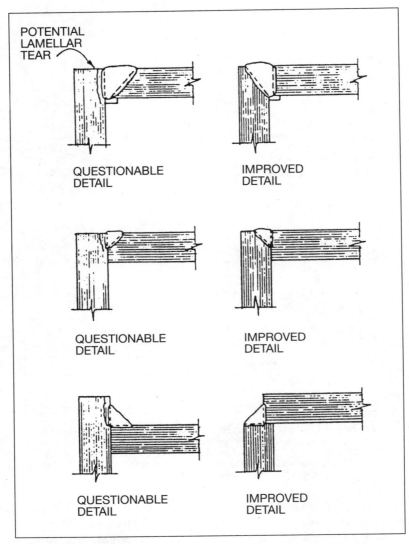

Figure 2.16 Lamellar tearing: (c) details prone to development of lamellar tearing and corresponding improved details.

tendency of the flanges to rotate (bow) away from the web. Therefore, the web-flange region at the end of the hole is stressed in tension, and any initial crack or microcracks in this region could propagate further. Assuming that a crack has formed, crack propagation will reduce the flange curvature, which in turn should lower the tensile force at the end of the hole, and the crack will stop propagating once it reaches a finite length. The final path followed by the crack will depend on the magnitude of all the components of the triaxial stress state shown on Figure 2.17a. A crack may rest in a stable position until external loads

Structural Steel 39

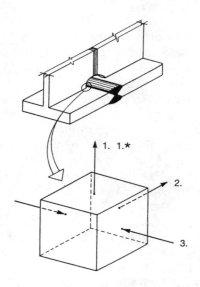

1. Tension due to tendency of flange to distort

1.★ Tension due to high restraint against longitudinal shrinkage of web welds (may not be present if cope holes large enough)

2. Tension: Transverse welding restraint of flange

3. Compression due to longitudinal shrinkage of flange welds

(a)

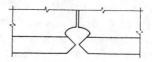

1. BEFORE WELDING THE FLANGES

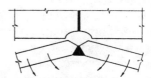

2. WELDING OF THE OUTSIDE OF FLANGES CAUSE TENDENCY TO LIFT

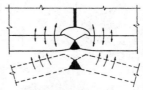

3. BY PULLING ACTION OF THE WEB, THE FLANGE STAYS IN PLACE BUT TENSION RESIDUAL STRESSES ARE INDUCED

(b)

Figure 2.17 Qualitative illustration of (a) triaxial state of stress at the point of cracking initiation and (b) effect of weld shrinkage on internal stresses. *(Bruneau and Mahin 1987.)*

applied to the structure raise the stresses to the threshold whereupon crack propagation will resume. Whether total fracture of flange will occur can be determined through use of fracture mechanics.

2.5.8 Fracture mechanics

Fracture mechanics (i.e., the study of the behavior and strength of solids having crack discontinuities) is a field of engineering unto itself. A comprehensive presentation of this topic is beyond the scope of this book, and the interested reader is referred to the literature (e.g., Rolfe and Barsom 1987). However, although fracture mechanics has become part of the design process in some applications of engineering (for example, in ships and airplane hull design and pressure vessels design), it is seldom used at this time for the design of mainstream civil engineering structures such as buildings and bridges (although fatigue, which is an extension of fracture mechanics, is obviously considered in many design situations). However, once brittle failures are encountered, investigators frequently resort to fracture mechanics to explain the observed behavior.

This state-of-practice is partly a consequence of the difficulty and limitations inherent in the currently available fracture mechanics analysis methods. For example, one simple and effective fracture mechanics approach requires comparison of a stress-intensity-factor, K_I, with the fracture toughness, K_{IC}, of the material (also termed critical stress-intensity-factor). Calculation of K_I for a given specimen and crack geometry requires resorting to empirically developed available solutions (Rooke and Cartwright 1976). K_{IC} must be determined by special fracture toughness tests (ASTM 1985a).

For cracks that propagate through base steel, fracture mechanics has proven to be an effective analytical tool. However, for fractures through welds, K_{IC} of the adjacent steel is irrelevant; K_{IC} of the weld itself is needed, a property sensitive to workmanship. This is a problem in postfailure investigations because the K_{IC} of a weld that has failed cannot be inferred from the other welds that have not fractured. Furthermore, K_{IC} values obtained on small standard welded specimens may be difficult to reliably extrapolate to large multipass welds; cracks tend to simply propagate through the weakest link between the passes in those welds. An example application of fracture mechanics principles to describe the brittle fracture of a high-strength heavy steel section splice with partial penetration welds is presented by Bruneau and Mahin (1987).

2.5.9 Partial penetration welds

Partial penetration welds should be avoided whenever ductile response is required (or implied) by design. When structural members connected by partial penetration welds are loaded to their ultimate strength in tension or flexure, only yielding of the weld is possible at best because the cross-sectional area of the weld is smaller than that of the adjacent base metal. As a result, because the weld's length is rather small, particularly at the root, large ductility of the weld metal translates into small or no ductility at the component level. Moreover, from a fracture mechanics perspective, the unwelded part of the connected metal can act as an initial crack; this cracklike effect at the toe of a typical partial penetration weld can be clearly seen in Figure 2.18 for a partial penetration weld executed on a 89 mm (3.5 inches) thick flange (cut, polished, and etched to allow visual observation of the weld).

The discontinuity of the base metal produced by the partial penetration weld also introduces a significant stress concentration at the toe of the weld. This is a problem particularly in heavy rolled steel sections

Figure 2.18 Polished and etched partial penetration weld on a 90 mm (3.5-in) thick flange.

because their metallurgy suffers from a reduced fracture toughness (K_{IC}), especially in the web-flange core area where the higher stresses occur as a result of this stress concentration. The web-flange core area of a heavy steel section does not get worked to the degree that thinner sections do during the rolling process. It has a "castlike" structure with large grains and inherently less fracture toughness. A higher fracture toughness is obtained away from this core area, closer to the outer surface of the flanges, the flange tips, and the bottom surface of flanges.

This is in spite of adequate and constant yield stress properties throughout the flange.

2.6 Low-cycle versus high-cycle fatigue

2.6.1 High-cycle fatigue

Crystal imperfection, dislocations, and other microcracks in steel can grow into significant cracks in structural components subjected to the action of repeated loads. For example, various components in bridges subjected to heavy traffic loading and off-shore structures subjected to wave loading will be subjected to millions of cycles of load during their service life. These components must be detailed with proper care to provide adequate resistance against crack initiation and crack propagation. The ability of metals and specific welded details to resist many hundreds of thousand cycles of strain below the yield level is termed "high-cycle fatigue resistance." One accomplishes design to ensure this resistance by limiting the maximum stress range due to cyclic loading to values that are usually much below the yield stress and selecting details that minimize stress concentrations. The design of these members can be accomplished according to traditional steel design methods, with stress ranges presented in various standards, and is therefore beyond the scope of this book.

2.6.2 Low-cycle fatigue

Anybody who has ever destroyed paper clips knows that these can be bent back and forth only a limited number of times. Some may even have noticed that the maximum number of cycles these paper clips can resist is a function of the severity of their postyield plastic deformations. The ability of metals to resist a limited and quantifiable number of strain cycles above the yield level is termed "low-cycle fatigue resistance" or "resistance to alternating plasticity." Although both low-cycle fatigue and high-cycle fatigue failures result from similar crack initiation and crack growth mechanisms, only low-cycle fatigue resistance characterizes the cyclic response of components and connections deformed beyond the yield limit and is thus related to ductile design.

There is also a significant difference between the low-cycle fatigue resistance of the material obtained by tests on coupons of the base metal and that of actual structural members. This difference principally arises as a result of the rather large inelastic strains that develop upon local buckling. For example, although tests of coupons indicate that mild steel can resist more than 400 cycles at a strain of 0.025 (ASM 1986), Bertero and Popov (1965) showed that local buckling in a steel beam of compact cross-section in bending started after half a cycle at that strain and that fracture occurred after the 16th cycle. Indeed, in steel sections that can develop their plastic moment capacity (see Chapter 3), local buckling due to cyclic loading will occur at smaller strains than predicted for monotonic loads, because the inelastic curvatures introduced in the flanges of beams subjected to flexural moments above the yield moment will typically act as initial imperfections and help trigger local buckling upon reversal of loading. This explains why smaller maximum width-to-thickness ratios are required by codes for members subjected to cyclic loading (as will be shown later). Fortunately, low-cycle fatigue resistance increases rapidly with reductions in the maximum cyclic strain (Figure 2.19), and steel structures detailed for ductile response will usually be able to resist the expected cyclic inelastic deformation demands without failing by low-cycle fatigue.

2.7 Material models

Once an appropriately ductile steel material has been chosen for a specific application, suitable stress-strain or moment-curvature models must be adopted for the purpose of calculations. As increases in model complexity often translate into additional computational difficulties, simple models may provide, in many cases, sufficiently accurate representations of behavior. Some of the simpler models commonly used are described below.

2.7.1 Rigid plastic model

As the name implies, the rigid plastic model neglects elastic deformations. The material is assumed to experience no strain until the yield stress is reached, and flexural components modeled with rigid plastic behavior undergo no curvature until the plastic moment is reached. This rudimentary model, illustrated in Figure 2.20a, can be useful, for example, when the plastic collapse load capacity of a structure is sought, as will be seen in Chapter 4. It has been used to model nonlinear friction hysteretic energy dissipation, in terms of forces versus displacement, in structures wherein this behavior could be relied upon to resist extreme dynamic lateral loads (e.g., Dicleli and Bruneau 1995). However, because this model implies infinite stiffness until plastification is reached and null stiffness thereafter, it is best suited for hand calculations and can be difficult to implement in structural analysis programs based on the stiffness-matrix method.

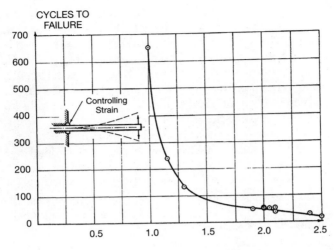

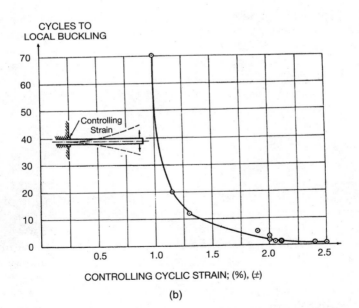

Figure 2.19 Low-cycle fatigue of structural shapes: (a) number of cycles required to attain fracture as a function of the controlling cyclic strain, (b) number of cycles required after which local buckling of flanges was detected as a function of the controlling cyclic strain. *(Figures a – d reprinted from "Effect of Large Alternating Strains of Steel Beams,"* Journal of Structural Engineering, *February 1965, V.V. Bertero and E.G. Popov, with permission of the American Society of Civil Engineers.)*

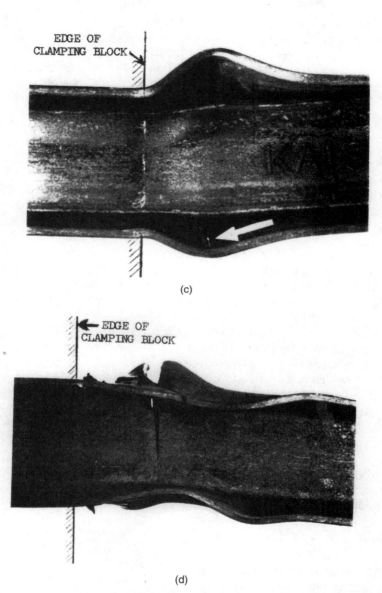

Figure 2.19 Low-cycle fatigue of structural shapes: (c) typical initiation of fracture, (d) typical complete fracture.

2.7.2 Elasto-plastic models

When response under progressive loading until collapse is desired, or when an accurate calculation of nonlinear deflections is needed, an elasto-perfectly plastic model is generally used. This hysteretic model considers two possible stiffness states: elastic or plastic, as shown in Figure 2.20b. The choice of zero post-yield stiffness is suitable for many applications in which strain hardening is not anticipated, and conservative for predicting the plastic collapse load and deformations whenever strain hardening would be expected to develop.

A computer is generally needed to take advantage of the features of this bilinear model, so one can easily consider the effect of strain hardening by assigning a nonzero postyielding stiffness; in fact, for numerical analysis reasons, nonzero positive stiffness is often desirable. Such an elasto-plastic model with nonzero post-yield stiffness to account for strain hardening is illustrated in Figure 2.20c. Various methods can be used to determine an appropriate post-yield stiffness to model material or member behavior. One such reasonable approach, shown in Figure 2.20f, consists of finding the slope of the straight line that would bisect the true stress-strain diagram in the plastic range such that an equal amount of plastic energy is dissipated before failure. To equate the energy, however, iterative procedures are generally needed to determine the maximum strains, ε_{max}, of all the yielded members in the structure, which is by no means a simple task. Therefore, values ranging between 0.005 and 0.05 times elastic stiffness are commonly used for the strain-hardening (post-yield) stiffness in bilinear models. Lower strain-hardening stiffness values will generally produce larger maximum plastic strains and curvatures, while higher values will translate into larger stresses and moments.

As reflected in the available literature, the bilinear elasto-plastic model has been widely used to model the cyclic hysteretic behavior of steel frame structures (although in some ways, this has often been constrained by the limitations of computer programs capable of cyclic hysteretic analysis). More complex piece-wise linear stress-strain models (Figure 2.20d) or other models (Figure 2.20e) have sometimes been used for analysis. However, the presumed additional benefit gained from the use of such models must always be weighed against all of the uncertainties associated with the design process, particularly when earthquakes are responsible for these hysteretic cycles.

2.7.3 Power, Ramberg-Osgood, and Menegotto-Pinto functions

Another class of material models, known as power functions, is particularly useful to analytically describe the experimentally obtained stress-strain relationships of various alloys. Although these functions complicate the characterization of the stress-strain relationship of low-grade steels having a well-defined yield plateau, they can effectively describe the inelastic moment-curvature relationships of members made with these steels, and even the inelastic shear force versus shear

Structural Steel 47

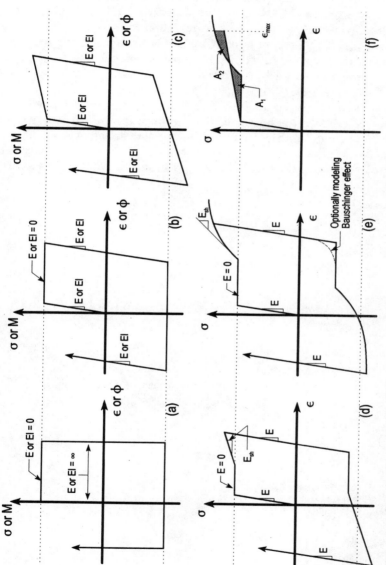

Figure 2.20 Some cyclic material models for structural steel: (a) bilinear rigid-perfectly plastic, (b) bilinear elasto-perfectly plastic, (c) bilinear elasto-plastic with strain-hardening, (d) trilinear elasto-plastic with strain-hardening, (e) cyclic model from monotonic-loading stress-strain curve, and (f) bilinear elasto-plastic with strain-hardening slope based on equal energy dissipation.

strain relationships in some other applications (Nakashima 1995). Therefore, it is worthwhile to be familiar with these power function models.

To remove some of the mathematical abstraction associated with power function models, the discussion is presented within the context of a numerical example. The data tabulated in Table 2.4, obtained from a standard tensile test of an aluminum specimen, is used for this purpose. Although this example is expressed in terms of stresses and strains, note that power functions can also be developed to model other behaviors expressed by a pair of related parameters, such as moments and curvatures, with only minor modifications.

Before a demonstration of how power functions can be used to model the stress-strain behavior of this aluminum alloy, it is worthwhile to note that, in this particular example, an elasto-perfectly plastic model of this material would provide a reasonably accurate representation of the material behavior. Through a graphical method

TABLE 2.4 Stress-strain data for aluminum alloy tensile test and results from various model.

Experimental Data		Power Function	Ramberg-Osgood	Menegotto-Pinto
σ (ksi)	ε	ε	ε	σ (ksi)
0	0	0.0	0.0	0.0
5	0.0005		0.0005	5.0
10	0.00102	0.0010	0.0010	10.2
15	0.00151		0.0015	15.1
20	0.00202	0.0020	0.0020	20.1
25	0.00251		0.0025	24.8
30	0.00301	0.0030	0.0030	29.2
31	0.00311		0.0031	30.0
32	0.00321		0.00321	30.7
33	0.00333		0.00333	31.6
34	0.00345		0.00348	32.4
35	0.0037		0.00371	34.0
36	0.0042	0.0041	0.00416	36.3
37	0.0052	0.0051	0.00516	38.8
38	0.008	0.0074	0.00748	40.1
39	0.015		0.01297	40.4
40	0.0265		0.0258	40.5
41	0.056	0.056	0.0555	41.0
42	0.122		0.123	42.0

(1 ksi = 6.895 μPa)

and the traditional 0.2 percent offset rule, the yield stress, σ_y, is calculated to be 37 ksi (255MPa), and the corresponding yield strain, ε_y, is 0.0037. The resulting elasto-perfectly plastic material curve could be drawn by joining the following three stress-strain (σ, ε) points: (0, 0), (37, 0.0037), and (37, ∞). Likewise, an elasto-plastic strain hardening model could be developed after some trial and error to define an appropriate strain-hardening stiffness.

2.7.3.1 Power functions The fundamental tenet of a power function is that the strain-stress relationship can be divided into an elastic part and a plastic part as shown below.

$$\varepsilon_{total} = \varepsilon_{elastic} + \varepsilon_{plastic} = \frac{\sigma}{E} + a \left(\frac{\sigma}{E}\right)^n \tag{2.6}$$

Terms for this equation are defined in Figure 2.21a. The task therefore lies in determining the parameters a and n that will provide best fit to the data. From the above expression, it transpires that the plastic strain can be expressed as a function of the elastic strain if Equation 2.6 is manipulated, to obtain:

$$\varepsilon_{plastic} = a \left(\frac{\sigma}{E}\right)^n = a \, \varepsilon_{elastic}^n \tag{2.7}$$

or, taking the logarithm of both sides of this equation,

$$\log \varepsilon_{plastic} = \log \left(a \, \varepsilon_{elastic}^n\right) = \log a + n \log \left(\varepsilon_{elastic}\right) \tag{2.8}$$

As illustrated in Figure 2.21b, Equation 2.8 is the equation of a straight line of slope n in a logarithmic space—that is, a straight line when plotted on log-log paper. Because all experimentally obtained data points will not always fall on a straight line, a least-squares solution could be used to find the best approximation to the true stress-strain curve. However, in most practical cases, this level of refinement is not necessary. A reasonably accurate result can be obtained if one takes two strain data points as far apart as possible, but not on the linear branch of the stress-strain curve where $\varepsilon_{plastic}$ would be zero, and calculates the values of a and n using the following relationships:

$$n = \left(\frac{\log \varepsilon_{plastic-2} - \log \varepsilon_{plastic-1}}{\log \varepsilon_{elastic-2} - \log \varepsilon_{elastic-1}}\right) \tag{2.9}$$

and

$$\log a = \log \varepsilon_{plastic-1} - n \log \varepsilon_{elastic-1} \tag{2.10}$$

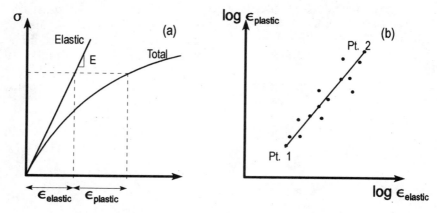

Figure 2.21 Power function material models for structural steel.

For the current example, assuming an elastic modulus of 10,000 ksi and arbitrarily taking for the two strain data points $\varepsilon=0.0037$ (at $\sigma = 35$ ksi, for which $\varepsilon_{elastic}=0.0035$ and $\varepsilon_{plastic}=0.0002$) and $\varepsilon=0.1220$ (at $\sigma = 42$ ksi, for which $\varepsilon_{elastic}=0.0042$ and $\varepsilon_{plastic}=0.1178$), Equations 2.9 and 2.10 give $n = 35.0$ and $a = 1.66 \times 10^{82}$, respectively, which can be substituted in the general power function expression. This gives the following result:

$$\varepsilon = 0.0001\sigma + 1.66 \times 10^{82} \left(\frac{\sigma}{10000}\right)^{35.0} \qquad (2.11)$$

The accuracy of this solution could be verified if it were plotted against the experimental data. However, because the resulting difference is rather small, results are instead compared through use of Table 2.4, and judged to be satisfactory.

Although the resulting expression may appear awkward to manipulate for hand calculations, the ability to express the entire stress-strain relationship by a single continuous function can be advantageous, particularly for computer programming purposes.

2.7.3.2 Ramberg-Osgood functions In earthquake engineering, Ramberg-Osgood functions are often used to model the behavior of structural steel materials and components. These functions are obtained when the above power function is normalized by an arbitrary strain, ε_o, for which the plastic component of strain, $\varepsilon_{plastic}$, is not zero. Generally, the yield strain, ε_y, provides a good choice for this normalizing strain, but other choices are also possible and equally effective. Therefore, the Ramberg-Osgood function is expressed as:

$$\frac{\varepsilon}{\varepsilon_0} = \frac{\sigma}{\sigma_0} + \frac{a}{\varepsilon_0}\left(\frac{\sigma}{E}\right)^n \qquad (2.12)$$

where E is the initial elastic modulus, and σ_o is equal to $E\varepsilon_o$. To define a Ramberg-Osgood model, one must evaluate a, n, and ε_o. Note that, when plotted in a parametric manner, normalized stresses as a function of the normalized strain give curves that meet at the point $(1+a, 1)$, as shown in Figure 2.22a. In that figure, the physical significance of increasing n is also evident: curves with progressively larger values of n will gradually converge toward the elasto-perfectly-plastic curve.

Because the Ramberg-Osgood function is a power function, the calculation of n and a proceeds as described previously: two points must be selected to calculate the slope n of a straight line in the log-log space, from which the constant a can then be calculated. However, taking advantage of the normalized form of the Ramberg-Osgood function, it is possible to derive close-formed expressions for this purpose.

If point A is defined as the point on the real data curve where $\sigma = \sigma_o$ (Figure 2.22b), substituting known values in the above equation at point A and defining m_A as the slope of the secant joining point A with the origin, produces the following expression:

$$\frac{\varepsilon_A}{\varepsilon_0} = 1 + \frac{a}{\varepsilon_0}\left(\frac{\sigma_0}{E}\right)^n = \frac{1}{m_A} \quad \Rightarrow \quad a = \frac{(1-m_A)/m_A}{\left(\frac{\sigma_0}{E}\right)^{n-1}} \qquad (2.13)$$

Similarly, taking a point B sufficiently far removed from point A along the experimentally obtained stress-strain curve (and preferably the last data point of increasing stress), Equation 2.13 becomes:

$$\frac{\left(\frac{\varepsilon_B}{\varepsilon_0}\right)}{\left(\frac{\sigma_B}{\sigma_0}\right)} = 1 + \frac{\frac{a}{\varepsilon_0}\left(\frac{\sigma_B}{E}\right)^n}{\left(\frac{\sigma_B}{\sigma_0}\right)} = \frac{1}{m_B} \qquad (2.14)$$

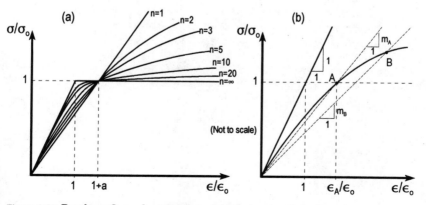

Figure 2.22 Ramberg-Osgood material model for structural steel.

Substituting a from Equation 2.13 into Equation 2.14 and solving for n, gives:

$$n = 1 + \frac{\log \frac{(1-m_A)\, m_B}{(1-m_B)\, m_A}}{\log(\sigma_0/\sigma_B)} \qquad (2.15)$$

where σ_o and σ_A have been constrained to be equal by this procedure. Once the slope n is known, the value a can be directly obtained from Equation 2.13, and the Ramberg-Osgood function is completely defined.

To summarize, a systematic procedure to define a Ramberg-Osgood function can be outlined as follows.

1. Draw the stress-strain diagram (or moment-curvature diagram, or whatever other force-deformation diagrams of interest) for the data and determine the initial elastic modulus, E (or EI, or other parameter relating the selected force and deformation terms per an expression similar to Equation 2.6). If good confidence exists in the accuracy and reliability of the first data points, this value can be determined analytically as the slope to the first data point from the origin. (For the example here, $E=10,000$.)

2. The yield strain can be selected as ε_o. Therefore, use the 0.2 percent offset method to first determine σ_o, and then calculate $\varepsilon_o = \sigma_o/E$. For the example here, the yield stress of 37 ksi and strains of 0.0037 as previously calculated are used.

3. Select point A as the point on the data curve for which $\sigma = \sigma_o$, and calculate the slope, m_A, of the secant joining point A with the origin. Note that for this secant modulus $m_A E$, m_A is always less than unity. Also, when σ_o is graphically obtained through the 0.2 percent offset method, the intersect of the data curve and the 0.2 percent curve directly gives point A. For the example at hand, $\varepsilon_A = 0.0052$ and $m_A = 0.711$.

4. Select a second point (point B) along the experimentally obtained stress-strain curve as one of the last points for which an increase in stress was measured and calculate the secant modulus $m_B E$.

5. Solve for n using Equation 2.15.

6. Solve for a using Equation 2.13.

For this example, $m_B = 0.151$, $n = 34.7$, and $a = 3.53 \times 10^{81}$. The resulting Ramberg-Osgood equation becomes:

$$\frac{\varepsilon}{0.0037} = \frac{\sigma}{37} + \frac{3.53 \times 10^{81}}{0.0037} \left(\frac{\sigma}{10000}\right)^{34.7} \qquad (2.16)$$

which can be simplified to:

$$\varepsilon = 0.0001\sigma + 3.53 \times 10^{81} \left(\frac{\sigma}{10000}\right)^{34.7} \qquad (2.17)$$

The slight difference between this result and that obtained using power functions (Equation 2.11) can be attributed to the choice of different data points to derive the expressions. However, for all intents and purposes, the two equations give identical results, as shown in Table 2.4.

2.7.3.3 Menegotto-Pinto functions One can derive a class of normalized equations, known as Menegotto-Pinto functions, following an approach similar to that presented above, to express stresses as a function of strains, instead of strains as a function of stresses per the Ramberg-Osgood functions. The general form of the Menegotto-Pinto functions is as follows:

$$\frac{\sigma}{\sigma_0} = b\left(\frac{\varepsilon}{\varepsilon_0}\right) + d = b\left(\frac{\varepsilon}{\varepsilon_0}\right) + \frac{(1-b)\left(\frac{\varepsilon}{\varepsilon_0}\right)}{\left[1 + \left(\frac{\varepsilon}{\varepsilon_0}\right)^n\right]^{1/n}} \quad (2.18)$$

where b is the ratio of the final to initial tangent stiffnesses, and d is a value that is graphically defined in Figure 2.23. In the normalized space of stress and strain, the initial stiffness has a slope of 1, the slope of the final tangent stiffness is b, and d varies from zero to $(1-b)$ as $\varepsilon/\varepsilon_0$ progressively increases from zero to a maximum value at the last data point.

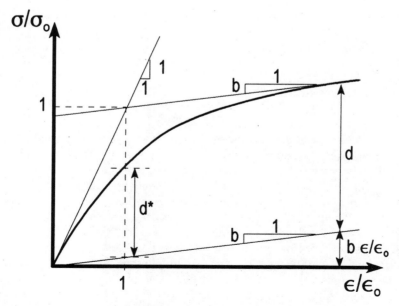

Figure 2.23 Menegotto-Pinto material model for structural steel.

To use Equation 2.18, one must evaluate the values of ε_o, σ_o, b, and n. A simple procedure to determine this needed information is outlined below:

1. Draw the initial and final tangent slopes to the data curves. The intersection of these two curves gives ε_o and σ_o, which should not be confused with the same variables used earlier for the Ramberg-Osgood case because these have a rather different meaning and cannot be chosen arbitrarily here. For this example, $\varepsilon_o = 0.004$ and $\sigma_o = 40.21$. In this case, it is possible for one to calculate these values using secants to the first and last pairs of data points, but normally, the engineer should make sure that these data points are reliable and representative.

2. Calculate the ratio of the tangent slopes. Here, this result is simply the slope of the last pair of data points, and therefore $b = 0.00152$.

3. At the point $\varepsilon/\varepsilon_o = 1$, note that Equation 2.18 simplifies to:

$$\frac{\sigma}{\sigma_0} = b\,(1) + d^* = b\,(1) + \frac{(1-b)(1)}{[1 + (1)^n]^{1/n}} = b + \frac{(1-b)}{2^{1/n}} \tag{2.19}$$

The stress σ at strain ε_o is obtained directly from the data (although some interpolation may be needed), and the value of d^* is calculated directly from Equation 2.19. In this example, σ is found to be 35.6, and therefore d^* is equal to 0.884. With knowledge of the value of d^*, one can compute n from the above expression as:

$$n = \frac{\log 2}{\log (1-b) - \log d^*} \tag{2.20}$$

Here, this gives $n = 5.68$. The resulting Menegotto-Pinto expression becomes:

$$\sigma = 15.17\varepsilon + 9985\,\frac{\varepsilon}{[1 + 249\varepsilon^{5.68}]^{1/5.68}} \tag{2.21}$$

This model is slightly less accurate than the power models presented earlier, as evidenced by the results summarized in Table 2.4, but differences are not of practical significance.

All the multilinear and continuous models presented above to describe monotonic response can be extended to capture cyclic behavior. The concepts and equations are similar, with the difference that "memory" must be introduced in the models to keep track of the strength level and extent of yielding (or accumulated plastic deformation history) at each point of load-reversal. Alternatively, for more complex structural elements, customized models of cyclic inelastic element behavior have been developed to account for multiple phases of plastic behavior.

2.8 Advantages of plastic material behavior

To illustrate some of the advantages of plastic material behavior, the three-bar truss of Figure 2.24a is subjected to an increasing statically applied load, F, up to its ultimate capacity. Incidentally, two- or three-bar trusses have been used by many authors to explain simple plasticity concepts. The simple truss of Figure 2.24 is one such classical example (e.g., Prager and Hodge 1951, Massonnet and Save 1967, Mrazik et al. 1987) that has been treated mostly from the perspective of noncyclic loading. However, here, behavior throughout at least one full plastic cycle is discussed, considering a simple elasto-perfectly plastic model. The section and material properties needed for this problem are indicated in Figure 2.24. All three bars are of same material (E and σ_y) and cross-sectional area (A). Moreover, the bars are assumed to be sufficiently stocky such that buckling will not occur prior to yielding in compression.

Although this structure is statically indeterminate, the force and elongation in member 3 will always be equal to that in member 1 as a result of structural and loading symmetry. Therefore, using small deformation theory, one can directly write the equation of static equilibrium for this problem as:

$$F = P_2 + 2P_1 \cos 45° = P_2 + 1.414\, P_1 \tag{2.22}$$

Likewise, geometric compatibility here gives:

$$\Delta_1 = \Delta_3 = \Delta_2 \cos 45° = \Delta \cos 45° = 0.707\, \Delta \tag{2.23}$$

where F and Δ are defined in Figure 2.24. These two relationships must always remain true as long as deformations do not reach a magnitude that would violate the assumptions of small deformation theory. This will be reassessed at the end of this example. What will change as plastic behavior develops is the stress-strain or force-deformation relationship for each structural member. As long as structural members remain elastic, the following relationships are valid:

$$P_1 = \frac{AE}{L_1}\Delta_1 = \frac{AE}{L/\cos 45°}\Delta \cos 45° = \frac{AE}{2L}\Delta \tag{2.24}$$

and

$$P_2 = \frac{AE}{L_2}\Delta_2 = \frac{AE}{L}\Delta \tag{2.25}$$

Therefore, until first yielding of any member:

$$F = P_2 + 2P_1 \cos 45° = \Delta + 2\frac{AE}{L}\Delta \cos 45°$$

$$= \frac{AE}{L}(1 + \cos 45°)\Delta = K\Delta \tag{2.26}$$

56 Chapter Two

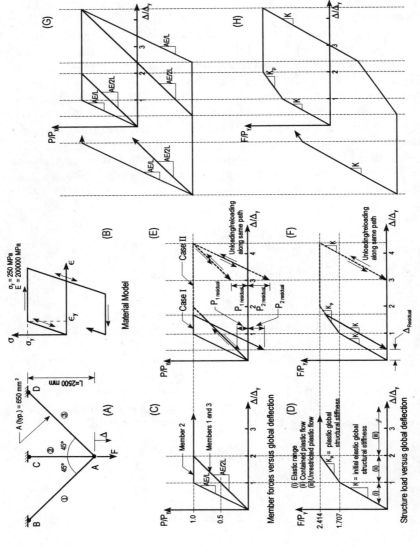

Figure 2.24 Simple example to illustrate cyclic plastic behavior of indeterminate structure.

where K is the elastic stiffness for this structure. The equation above shows that member 2 reaches its yield strength (P_y) first, when members 1 and 3 are stressed to only half their capacity (i.e., 0.5 P_y). The yield displacement of the structure (Δ_y) could be defined in terms of the truss bar's properties and geometry:

$$\Delta_y = \frac{P_y L}{AE} \tag{2.27}$$

where all terms have been previously defined. The state of member forces and applied load as a function of the vertical deflection at point A are presented in a normalized format in Figures 2.24c and 2.24d. At $\Delta=\Delta_y$, the structure has been loaded to $F = 1.707\, P_y$. This case reflects the current standard structural engineering practice worldwide, in which linear elastic analyses are conducted to determine the maximum load that can be applied to a structure until the capacity of a single member is reached. Here, the capacity of member 2 is reached first, and members 1 and 3 are underutilized. Note that although one could conclude from elastic analysis that having bars of the same size is inefficient and likely propose a solution with bars of different cross-sections, this type of optimization based on minimum weight is often at odds with practical considerations, where architectural, manufacturing, or construction concerns may dictate equal size bars. Hence, in this problem, it is reasonable to constrain the bars to be of identical cross-section.

Even though truss bar 2 has yielded, the load applied to the structure can be increased beyond 1.707 P_y because members 1 and 3 have not reached their ultimate capacity. In the process, member 2 will undergo plastic elongation while sustaining its capacity. Therefore, for $\Delta > \Delta_y$:

$$P_1 = \frac{AE}{2L}\Delta \quad \text{and} \quad P_2 = P_y \tag{2.28}$$

and

$$F = P_2 + 2P_1 \cos 45° = P_y + \frac{AE}{L} \Delta \cos 45° \tag{2.29}$$

This condition of further loading beyond the elastic range, called contained plastic flow, is illustrated in Figures 2.24c and 2.24d. Note that the stiffness of the structure has decreased as a result of this partial yielding. Eventually, when the applied load reaches 2.414 P_y, all three truss bars yield, and a condition of unrestrained plastic flow develops. In this example, the ultimate capacity calculated by tolerating plastification of the structure is 41 percent larger than the elastic limit, providing some evidence that it is generally worthwhile to take advantage of the existing plastic strength available beyond the elastic range.

58   Chapter Two

It is also important to realize that the maximum plastic capacity of a structure can be calculated directly through use of static equilibrium alone, without knowledge of prior elastic or plastic loading history. A logical conclusion from this observation is that lack-of-fit and other displacement-induced forces have no influence on the magnitude of the maximum load a structure can support, termed the plastic capacity. No supporting calculations are presented here, but the reader may wish to verify this remark by redoing the example assuming that member 2 was constructed 25 mm too short and forced to fit into the frame during construction before any loads were applied. Although doing so would reduce the value of the maximum load that can be applied if the structure is to remain elastic, it does not change the plastic capacity of the structure.

The behavior of this plastified truss-bar structure upon unloading is investigated next. As described in an earlier section of this chapter, steel unloads at a stiffness equal to the original elastic stiffness (as also shown in Figure 2.24b). However, in this example, two scenarios are evaluated, considering various states of plastification prior to unloading. In both cases, because the principle of superposition is not valid beyond the elastic range (as evidenced by Figure 2.24d), calculations must proceed following basic principles.

First, if unloading to $F = 0$ starts before all three truss bars have yielded (shown as Case I in Figures 2.24e and 2.24f), it can be shown that:

$$P_1 = \frac{AE}{2L}\Delta_{MAX} - \frac{AE}{2L}\delta\Delta$$

$$P_2 = P_y - \frac{AE}{L}\delta\Delta \qquad (2.30)$$

$$\delta\Delta = \Delta_{MAX} - \Delta$$

where $\delta\Delta$ is the incremental displacement imposed by the unloading, and Δ is the resulting displacement after unloading. The above approach also recognizes that only member 2 had plastified prior to unloading. Because the objective is to unload to $F=0$, the equilibrium condition developed earlier can be used to solve for all other variables. As a result:

$$F = 0 = 2\left[\frac{AE}{2L}\Delta_{MAX} - \frac{AE}{2L}(\Delta_{MAX} - \Delta)\right]\cos 45° + \left[P_y - \frac{AE}{L}(\Delta_{MAX} - \Delta)\right] \qquad (2.31)$$

and solving for Δ:

$$\Delta = \frac{\frac{AE}{L}\Delta_{MAX} - P_y}{\frac{AE}{L}(1 + \cos 45°)} = \frac{\Delta_{MAX} - \Delta_y}{(1 + \cos 45°)} = 0.586\,(\Delta_{MAX} - \Delta_y) \qquad (2.32)$$

This residual displacement after unloading is shown in Figures 2.24e and 2.24f. Based on Figure 2.24g, it could be demonstrated that the residual displacement is also equal to:

$$\Delta = \Delta_{MAX} - \frac{F_{MAX}}{K} \tag{2.33}$$

where K is the global elastic structural stiffness defined earlier. Numerical examples would confirm the expediency of this alternative procedure.

Moreover, as shown in Figure 2.24e, residual member forces remain after unloading and exist simultaneously to that state of residual deformations. Because no external loads are applied to the structure, these member forces must be self-equilibrating. Numerical values can be obtained through simple substitution in the above equations.

In the second scenario considered here, unloading to $F = 0$ starts after all members have yielded (shown as Case II in Figures 2.24e and 2.24f). The approach remains the same, with slightly different results, as:

$$F = 0 = 2\left[P_y - \frac{AE}{2L}(\Delta_{MAX} - \Delta)\right]\cos 45° + \left[P_y - \frac{AE}{L}(\Delta_{MAX} - \Delta)\right] \tag{2.34}$$

and:

$$\Delta = \frac{\Delta_{MAX}(1 + \cos 45°) - \Delta_y(2\cos 45° + 1)}{(1 + \cos 45°)} = \Delta_{MAX} - 1.414\Delta_y \tag{2.35}$$

Again, this result could be obtained directly (and more easily) through use of the alternative method suggested above.

Note that substantial residual member forces develop in this second example (Figure 2.24e). Calculations would show that these reach approximately 30 percent and 40 percent of the yield force, P_y, in members 1 and 2 respectively.

As a result of the residual stresses left after unloading in Cases I and II, the truss bars would follow their elastic path until the previously applied maximum load is reached anew if the structure is reloaded in the same direction (Figures 2.24e and 2.24f). This has a number of practical implications, as shown in the following chapters. However, this extended elastic range exists only as long as the state of residual stresses in the bars at $F = 0$ remains unchanged; cyclic loading is one factor that can alter these residual stresses.

When the loading cycle for this example structure is completed, the elastic range for the reversed loading is reduced because of the residual compression force in member 2. Formal analytical expressions to demonstrate this subsequent behavior can be derived from the same basic principles presented above, but it should be obvious that complex calculations and formulations are not necessary given the clear graphical relationship already established in this example between the load-deformation relation of the structure and that of the members, as shown in Figure 2.24. Therefore, Figures 2.24g and 2.24h can be drawn directly as extensions of parts e and f, through use of simple trigonometric calculations to obtain numerical values. Note that the same structural plastic capacity is reached at $F = 2.414 P_y$ in the reverse direction despite the earlier threshold of yielding, but this required greater contained plastic deformations.

More in-depth observations on the behavior of structural members subjected to inelastic cyclic loading are reserved for later chapters. Nonetheless, it is worthwhile to emphasize at this point that the plastic analysis presented above is possible only if connections of the structural members are stronger than the members themselves. Although this may sound obvious in the context of this simple example, this logical requirement sometimes become controversial when the severity of the expected ultimate loads is hidden in more simple static loading models or design philosophies (such as working stress design).

Finally, it is instructive to review the assumption of small deformations. For mild steels typically used in structural engineering applications, strains of up to 20 percent are possible. However, in this problem, calculations show that near this extreme level of deformations (corresponding to $100\varepsilon_y$), the angle of members 1 and 3 to the vertical is approximately 40°. Equilibrium in the deformed configuration gives a maximum value of F barely 5 percent greater. Such a gain in strength is hardly sufficient to warrant the consideration of large deformations theory, especially in the more practical range of 2 percent strains. A more important issue is the development of strain hardening in the truss bars and its tangible impact on member strength. The effect of strain hardening is addressed as necessary in this book.

2.9 References

1. AISC. 1973. "Commentary on Highly Restrained Welded Connections." *Engineering Journal*. Vol. 10, No. 3: 61–73.
2. AISC. 1994. "Load and Resistance Factor Design. Manual of Steel Construction, 2nd ed.," American Institute of Steel Construction. Chicago.
3. AISC. 1997a. "Steel Industry Announces Improved Structural Grade for Buildings." *Modern Steel Construction*. Vol. 37, No. 4: 18.
4. AISC. 1997b. *Technical Bulletin #3—New Shape Material*. American Institute of Steel Construction.
5. ASCE. 1992. "Structural Fire Protection," ASCE Manuals and Reports on Engineering Practice No. 78. T.T. Lie, Editor. ASCE Committee on Fire Protection. American Society of Civil Engineers. New York.
6. ASM. 1986. *Atlas of Fatigue Curves*. Edited by H.E. Boyer, Materials Park, Ohio: American Society for Metals.

7. ASTM. 1985a. "Standard Method of Test for Plane-Strain Fracture Toughness of Metallic Materials—ASTM Designation E399-83," Vol. 03.01, ASTM Annual Standards, American Society for Testing and Materials.
8. ASTM. 1985b. "Standard Method for Conducting Drop Weight Test to Determine Nil-Ductile Transition Temperature of Ferric Steels—ASTM Designation E208," ASTM Annual Standards, American Society for Testing and Materials.
9. AWS. 1990. "Welding Specifications," AWS D1.1-90> Miami: American Welding Society.
10. AWS. 1976, "Fundamentals of Welding," *Welding Handbook:* Vol. 1, 7th Edition. Miami: American Welding Society.
11. Bertero, V.V., and Popov, E.P. 1965. "Effect of Large Alternating Strains of Steel Beams," *Journal of the Structural Division.* American Society of Civil Engineers, Vol. 91, No. ST1.
12. Bjorhovde, R. 1987. "Welded Splices in Heavy Wide-Flange Shapes," National Engineering Conference & Conference of Operating Personnel. New Orleans.
13. Blodgett, O.W. 1977a. "Distortion ... How to Minimize It with Sound Design Practices and Controlled Welding Procedures plus Proven Methods for Straightening Distorted Members," Publication G261. Cleveland: The Lincoln Electric Co.
14. Blodgett, O.W. 1977b. "Why Welds Crack—How Cracks Can Be Prevented," Publication G230. Cleveland: The Lincoln Electric Co.
15. Bruneau, M.; Mahin, S.A.; and Popov, E.P. 1987. "Ultimate Behavior of Butt Welded Splices in Heavy Rolled Steel Sections," EERC Report No. 87-10. Earthquake Engineering Research Center. California: University of California at Berkeley.
16. Dafalias, Y.F. 1992. "Bounding Surface Plasticity Models for Steel under Cyclic Loading," *Stability and Ductility of Steel Structures Under Cyclic Loading.* Fukomoto and Lee, Editors. Boca Raton, Florida: CRC Press.
17. Dicleli, M., and Bruneau, M. 1995. "An Energy Approach to Sliding of Single-Span Simply Supported Slab-on-Girder Steel Highway Bridges with Damaged Bearings," *Earthquake Engineering and Structural Dynamics,* Vol. 24: 395–409.
18. Englehardt, M.D., and Husain, A.S. 1993. "Cyclic Loading Performance of Welded Flange Bolted Web Connections," *ASCE Structural Engineering Journal,* Vol. 119, No. 12: 3537–3550.
19. Fisher, J.W., and Pense, A.W. 1987. "Experience with Use of Heavy W Shapes in Tension," *A.I.S.C. Engineering Journal,* Vol. 24, No. 2: 63–77.
20. Galambos, T.V., and Ravindra, M.K. 1978. "Properties for Steel Used in LRFD," *Journal of the Structural Division, ASCE,* Vol. 104, No. 9: 1459–1468.
21. Lay, M.G. 1965. "Yielding of Uniformly Loaded Steel Members," *Journal of the Structural Division, Proceedings of the ASCE.* Vol. 91, No. ST6: 49–66.
22. Lee, G.C.; Chang, K.C.; and Sugiura, K. 1992. "The Experimental Basis of Material Constitutive Laws of Structural Steel under Cyclic and Nonproportional Loading," *Stability and Ductility of Steel Structures Under Cyclic Loading.* Fukomoto and Lee, Editors. Boca Raton, Florida: CRC Press .
23. Massonnet, Ch., and Save, M. 1967. "Calcul Plastique des Constructions—Volume 1: Structures Dépendant d'un Paramètre," Deuxième Édition. Centre Belgo-Luxembourgeois d'information de l'Acier, Bruxelles, 547 pp.
24. Mizuma, T.; Shen, C.; Tanaka, Y.; and Usami, T. 1992. "A Uniaxial Stress-Strain Model for Structural Steels under Cyclic Loading." *Stability and Ductility of Steel Structures Under Cyclic Loading.* Fukomoto and Lee, Editors. Boca Raton, Florida: CRC Press.
25. Mrazik, A.; Skaloud, M.; and Tochacek, M. 1987. "Plastic Design of Steel Structures," New York: Halsted Press, John Wiley & Sons, p. 637.
26. Nakashima, M.; Iwai, S.; Iwata, M.; Takeuchi, T.; Konomi, S.; Akazawa, T.; and Saburi, K. 1994. "Energy Dissipation Behavior of Shear Panels Made of Low Yield Steel." *Earthquake Engineering and Structural Dynamics,* Vol. 23: 1299–1313.
27. NBS. 1980. "Development of a Probability-Based Criterion for American National Standard A58 Building Code Requirements for Minimum Design Loads in Buildings and Other Structures," NBS Special Publication 577, National Bureau of Standards, U.S. Department of Commerce.

28. Popov, E.P. 1968. "Introduction to Mechanics of Solids." Englewood Cliffs, New Jersey: Prentice Hall.
29. Popov, E.P., and Stephen, R.M. 1971. "Cyclic Loading of Full Size Steel Connections," Bulletin No. 21, Washington, DC: American Iron and Steel Institute.
30. Popov, E.P., and Stephen, R.M. 1976. "Tensile Capacity of Partial Penetration Welds," EERC Report No. 70-3, Earthquake Engineering Research Center. California: University of California at Berkeley.
31. Prager, W., and Hodge, P.G. 1951. "Theory of Perfectly Plastic Solids." New York: John Wiley & Sons: 265.
32. Rolfe, S. T., and Barsom, J.M. 1987. "Fracture and Fatigue Control in Structures—Application of Fracture Mechanics." Englewood Cliffs, New Jersey: Prentice Hall.
33. Rooke, D.P., and Cartwright, D.J. 1976. "Compendium of Stress Intensity Factors," Procurement Executive, Ministry of Defense, Her Majesty's Stationery Office. London.
34. SAC. 1995. "Advisory No. 3—Steel Moment Frame Connections," Report SAC-95-01, SAC Joint Venture, Sacramento, California.
35. Timoshenko, S.P. 1983. "History of Strength of Materials." New York: Dover Publications.
36. Tuchman, J.L. 1986. "Cracks, Fractures Spur Study." Engineering News Record. August 21.
37. Van Vlack, L.H. 1980. "Elements of Materials Science and Engineering." Reading, Massachusetts: Addison-Wesley.
38. Wright, W. 1996. "High Performance Steels for Highway Bridges," Proc. 4th National Workshop on Bridge Research in Progress. Buffalo, New York: 191–194.

Chapter 3

Plastic Behavior at the Cross-Section Level

Once material properties have been obtained and an appropriate stress-strain analytical model has been formulated, plastic capacities can be calculated at the cross-section level. This is a crucial phase of plastic analysis. The level of sophistication embraced in the calculation of these capacities will have the foremost impact on the resulting member and structural plastic strengths. As plastic analysis is generally used to compute ultimate structural capacities, erroneous and potentially dangerous conclusions can be reached if an overly simplistic cross-sectional model is used, thus luring the engineer into a false sense of security. Hence, it is worthwhile to review how various expressions can be derived for these cross-sectional properties.

3.1 Pure flexural yielding

The basic case of pure flexural yielding should be familiar to engineers who have used Limit States Design or Ultimate Strength Design. Nonetheless, it is reviewed here because it provides some of the building blocks necessary to understand the more complex models presented later in the chapter.

A number of simplifying assumptions are made to calculate the plastic moment capacity as follows:

- Plane sections remain plane; even though plastic deformations are typically larger than elastic deformations, their overall magnitudes are still sufficiently small to satisfy this condition.
- Structural members must be prismatic and have at least one axis of symmetry parallel to the direction of loading.
- Members are subjected to uniaxial bending under monotonically increasing loading (neutral axis perpendicular to axis of symmetry).

- Shearing deformations are negligible.
- Member instability (flange local buckling, web local buckling, lateral-torsional buckling) is avoided.
- Structural members must not be subjected to axial, torsional, or shear forces. This assumption will obviously be relaxed in the later sections of this chapter.

The elasto-perfectly plastic model is generally used to calculate plastic moment capacities and leads to acceptable results for most practical problems in structural engineering (ASCE 1971). It is therefore used in the following derivations. However, nothing precludes the consideration of more complex models (or a smaller number of fundamental assumptions for that matter) as the corresponding cross-sectional strengths and other relevant plastic properties could be determined from the same basic concepts presented in this book.

3.1.1 Doubly symmetric sections

Doubly symmetric sections constitute the simplest cross sections for which analytical expressions of plastic strength can be developed, because their neutral axis will always be located at their geometric centroid. With an elasto-perfectly plastic material model, for a given cross section, the stress-diagrams can be constructed directly from the strain diagrams, as shown in Figure 3.1, for various levels of increasing strains on an arbitrary doubly symmetric structural shape. At any point along a given loading history (such as that shown in Figure 3.1), the cross-section curvature, ϕ, can be calculated by simple geometry, as:

$$\phi = \frac{\varepsilon_{max}}{(h/2)} \tag{3.1}$$

where ε_{max} is the maximum strain acting over the doubly symmetric cross section of depth h. If at a given curvature the yield strain, ε_y, is located at a distance $y*$ from the neutral axis, then, as shown in Figure 3.1, by similar triangles:

$$\phi = \frac{\varepsilon_{max}}{(h/2)} \frac{\varepsilon_y}{y^*} \quad ; \quad y^* = (h/2) \frac{\varepsilon_y}{\varepsilon_{max}} \tag{3.2}$$

Once $y*$ is known for a given curvature, it becomes a simple matter to compute the corresponding moment using simple mechanics of materials principles:

$$M = \int_{-h/2}^{h/2} \sigma y \, dA = \int_{-h/2}^{h/2} \sigma y \, b(y) \, dy \tag{3.3}$$

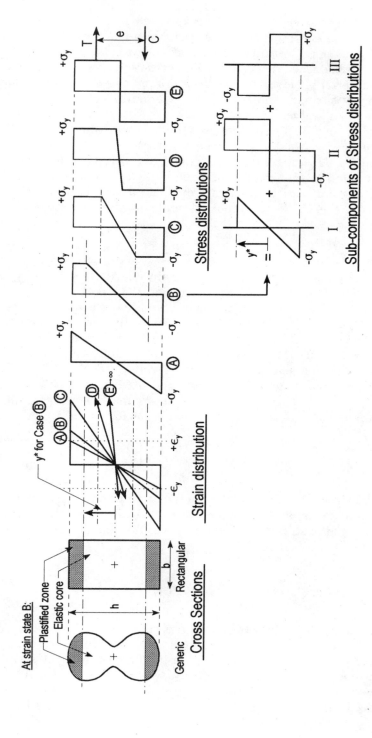

Figure 3.1 Strain and stress distributions in the plastic range for members of symmetric cross section.

where b is the cross-sectional width, expressed as a function of y.

For any given curvature, the cross section can be divided into elastic and plastic zones. Within $\pm y^*$ of the neutral axis, the section is elastic. Outside this zone, also termed the elastic core, strains exceed ε_y, and the material is plastified. As curvature increases, the elastic core progressively shrinks, and plastification progressively spreads over the entire cross section.

The impact of this progressive growth of the plastified zone on the moment-curvature relationship is best understood from a case study. For simplicity, a rectangular cross section is used. The calculations are directly related to the strain profiles identified in Figure 3.1, which use a rectangular section of height h and base b. For that example, in the elastic range, the flexural moment and curvature are given by:

$$M = S\sigma = \left(\frac{bh^2}{6}\right)\sigma \quad ; \quad \phi = \frac{M}{EI} = \left(\frac{12}{bh^3}\right)\frac{M}{E} \quad (3.4)$$

where S is the section modulus. For strain distribution A in Figure 3.1, the yield strain has just been reached at the top fiber of the cross section. Replacing σ by σ_y in the above equation will give the yield moment M_y and the yield curvature ϕ_y: the transition point between purely elastic and elasto-plastic behavior. From that point onward, any increase in curvature introduces partial plastification of the cross section, a condition also sometimes called contained plastic flow. For example, for an arbitrary strain profile represented by B in Figure 3.1, the curvature is given by equation 3.1 above, and the corresponding moment can be expressed as a function of y^* as follows:

$$M = 2\int_0^{y^*} \sigma\, y\, b\, dy + 2\int_{y^*}^{h/2} \sigma_y\, y\, b\, dy \quad (3.5)$$

where the first term reflects the contribution of the elastic core to moment resistance, and the second term, that of the plastified zone. Using the relationship $\sigma/\sigma_y = y/y^*$ valid over the elastic core, equation 3.5 becomes:

$$M = \left[\frac{2b}{y^*}\int_0^{y^*} y^2\, dy + 2b\int_{y^*}^{h/2} y\, dy\right]\sigma_y = \frac{2}{3}b(y^*)^2\sigma_y + \frac{bh^2}{4}\sigma_y - b(y^*)^2\sigma_y \quad (3.6)$$

Interestingly, the three terms on the right side of this equation correspond to the contributions to the moment that can be obtained by a piecewise decomposition of the stress diagram into the simpler subdiagrams I, II, and III, respectively, shown in Figure 3.1. Adding and subtracting stress-diagrams in this manner is a statically correct procedure that can prove useful to simplify complicated problems or to graphically verify the adequacy of analytically derived results.

To complete the above derivation, it is worthwhile to regroup the terms in $(y^*)^2$ and express the results in terms of curvature. Thus, when the following relationship is used (again, obtained through use of the properties of similar triangles):

$$\frac{y^*}{h/2} = \frac{\varepsilon_y}{\varepsilon_{max}} = \frac{\phi_y}{y}\frac{y}{\phi} = \frac{\phi_y}{\phi} \qquad (3.7)$$

the expression for the flexural moment at a given magnitude of curvature (i.e., the moment-curvature relationship) becomes:

$$M = \left[\frac{bh^2}{4} - \frac{bh^2}{12}\left(\frac{\phi_y}{\phi}\right)^2\right]\sigma_y \qquad (3.8)$$

Whenever possible, a solution expressed in a normalized manner that is, in terms of both (M/M_y) and (ϕ/ϕ_y), should be sought. For example, when Equation 3.4 is used to obtain an expression for the yield moment, the final result becomes:

$$\frac{M}{M_y} = \frac{3}{2}\left[1 - \frac{1}{3}\left(\frac{\phi_y}{\phi}\right)^2\right] = \frac{3}{2}\left[1 - \frac{1}{3}\left(\frac{\varepsilon_y}{\varepsilon_{max}}\right)^2\right] \qquad (3.9)$$

The latter part of that equation, expressed in terms of the strain-ratio, is obtained if one takes advantage of the simple equivalence that exists between normalized curvatures and strains.

When Equations 3.4 and 3.9 are used over their respective range of validity, the entire moment-curvature relationship for this cross section can be calculated and plotted. This has been done in Figure 3.2 for different sections. Figure 3.1 and Equation 3.9 show that, theoretically, for any cross section, full plastification and maximum flexural moment will be reached only at an infinite curvature (Case E in Figure 3.1). For the rectangular cross section, this maximum moment is 1.5 times the yield moment, as can be easily deducted from Equation 3.9. However, for practical purposes, as seen in Figure 3.2, this maximum moment is rapidly approached and nearly reached at only three or four times the yield curvature. In fact, when the maximum strain over the cross section approaches the onset of strain hardening of the steel material, at approximately 10 times the yield strain value (as mentioned in the previous chapter), Equation 3.9 indicates that 99.7 percent of the maximum moment has been reached. This demonstrates that a fully plastified cross section can reliably be used to calculate the maximum moment, referred to hereafter as the "plastic moment."

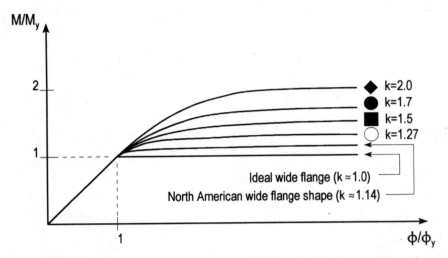

Figure 3.2 Normalized moment curvature relationship and flexural shape factor, k, for different cross sections.

For example, for the rectangular cross section, one can calculate the plastic moment directly using the resulting forces and lever-arms corresponding to the stress distribution E of Figure 3.1. This gives:

$$M_p = 2\,T(e/2) = 2\big[\sigma_y\,(h/2)\,b\big](h/4) = \frac{bh^2}{4}\sigma_y = Z\sigma_y = 1.5\,M_y \qquad (3.10)$$

where Z is the plastic section modulus, a geometrical property of any given cross section. Another useful sectional property is given by the shape factor, k, which is the ratio between the plastic section modulus and the elastic section modulus. This factor, expressed by:

$$k = \frac{M_p}{M_y} = \frac{Z}{S} \qquad (3.11)$$

provides information on the additional cross-sectional strength available beyond first yielding. The shape factors for various cross sections are shown in Figure 3.2.

For the wide-flange sections typically used in North American steel construction (AISC 1992, CISC 1995), the shape factors typically vary from 1.12 to 1.16, with an average of 1.14. Moreover, the following normalized moment curvature expressions could be derived:

1. When y^* is located in the flange:

$$\frac{M}{M_y} = \frac{\phi}{\phi_y}\left(1 - \frac{bd^2}{6S}\right) + \frac{bh^2}{4S}\left[1 - \frac{1}{3}\left(\frac{\phi}{\phi_y}\right)^2\right] \qquad (3.12)$$

2. When y^* is located in the web:

$$\frac{M}{M_y} = \frac{M_p}{M_y} - \left(\frac{wd^2}{12S}\right)\left(\frac{\phi_y}{\phi}\right)^2 \qquad (3.13)$$

where b is the flange width, d is the total depth of the structural section, h is the distance between flanges (or web length), w is the web thickness, and all other terms have been defined previously.

Basic principles of mechanics of materials indicate that an ideal wide-flange section for flexural resistance would have all its material concentrated in flanges of infinitely small thicknesses (obviously an impractical theoretical case). The shape factor of such an ideal section would be unity, because the entire cross section would reach the yield strain simultaneously, without any possible spread of plasticity given that no material would exist between these flanges. Hence, both the plastic and the elastic section moduli in that case would be equal to the area of one flange times the distance between the two flanges.

3.1.2 Sections having a single axis of symmetry

The procedure developed above is applicable for sections having only a single axis of symmetry parallel to the applied load, with the essential difference that the location of the neutral axis must now be determined explicitly. Therefore, for any given curvature, the calculations must start by a determination of the neutral axis. There are two instances when this calculation is relatively simple. First, as long as the material is entirely linear-elastic, the location of the neutral axis remains located at the center of geometry of the cross section. Second, when this same material is fully plastic, the neutral axis must be located such that it evenly divides the cross-sectional area. This latter case is simply a consequence of having the entire cross section subjected to the yield stress, in either tension or compression. Summing the axial forces on the cross section:

$$P = \int_{Area} \sigma \, dA = A_{tension}\, \sigma_y + A_{compression}\,(-\sigma_y) = 0 \quad \Rightarrow \quad A_{tension} = A_{compression} \qquad (3.14)$$

which defines the location of the neutral axis in the fully plastic condition.

In the transition phase between the fully elastic and fully plastic conditions, where plastification progresses through the cross section as the applied moment is increased, the neutral axis progressively migrates. This is schematically illustrated in Figure 3.3. For these

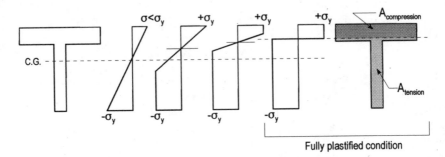

Figure 3.3 Migration of neutral axis in singly symmetric cross section as applied moment increases during progressive plastification.

intermediate cases, the development of analytical moment-curvature expressions, although possible, can be rather complex, except for the simplest cases.

Therefore, even though explicit moment-curvature relations can be derived, as done in Section 3.1.1, for many doubly or singly symmetric cross-sectional shapes, it is sometimes more convenient to simply calculate the moments corresponding to a large number of curvatures and fit the data using a Ramberg-Osgood or Menegotto-Pinto function. When using hand calculations, one can obtain each moment-curvature point by selecting a value for the strain at the top of the cross section, arbitrarily choosing a location for the neutral axis, calculating the axial force resulting from that assumed strain diagram, and iterating by changing the position of that neutral axis until zero axial force is obtained. By repeating that process for various strain values, one can plot the moment-curvature relationship for a given cross section. Following the same logic, moment-curvature points can be generated through the use of computer programs. The approximate moment-curvature relation obtained in that manner would still be sufficiently accurate to allow precise computation of member stiffnesses or deflections.

There exist numerous computer programs capable of developing moment-curvature relations for arbitrary cross sections under uniaxial or biaxial bending, and these are often capable of considering axial and shear forces and other factors. Such programs can also easily be written because their structure is generally rather simple. For example, for an arbitrary section having a single axis of symmetry (parallel to the applied load) and subjected to uniaxial bending, it is sufficient to use a layered model of the cross section. This consists simply of "slicing" the cross section into a large number of layers, say 1000, and calculating and integrating the contributions of all layers to the flexural moment at a given curvature. To structure such a program, one could write subroutines to accomplish the following tasks:

- Perform an automatic layering of the cross section (based on simple input of geometry characteristics)
- Initialize the stress values for all layers and establish other initial parameters
- Set up controls for the iteration strategy
- Increment the curvature for calculation of a given moment-curvature data point
- Estimate the location of the neutral axis, adjusting this estimate in accordance with an iteration strategy (i.e., considering the results from previous iterations)
- For given curvature and neutral axis location, calculate strains for all layers
- Calculate stresses for all layers per the assumed material model
- Calculate the resulting moment by summing the contribution of all layers about the neutral axis
- Calculate the resulting axial force on the cross section by summing the contribution of all layers
- Check for convergence using, for example, a user-specified tolerance on the axial force
- Iterate until the axial force is equal to zero, within the specified tolerance. Convergence gives a single M and ϕ point and corresponding stresses at all layers of the cross section (i.e., stress distribution). Repeat calculations at other curvatures to obtain the entire M-ϕ curve.

Note that any material model could be implemented in such a computer program, although the simple elasto-perfectly plastic model is frequently sufficient. Furthermore, the above algorithm can easily be modified to allow consideration of cyclic loading, biaxial bending, non-zero axial forces, residual stresses, and nonsymmetric cross-sectional shapes.

3.1.3 Impact of some factors on inelastic flexural behavior

A large body of experimental research has confirmed that the plastic flexural moment can indeed be developed in beams. A summary of some of that earlier experimental work is presented in ASCE (1971). It is understood that, for this plastic moment to develop, the constraints set by the assumptions listed at the beginning of Section 3.1 must be respected. However, a few additional factors that may have an impact on this inelastic flexural behavior deserve the following brief review.

3.1.3.1 Variability in material properties As the plastic moment directly depends on the yield stress of the steel, it is worthwhile to question what value can be reliably used for its calculation.

A potentially fatal mistake would consist of using the mill test certificate value. Steel mills typically perform a single coupon test per batch of steel produced (the size of a batch will vary depending on the mills, but typically consists of many tons of steel). This coupon is tested at strain rates that usually raise the yield strength by 30 MPa (4.4 ksi) or more and are intended only to provide confidence that the entire batch will meet the applicable specifications. Fluctuations in steel properties will exist within a given batch of steel produced from the same heat, and the mill test certificate can provide, at best, only one value for the entire tonnage of steel produced by that heat. Therefore, using the value reported on the mill test certificate is improper and fraught with danger; building failures have been documented wherein such a mistake has been identified as a major reason contributing to collapse (e.g., Closkey 1988).

For the design of new structures, the specified yield strength is the value that should be used. Engineers familiar with limit states design concepts (LSD, LRFD, or others) appreciate the variability inherent in any engineering parameter, including material properties. These variabilities are generally included in the load and resistance factors.

However, when one is evaluating existing structures, it is important to appreciate the cross-sectional variability of yield strength. Extracting material from a steel member to obtain its yield stress using a standard coupon test (such as the standard ASTM E6 test) could give quite different results depending on whether the coupon is taken from the flanges or the web. Galambos and Ravindra (1976) reported that mean values of yield stress for coupons taken from flanges and webs were respectively 5 percent and 10 percent larger than the specified values, with coefficients of variation of 0.11 and 0.10, respectively. This is largely a consequence of the different treatment the web and flanges receive during the rolling process: thicker plates and members are less worked and cool more slowly, resulting in a slightly different grain structure of the metal and weaker strength properties. Likewise, variations exist depending on the thickness of the rolled shapes. In fact, special alloys are added to very thick steel section (such as AISC Group 4 and 5 shapes) to provide the same yield strength as thinner sections for a given metallurgical composition. For some types of steels, when mills do not modify the chemical composition of the steel to compensate for this loss of strength, lower specified yield strengths are provided for use in design (e.g., CISC 1995).

3.1.3.2 Residual stresses Contrary to what is commonly assumed in design, a steel member is not stress-free prior to the application of external loads. In fact, large internal stresses exist there in a state of self-equilibrium. These are generally "locked in" during the rolling process and are also affected by welding or any other heat-imparting or cold-working operations. To understand the origin of these residual stresses, one must visualize the cooling process of a rolled steel section, keeping in mind that the modulus of elasticity of steel at high temperatures is very low, and increases rapidly as the steel progressively cools down below 1000°F, and that steel (like other materials) shrinks as it cools.

Thus, when a rolled section is cooled by the surrounding air, the tips of the flanges that are surrounded by air on three sides cool and gain stiffness first. Shrinkage is essentially unrestrained by the adjacent softer steel. However, as cooling progresses along the flanges, the tips of the flanges that have already partly cooled and acquired some stiffness provide partial restraint against shrinkage of the adjacent flange material. Hence, the tip of the flange is placed in compression, and the adjacent cooling material, in tension. As cooling progresses further along the flanges, the process repeats itself, and all previously partly cooled and stiffer material is compressed by the adjacent material that is beginning to cool. Consequently, the tips of the flanges that have cooled first will be the most compressed, and the flange-web core material, which is surrounded on all sides by steel and cools the slowest, will be subjected to the largest internal tensile stresses.

Members that can cool rapidly, such as thin steel plates, will be subjected to the largest magnitude of residual stresses, with values occasionally reaching up to the yield stress. However, in most rolled steel sections, the maximum residual stresses are approximately 33 percent of the yield stress.

The same logic also illustrates how welding introduces residual stresses. As welding is accomplished, the cooling and shrinking of both the weld metal and the steel in the heat-affected zone is restrained by the adjacent steel material. Therefore, following welding, the welds will be in tension. The existing residual stress pattern in the base metal is also locally affected by the welding operations.

A schematic representation of these self-equilibrating residual stresses is shown in Figure 3.4, using linear variations of stresses along the flanges and web. For comparison, an actual residual stress distribution in a steel section is illustrated in Figure 3.5. These internal stresses are in self-equilibrium as the integral of their effects produce no resulting axial force or moment on the cross section.

Although residual stresses can be large, they have no impact on the plastic moment of a cross section. For example, consider the section of Figure 3.4 for which residual stresses are assumed to be of a magnitude equal to half the yield stress. When that section is subjected to pure axial loading in compression, the tips of the flanges will reach their yield stress at only half the applied axial load that

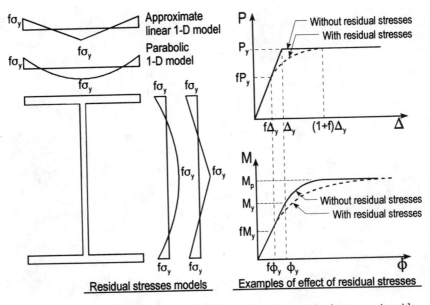

Figure 3.4 Schematic representation of self-equilibrating residual stresses in wide-flange structural shape.

would produce full yielding of the cross section if there were no residual stresses. From that point onward, plastification will start spreading over the cross section. Essentially, because every individual point along the cross section starts from a different initial stress, it must be subjected to a different magnitude of strain prior to reaching the yield stress. Eventually, the flange-web core zones will be the last points to yield, when the applied strain will be 50 percent larger than would have otherwise been necessary to plastify a section free of residual stresses. Indeed, a force-elongation diagram similar to that shown in Figure 3.4 is usually obtained when one tests full cross-section stub-column specimens in axial compression (or tension) instead of standard material coupons.

If the same cross section were subjected to flexure instead of axial force, it would start yielding at a moment equal to half of M_y, with plasticity spreading from the tips of the flanges inward for the flange in compression, and from the flange-core outward for the flange in tension, as the respective flanges would be subjected to larger compression and tension as the flexural moment increased. However, the plasticity moment would be unchanged and remain M_p. This is demonstrated in more detail in the example below.

Although residual stresses do not impact the strength of members, the accelerated softening of the axial force versus axial deformation or moment versus curvature curves, as well as the earlier initiation of

Plastic Behavior at the Cross-Section Level 75

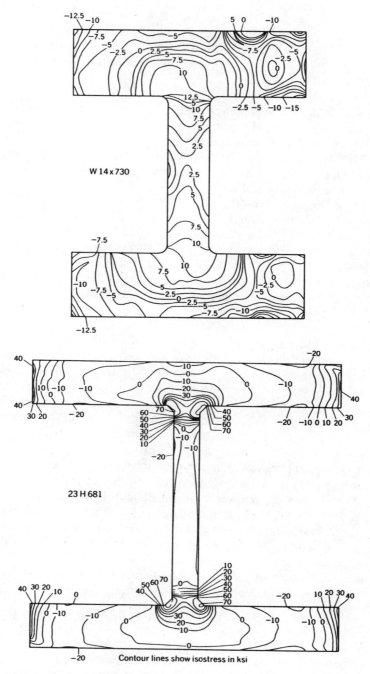

Figure 3.5 Two-dimensional distribution of residual stresses in rolled and welded wide-flange structural shapes. (*From L. Tall*, Structural Steel Design, *2nd edition, 1974.*)

the yielding process, will have an impact on members' deflections and buckling resistances. Incidentally, the analytical expressions included in steel-design codes and standards to calculate the stability and strength of structural members (such as for columns in compression) already take this into account.

Example 3.1.3.3: Ideal wide flange with residual stresses An "ideal" wide flange section for flexural resistance has flanges of negligible thickness and area A_f, a height d, and negligible web area (Figure 3.6). The shape factor for this section is 1.0, and $M_p = dA_f\sigma_y$. It is assumed that the initial residual stresses introduced by the rolling process (thus prior to the application of any external loads) have a peak value of $0.75\sigma_y$ (in compression) at the tips of the flanges and $0.75\sigma_y$ (in tension) at the intersection with the web and that they vary linearly in between. The material is assumed to be elasto-perfectly plastic.

The M-ϕ curves for the initially stress-free case, and for the case having residual stresses, are to be drawn. For this purpose, although accurate analytical expressions could be derived, for simplicity here, the solution will proceed by calculating representative points to accurately plot these curves.

Using the principle of stress superposition, and the knowledge that, at any point, the newly applied stresses and strains are related as per the elasto-perfectly plastic model, one can construct Figure 3.6. In the resulting moment-curvature plot in this figure, the dashed line is obtained when one considers the cross section free of residual stresses, using ϕ_y and ε_y corresponding respectively to the curvature and maximum strain at the onset of yielding M_y (which happens to be equal to M_p in this example). The resulting M-ϕ curve is bilinear for this ideal cross section having a shape factor of unity.

More interesting, however, is the calculation of the M-ϕ curve for the case having residual stresses. Stage A identifies the end of the elastic range, because the application of strains equal to $\varepsilon_y/4$ will uniformly add a stress of $\sigma_y/4$ to the flanges and bring to yield the points for which the magnitude of residual stresses was $0.75\sigma_y$. Values of moments and curvature at multiples of this curvature are calculated. The resulting strains, stresses, and M-ϕ (solid line) are shown in Figure 3.6. As an example of an intermediate calculation, for stage C, when the curvature is ϕ_y, the corresponding moment calculated from the stress diagram is:

$$M_{stage-C} = 2\left[2\left(\frac{A_f}{4}\sigma_y\right) + \frac{1}{2}\left(\frac{\sigma_y}{4} + \sigma_y\right)\frac{A_f}{4}\right]\frac{d}{2} = 0.8125\, A_f d\, \sigma_y = 0.8125\, M_p \quad (3.15)$$

3.1.3.4 Local instabilities Plastic curvature cannot increase indefinitely, and eventually local buckling, or lateral torsional buckling, will occur. The problem of member instability receives a comprehensive treatment in Chapter 10. However, at this point, it is important to realize

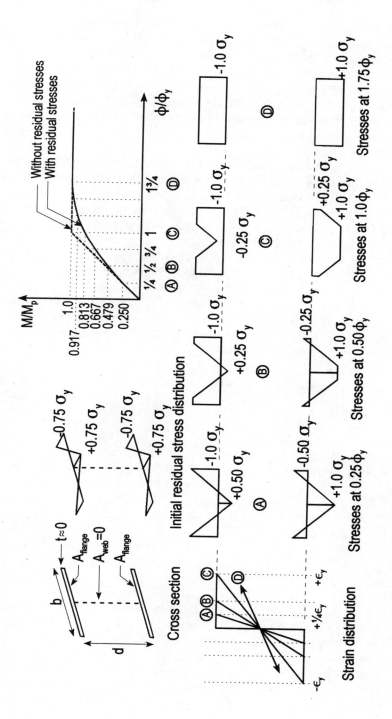

Figure 3.6 Plastic flexural behavior of ideal structural shape with residual stresses.

that local buckling must eventually occur in any structural member compressed far into the plastic range (such as the flange of wide-flange beams commonly used in structures) as the strain-hardening tangent section modulus progressively decreases when plastic strains increase (as demonstrated in Chapter 2). More stringent width-thickness limits imposed by plastic design procedures simply delay local buckling to ensure the development of large plastic deformations within the limits expected in normal applications.

3.1.3.5 Strain hardening Models that neglect strain hardening, such as the elasto-perfectly plastic model, are effectively much simpler to use and particularly useful for hand-calculation. Although the consideration of strain-hardening is always possible, at the cost of additional computational efforts, it has been proven to be generally conservative to neglect the effect of strain hardening in beams subjected to moment gradient, as long as strength is the primary concern. The influence of strain hardening on plastic rotation calculations can be more significant. For the special case of beams subjected to uniform moments (in which plastification occurs simultaneously over the entire uniform moment region), the consideration of strain hardening would bring few benefits as local buckling typically promptly occurs at the onset of strain hardening in such beams.

3.1.4 Behavior during cyclic loading

A major application of plastic analysis concepts is found in the design of earthquake-resistant structures. Consequently, it is important to examine the effect of cyclic loading on a partly or fully plastified cross section. The key to understanding the plastic behavior of a cross section is to consider it as a series of layers of material. All layers of a cross section must abide by the same material model rules, but each layer is strained differently, so its stress history will also differ. For example, in Figure 3.7, the points A to F are assigned to various layers along the height of a rectangular cross section that has been subjected to a moment larger than its yield moment, M_y, but less than its plastic moment, M_p. When the applied moment is removed, the cross section is unloaded to its point of zero applied moment on the M-ϕ diagram, and all layers unload elastically according to the elasto-plastic element model. In fact, a layer will always unload elastically as long as its stress doesn't reach the opposite yield level of $-\sigma_y$. Therefore, one must first try to remove the applied moment elastically, as shown in Case A of Figure 3.7.

If none of the stresses in the resulting stress diagram exceed the yield stress, the solution is deemed acceptable. Note that although the externally applied load has been removed, an internal residual-stress diagram in self-equilibrium has been created. Moreover, as can be

Plastic Behavior at the Cross-Section Level 79

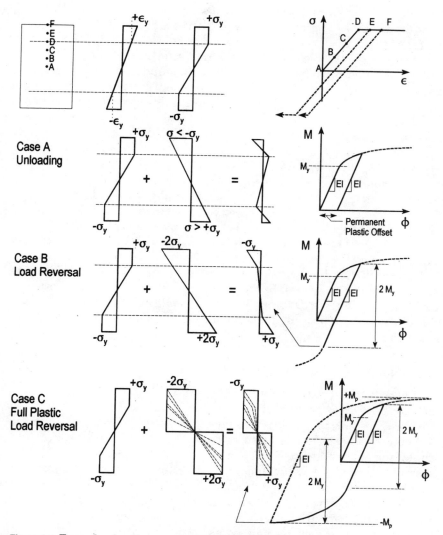

Figure 3.7 Example of cyclic cross-sectional plastic behavior.

seen from the M-ϕ diagram, a residual curvature remains. One can obtain the magnitude of this curvature by first calculating the maximum curvature that was reached during the initial loading phase (for the rectangular section used, Equation 3.9 derived previously can be used for this purpose) and subtracting from this value the curvature that would correspond to the removal of this moment, assuming elastic response (using Equation 3.4, that is $\phi = M/EI$). Because the two values are quite different, a residual curvature must remain.

Following the above logic, one can observe, as in Case B of Figure 3.7, that for any cross section that was first stressed above M_y, it is possible to remove elastically a moment as large as $2M_y$ without having any stresses exceed the yield value in the resulting stress diagram. In fact, the section can be subjected to reversed loading histories within this range of $2M_y$ without producing any new plastification. This is equivalent to having a new elastic range shifted upward by the difference between M and M_y. However, as soon as loading exceeds these bounds, new yielding occurs, and the stress-strain history of each layer must be followed to determine the actual stress distribution throughout the cross section. To determine this stress profile, the procedure followed in the example on residual stresses in the previous section can be used, but one must account for the offset produced by the residual curvature at zero moment. As may be expected by now, a maximum moment of $-M_p$ will eventually be reached, as shown in Case C of Figure 3.7. The procedure can be repeated by reversing the loading and cycling repeatedly, thus producing M-ϕ hysteretic curves.

Generally, the ductile behavior of steel members will develop until a local instability occurs, because of either excessive straining under noncyclic loading or fracture under alternating plasticity (e.g., low-cycle fatigue under cyclic inelastic loading) as discussed in the previous chapter.

3.2 Combined flexural and axial loading

In Section 3.1, it was assumed that no externally applied axial load acted on the cross section. However, in many instances this will not be the case, and it is necessary to investigate how the plastic moment is affected by the presence of axial force. The same fundamental concepts and modeling previously presented still apply: strains linearly distributed across the member's cross section are related to stresses using an elasto-plastic model, and the moments and axial forces are obtained by integrating the stresses acting on the cross section (or, more simply, using the stress-resultant forces). Figure 3.8 illustrates how the stress diagram changes as the applied moment is progressively increased for a given axial force.

Calculation of the reduced plastic moment, M_{pr}, in the fully plastified state for a given axial load, P, is a straightforward operation. Directly, equilibrium of the horizontal forces acting on the cross section gives:

$$P = C - T = A_{compression}\, \sigma_y - A_{tension}\, \sigma_y = (A_{compression} - A_{tension})\sigma_y \quad (3.16)$$

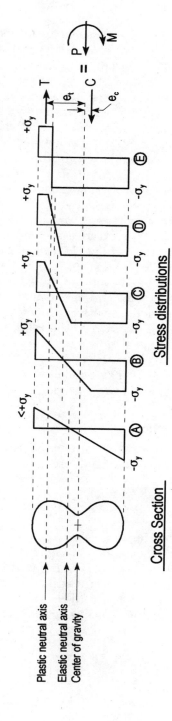

Figure 3.8 Stress diagrams as plasticity progresses in a cross section subjected to combined flexural and axial loading.

Given that the sum of $A_{compression}$ and $A_{tension}$ must equal the total cross-section area, A, one can directly solve for the location of the neutral axis; the stress-resultant forces, C and T; and the corresponding reduced plastic moment, M_{pr}. By repeating the process for axial forces varying from zero to the axial plastic load ($= A\sigma_y$), one can plot an interaction diagram for a given cross section. Alternatively, for the simplest cross sections, closed-form solutions may be developed.

Some of these closed-form solutions are provided here, taking advantage of the fact that the fully plastified stress diagram for combined flexural-axial response can be divided into a pure moment contribution and a pure axial contribution, as shown in Figure 3.9. For convenience, that figure is developed using the same arbitrary neutral axis location for various cross sections; the implication is that, for this generic state of full plasticity, different corresponding axial loads and moments would be obtained for each cross section. The basic principle, nonetheless, remains the same: when a location for the neutral axis is assumed, expressions for the corresponding applied axial force and reduced plastic moment can be developed, which are valid over all or some depths of the cross section. Through algebraic manipulations, it is possible to develop equations for interaction diagrams that express the applied axial force as a function of the reduced plastic moment, although this sometimes proves to be a tedious process.

These equations are developed hereunder for some simple doubly symmetric sections, for a neutral axis generically located at a distance y_o above the geometric center of these sections.

3.2.1 Rectangular cross section

If one divides the stress diagram of Figure 3.9 into pure flexural and axial contributions, the resulting axial force is:

$$\left[\frac{P}{P_y}\right] = \left[\frac{(P = 2\,b\,y_o\,\sigma_y)}{(P_y = h\,b\,\sigma_y)}\right] = \frac{2\,y_o}{h} \tag{3.17}$$

where P_y is the capacity of the cross section fully plastified axially, and all geometric parameters are defined in Figure 3.9.

The expression for the reduced plastic moment can be developed if one subtracts the plastic moment of a section of depth $2y_o$ (i.e., that portion of the cross section assumed to resist the axial load, P) from the plastic moment (i.e., the flexural strength in absence of axial load), making it possible to obtain:

$$M_{pr} = Z\,\sigma_y - Z\,y_o\,\sigma_y = \left[\left(\frac{bh^2}{4}\right) - \left(\frac{b(2y_o)^2}{4}\right)\right]\sigma_y = \frac{b(h^2 - 4y_o^2)}{4}\sigma_y \tag{3.18}$$

In a normalized format, and substituting the result of Equation 3.17:

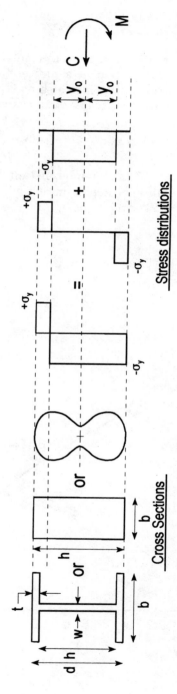

Figure 3.9 Fully plastified condition for an arbitrary location of neutral axis (corresponds to different set of P and M for the different cross sections shown).

$$\frac{M_{pr}}{M_p} = \frac{b(h^2 - 4y_o^2)\sigma_y}{4}\left(\frac{4}{b\,h^2\,\sigma_y}\right) = 1 - \frac{4y_o^2}{h^2} = 1 - \left(\frac{P}{P_y}\right)^2 \quad (3.19)$$

The resulting interaction diagram is plotted in Figure 3.10. Note that, for a rectangular cross section, a significant difference exists between the elastic interaction curve for which no strain is allowed to exceed the yield strain, and the plastic interaction curve derived above.

3.2.2 Wide-flange sections: strong axis bending

For wide flange sections (i.e., those having constant flange thickness), the closed-form solution will vary depending on whether the neutral axis falls in the web ($y_o \leq h/2$) or the flange ($h/2 < y_o \leq h/2$). For the former case:

$$\left[\frac{P}{P_y}\right] = \left[\frac{(P = 2 y_o\, w\, \sigma_y)}{(P_y = A\, \sigma_y)}\right] = \frac{2\, w\, y_o}{A} \leq \frac{A_w}{A} \quad (3.20)$$

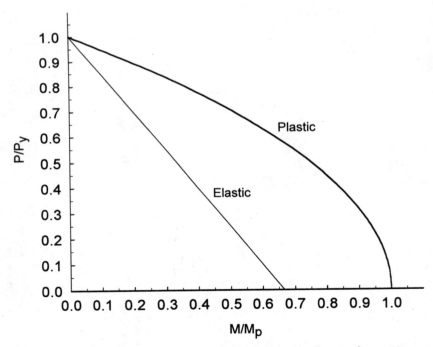

Figure 3.10 Elastic and plastic M-P normalized interaction diagram for a rectangular cross section.

Plastic Behavior at the Cross-Section Level 85

where A is the total cross-section area ($=2bt+wh$), A_w is the web area ($=wh$), and:

$$M_{pr} = Z\sigma_y - Zy_o\sigma_y = \left[Z - \left(\frac{w(2y_o)^2}{4}\right)\right]\sigma_y = (Z - wy_o^2)\sigma_y \qquad (3.21)$$

Then, to obtain a normalized M-P interaction curve, divide the above equation by M_p and substitute the result of Equation 3.20:

$$\frac{M_{pr}}{M_p} = 1 - \left(\frac{Zy_o}{Z}\right) = 1 - \left(\frac{wy_o^2}{Z}\right) = 1 - \left(\frac{P}{P_y}\right)^2 \frac{A^2}{4wZ} \quad \text{for} \quad \frac{P}{P_y} \leq \frac{A_w}{A} \qquad (3.22)$$

When the neutral axis falls in the flange, the following expression is obtained using a similar procedure.

$$\frac{M_{pr}}{M_p} = A\left(1 - \frac{P}{P_y}\right)\left[d - \frac{A}{2b}\left(1 - \frac{P}{P_y}\right)\right]\left(\frac{1}{2Z}\right) \quad \text{for} \quad \frac{P}{P_y} > \frac{A_w}{A} \qquad (3.23)$$

Note that, until the axial force exceeds 30 percent of the axial plastic value, the reduction in moment capacity is typically less than 10 percent for these structural shapes. For most commonly available wide-flange sections, the normalized M-P interaction curve is a function of the ratio of the web area to total cross-section area (which can be alternatively expressed as the web area to sum of the flange areas, or many other variations). This could be demonstrated, as an example, by expansion of the non-normalized term of Equation 3.22 as follows:

$$\frac{A^2}{4wZ_y} = \frac{A^2}{4w\left(\frac{wh^2}{4} + \frac{A_f(d-t)}{2}\right)}$$

$$= \frac{A^2}{A_w^2 + 2A_f(A_w + tw)} \qquad (3.24)$$

$$= \frac{1}{\left(\frac{A_w}{A}\right)^2 + \left(\frac{2A_fA_w}{A^2}\right) + \left(\frac{2A_ftw}{A^2}\right)}$$

where A_f is the total flange area (i.e., sum of the areas of both flanges). Simple observation, or trial calculation, reveals the very small significance of the third denominator term on the resulting normalized M-P interaction curve. This term is the only nonconstant value for a given ratio of web-to-flange area. As a result, normalized interaction curves can be conveniently expressed as a function of the flange-to-web area ratio, as shown in Figure 3.11a.

It can be observed from Figure 3.11a that the normalized M-P interaction curves of wide-flange sections that have the lowest ratio of

86   Chapter Three

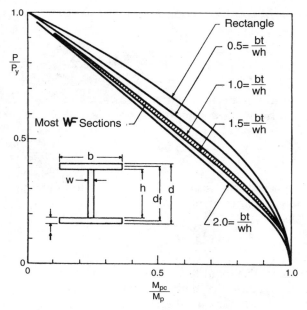

Figure 3.11 Plastic M-P normalized interaction diagrams: (a) for wide-flange structural shape, strong-axis bending. (*Figures a and c reprinted from ASCE Manual #41:* Plastic Design in Steel: A Guide and Commentary, *2nd edition, with permission of American Society of Civil Engineers.*)

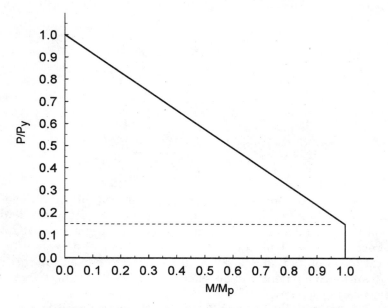

Figure 3.11 Plastic M-P normalized interaction diagrams: (b) simplified design interaction curve for wide-flange structural shape, strong-axis bending.

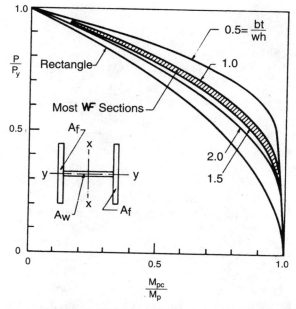

Figure 3.11 Plastic M-P normalized interaction diagrams: (c) for wide-flange structural shape, weak-axis bending.

flange-to-web areas are closest to the curve obtained previously for the rectangular cross section, as is logically expected. Although this observation is useful to demonstrate the relative physical behavior of various cross sections, it should be remembered that steel rectangular shapes inefficiently use material and are not desirable, in spite of their more extensive plastic range. Finally, note that wide-flange sections are usually rolled with a relatively constant ratio of flange-to-web area, normally between 2 and 3, approximately corresponding to the shaded area in Figure 3.11a. This has made possible the development of a convenient and reliable M-P design interaction curve for the plastic strength of wide-flange cross sections in strong axis bending, as shown in Figure 3.11b, and expressed by:

$$\frac{M}{M_p} = 1.18 \left(1 - \frac{P}{P_y}\right) \leq 1.0 \tag{3.25}$$

3.2.3 Wide-flange sections: weak axis bending

Normalized M-P interaction equations can be obtained for wide-flange sections in weak-axis bending following an approach similar to that presented above for strong axis bending. Two cases must be considered, depending on whether the neutral axis falls in the web ($y_o \leq w/2$) or outside the web ($w/2 < y_o \leq b/2$). For the former:

$$\frac{M_{pr}}{M_p} = 1 - \left(\frac{P}{P_y}\right)^2 \frac{A^2}{4dZ_y} \quad \text{for} \quad \frac{P}{P_y} \le \frac{wd}{A} \qquad (3.26)$$

whereas, for the latter case:

$$\frac{M_{pr}}{M_p} = \left[\frac{4bt}{A} - \left(1 - \frac{P}{P_y}\right)\right]\left(1 - \frac{P}{P_y}\right)\left(\frac{A^2}{8tZ_y}\right) \quad \text{for} \quad \frac{P}{P_y} > \frac{wd}{A} \qquad (3.27)$$

Again, for most commonly available wide-flange sections, the normalized M-P interaction curve is a function of the ratio of the flange area to the web area. Normalized interaction curves expressed as a function of that ratio are shown in Figure 3.11c. This time, because a wide-flange shape in weak-axis bending is simply two rectangular cross sections joined at their centroids by a thin member, the wide-flange sections that have the largest ratio of flange-to-web areas will have the interaction curves closest to those for a rectangular cross section, as can be observed from Figure 3.11c. Again, the relatively narrow shaded area in that figure represents the range of flange-to-web area ratios corresponding to most wide-flange sections rolled today. Based on this observation, the following design interaction curve expression for the plastic strength of wide-flange cross sections in weak axis bending has been proposed:

$$\frac{M}{M_p} = 1.19\left[1 - \left(\frac{P}{P_y}\right)^2\right] \le 1.0 \qquad (3.28)$$

In some instances, however, simpler but more conservative equations have also been used in some design codes (e.g., CISC 1995).

3.2.4 Moment-curvature relationships

If the complete moment-curvature relationship for a given axial load and cross section is desired, the use of numerical techniques is recommended, even though some closed-form solutions can be developed for the simplest cross sections. In fact, to account for combined flexural and axial loading, only minor changes are necessary to the moment-curvature algorithm described in the previous section.

3.3 Combined flexural and shear loading

The interaction between axial force and moment could easily be considered in the previous section because they both produce axial strains in a structural member. However, before one can consider the combined effect of flexural and shear on the plastic moment of a cross section, the interaction between axial and shear stresses when yielding is reached must first be described. Based on experimental observations, two mechanics-of-materials rules have been formulated to describe this fundamental material behavior: the Tresca and the Von Mises yield conditions (Popov 1968). The latter has been most widely used to describe aspects of the behavior of steel structures. The Von Mises criterion can be expressed as follows:

$$\sigma^2 + 3\tau^2 = \sigma_y^2 \tag{3.29}$$

where σ is the axial stress, τ is the shear stress, and σ_y is the yield stress in uniaxial tension.

According to that criterion: (1) no shear stress, τ, can be applied when the axial stresses reach yield, and (2) in absence of axial stresses, the yielding shear stress, τ_y, is equal to $0.577\sigma_y$. Such a model neglects strain hardening, which is conservative and consistent with what has been done so far, although it is well known that a dependable and slightly higher shear stress can usually easily be reached because of strain hardening. With this criterion as a starting point, many models have been proposed to determine the plastic moment as reduced by the presence of shear forces. The solutions presented here are essentially those based on equilibrium considerations; such solutions are generally lower-bound solutions and conservative, as shown in Chapter 4.

For the rectangular cross section used here to demonstrate flexure-shear interaction, a distribution of axial and shear stresses that respect the above Von Mises yield criterion is shown in Figure 3.12. The distribution of shear stresses is obtained considering only the remaining (non-yielded) elastic core and knowledge that shear stresses vary parabolically along a rectangular cross section. Thus, with the knowledge that the maximum shear stress on a rectangular cross section is 50 percent larger than the average shear stress, and through use of the Von Mises yield criterion, the depth of the elastic core, $2y_y$, required to resist the shear force, V, can be calculated as:

$$\tau_y = \frac{1.5\,V}{A_{effective}} = \frac{1.5\,V}{2y_y b} = \frac{\sigma_y}{\sqrt{3}} = 0.577\sigma_y \quad \text{thus} \quad y_y = \frac{1.3\,V}{b\sigma_y} \tag{3.30}$$

The resulting axial stress diagram, elastic within $\pm y_y$ from the neutral axis and plastic beyond that value, is used to calculate the reduced plastic moment capacity. Using the principle of stress superposition demonstrated earlier, and the procedure demonstrated in Figure 3.12, one obtains the following formula for the plastic moment reduced by shear, M_{prs}:

$$\frac{M_{prs}}{M_p} = \frac{M_p - 2\left[\left(\frac{by_y}{2}\right)\left(\frac{y_y}{3}\right)\right]\sigma_y}{\left(\frac{bd^2}{4}\right)\sigma_y} = 1 - \frac{4}{3}\left(\frac{y_y}{d}\right)^2 = 1 - \frac{3}{4}\left(\frac{V}{V_p}\right)^2 \quad \text{for} \quad \left(\frac{V}{V_p}\right) \leq \frac{2}{3} \tag{3.31}$$

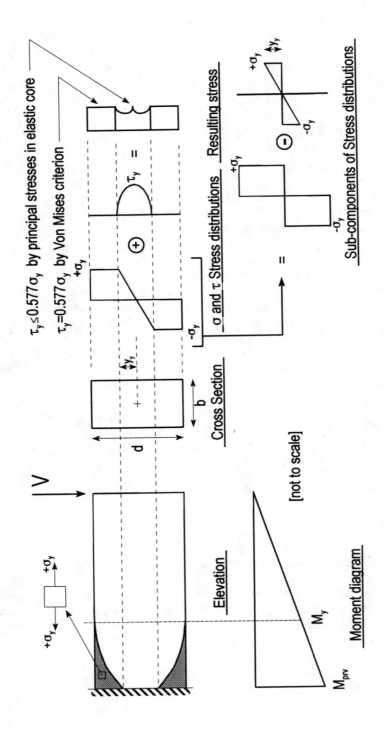

Figure 3.12 Axial and shear stress distribution on rectangular cross section in plastic range.

where the last equality in the above equation is obtained through substituting the value of y_y defined in the previous equation and by defining V_p as the shear corresponding to the fully plastified section under pure shear (a purely theoretical quantity because this would violate the important condition that shear stresses must become zero at the top and bottom of the cross section). For the rectangular cross section, V_p is equal to $bd\tau_y$. The limit of applicability of the above equation is simply a result of the parabolic shear stress distribution assumed; in other words, the maximum shear stress is limited to 1.5 times the maximum average stress that can act on the cross section in the absence of axial yield stresses; that is, $\tau_y = 1.5\,\tau_{average} = 1.5\,V_{max}/bd$.

Given that axial stresses and shear stresses vary linearly and parabolically, respectively, over the $\pm y_y$ region, the following Mohr's circle expression can be used to calculate the principal shear stresses at any point in that region:

$$\tau_{max} = \sqrt{\left(\frac{\sigma_y}{2}\right)^2 + \tau^2} = \sqrt{\left(\frac{\sigma_y y}{2 y_y}\right)^2 + \left[\frac{\sigma_y}{\sqrt{3}}\left(1 - \left(\frac{y}{y_y}\right)^2\right)\right]^2} = \sigma_y \sqrt{\frac{1}{3} - \frac{5}{12}\left(\frac{y}{y_y}\right)^2 + \frac{1}{3}\left(\frac{y}{y_y}\right)^4}$$

(3.32)

This equation demonstrates that a principal shear stress of τ_y is reached only at $y=0$, and at $\pm y_y$, and is not exceeded anywhere in between because $(5/12)(y/y_y)^2$ is always greater than $(1/3)(y/y_y)^4$ in the region of interest (i.e., $y \leq y_y$). Also, calculating the minimum shear stress in that region by making the first derivative of the above equation equal to zero gives a value of approximately 80 percent of τ_y; this illustrates how effectively the section is utilized. As for the part of the cross section yielded under axial strains, σ_y, no shear forces can be applied per the Von Mises criterion, but for sake of completing the principal stresses diagram, σ_y can be expressed as equivalent to $\sqrt{3}\tau_y$, or alternatively, $\tau_y = \sigma_y/\sqrt{3}$.

Although the same procedure could be followed to calculate the reduced moment capacity of wide-flange shapes commonly used in practice, a more expedient approach exists to provide a reliable estimate of this value. This alternative procedure is illustrated in Figure 3.13. First, a uniform shear stress is assumed to act on the web as a result of the applied shear force and is calculated as follows:

$$\tau_w = \frac{V}{hw} \qquad (3.33)$$

Then, the Von Mises yield condition is used to calculate the maximum axial stress that can be applied on the web (i.e., remaining axial stress capacity available):

$$\sigma_w = \sqrt{\left(\sigma_y^2 - 3\tau_w^2\right)} = \sigma_y \sqrt{\left[1 - \left(\frac{\tau_w}{\tau_y}\right)^2\right]} \qquad (3.34)$$

The maximum axial stress that can be applied to the flange remains σ_y. Then, one can determine the reduced plastic moment directly from the resulting axial stress diagram, again using the principle of superposition of stress diagrams to simplify calculations. It is equal to:

$$M_{prv} = M_p - M_{loss} = Z_{prv}\,\sigma_y$$

$$\text{where} \quad M_{loss} = \frac{h^2 w}{4}(\sigma_y - \sigma_w) = \frac{h^2 w}{4}\left[1 - \sqrt{1 - \left(\frac{\tau_w}{\tau_y}\right)^2}\right]\sigma_y = Z_{loss}\sigma_y \qquad (3.35)$$

and where Z_{PRV} is defined as the plastic section modulus reduced by the presence of shear. This procedure is valid only for shear forces smaller than or equal to what would produce shear yielding of the entire web. Furthermore, this procedure cannot be extended to other cross sections, unless these also have webs similarly capable of resisting the bulk of the applied shear force in a condition of near uniform shear stresses. For example, for wide-flange shapes in weak axis bending, the equations derived for the rectangular cross section are actually more suitable.

Although many other expressions have been proposed in the available literature to describe flexure-shear plastic interaction, the above approaches are conservative and simple for hand calculations. Experiments have demonstrated that the reduction in plastic moment capacity is in fact less than predicted by the above equations (e.g., ASCE 1971, Kasai and Popov 1986). Furthermore, in most practical cases, the impact of shear forces on the plastic moment capacity is insignificant.

3.4 Combined flexural, axial, and shear loading

The combined interaction of flexural, axial, and shear loading can be considered through the above principles. Following the same procedure used in the previous section, by apportioning some of the cross section to resist each load effect individually, one could develop close-form solutions, but in most cases, direct calculation of the needed result from the stress diagram is actually more expedient.

For example, the reduced plastic moment of a W920x238 shape (equivalent to a W36x160 in U.S. units) when a shear force, V, of 1334 kN (300 kips) and an axial force, P, of 2668 kN (600 kips) are simultaneously applied, can be calculated as follows, assuming that a mild

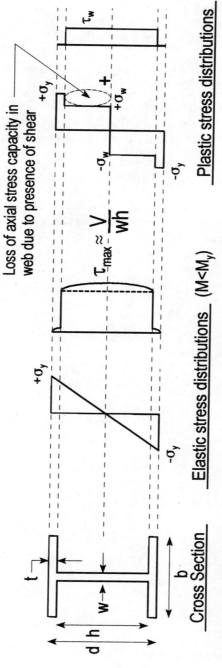

Figure 3.13 Elastic and plastic stress distributions in wide flange structural shape subjected to shear and flexure.

steel that yields at 250 MPa (36 ksi) is used (some of the steps of this example are illustrated in Figure 3.14).

First, the average shear stress on the web of that wide flange resulting from the applied shear force is calculated:

$$\tau_w = \frac{V}{wh} = \frac{1{,}334{,}000 \text{ N}}{(16.5 \text{ mm})(863.2 \text{ mm})} = 93.7 \text{ N/mm}^2 = 93.7 \text{ MPa} \quad (3.36)$$

Incidentally, this is about 65 percent of the yield strength of the web, which is $0.577\sigma_y$, or 144.3 MPa (20.9 ksi). Then, the Von Mises yield criterion is used to calculate the axial stress capacity available in the web:

$$\sigma_w = \sqrt{\sigma_y^2 - 3\tau_w^2} = \sqrt{(250)^2 - 3(93.7)^2} = 190.2 \text{ MPa} = 0.761\sigma_y \quad (3.37)$$

With that remaining web axial stress capacity of 190.2 MPa (27.6 ksi) and a maximum axial flange stress capacity of 250 MPa, one can calculate the portion of the cross section needed to resist the axial force using the procedure demonstrated earlier. In this case, it is found that the neutral axis is in the web:

$$\sigma_w = 190.2 \text{ MPa} = \frac{2{,}668{,}000 \text{ N}}{y_o (16.5)} \Rightarrow y_o = 425 \text{ mm} < \frac{h}{2} = 431.6 \text{ mm} \quad (3.38)$$

One can add the contributions of the web and flanges to the moment resistance as follows:

$$M_{pr\text{-}web}^{P,V} = \frac{wh^2}{4}\sigma_w - \frac{w(2y_o)^2}{4}\sigma_w$$

$$= \frac{(16.5)(863.2)^2}{4}(190.2) - \frac{(16.5)(850)^2}{4}(190.2)$$

$$= 584.5 - 567 \text{ kN–m} = 17.5 \text{ kN–m} \quad (3.39)$$

$$M_{pr\text{-}flange}^{P,V} = 2\left[bt\left(\frac{h-t}{2}\right)\right]\sigma_y$$

$$= (305)(25.9)\left(\frac{915 - 25.9}{2}\right)(250) \quad (3.40)$$

$$= 1755.9 \text{ kN–m}$$

$$M_{pr}^{P,V} = M_{pr\text{-}flange}^{P,V} + M_{pr\text{-}web}^{P,V} = (1755.9 + 17.5) \text{ kN–m} = 1773.4 \text{ kN–m} \quad (3.41)$$

where the superscript (P,V) indicates that this plastic moment is reduced to take into account the applied axial and shear forces (Figure 3.14). This structural shape has a plastic modulus, Z, of $10{,}200{,}000 \text{ mm}^3$

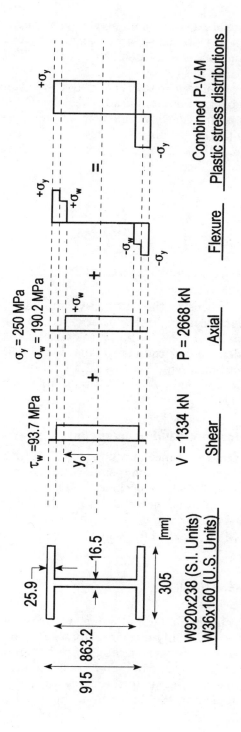

Figure 3.14 Example of plastic resistance of a wide-flange structural shape subjected to flexure, axial, and shear forces.

(622.4 in³) and therefore a corresponding plastic moment of 2550 kN-m (1882kip-ft) in the absence of shear or axial forces. Therefore, the reduced plastic moment capacity in this example is 70 percent of the unreduced value.

When one takes advantage of the readily available value of the plastic section modulus (e.g., tabulated in AISC 1992 or CISC 1995), a subtractive approach to the above problem is actually more expedient. Indeed, deducting the stress diagrams associated with the axial and shear contributions to the moment resistance from the plastic moment, Mp, gives:

$$M_{pr}^{P,V} = M_p - M_{pr}^V - M_{pr}^P$$

$$= M_p - \frac{wh^2}{4}(\sigma_y - \sigma_w) - \frac{w(2y_0)^2}{4}\sigma_w \quad (3.42)$$

$$= (2550 - 183.8 - 567) \text{ kN-m} = 1779.2 \text{ kN-m } (1313.1 \text{ kip-ft})$$

The results obtained using the subtractive and additive approaches are slightly different because in the latter case, the value of the plastic section modulus from a design handbook was used. Handbook values usually consider the true rounded shape of the cross section at the flange-web intersections, which are usually neglected when this section property is determined by hand calculation. Practically, this difference is insignificant.

3.5 Pure plastic torsion: sand-heap analogy

The plastic torsional resistance of a given cross section can be determined as a logical extension of the results obtained from the theory of elasticity. A complete derivation of the elastic torsion theory is beyond the scope of this book and is usually covered comprehensively in mechanics-of-material textbooks. However, some important results from that elastic theory needed to understand plastic torsion are summarized below.

3.5.1 Review of important elastic analysis results

For a structural member in pure-torsion, with its longitudinal axis parallel to the z-axis, the only nonzero stresses are the shear stresses acting in the cross section plane, τ_{zx} and τ_{zy}, and their equal reciprocal components, τ_{xz} and τ_{yz}. The first and second indices, respectively, indicate the axis perpendicular to the plane on which these stresses are acting and the axis in the direction of their action. These values can be defined through the generalized Hooke's Law constitutive relationship (i.e., stress-strain relationship). To satisfy equilibrium, expressed in differential equations as defined in the theory of elasticity, the following relation must be satisfied:

$$\frac{\partial \tau_{zx}}{\partial x} + \frac{\partial \tau_{zy}}{\partial y} = 0 \qquad (3.43)$$

An equation of compatibility for this problem is obtained by derivation of the shear stress expression obtained from the generalized Hooke's Law, with the following result:

$$\frac{\partial \tau_{zx}}{\partial y} + \frac{\partial \tau_{zy}}{\partial x} = \frac{\partial^2 \phi}{\partial y^2} + \frac{\partial^2 \phi}{\partial x^2} = -2G\theta \qquad (3.44)$$

where θ is the angle of twist per unit longitudinal length, G is the shear modulus, and ϕ, called the Prandtl stress function, is a convenient mathematical substitution that will satisfy the above relationship if the following equalities are true:

$$\tau_{zx} = \frac{\partial \phi}{\partial y} \quad ; \quad \tau_{zy} = -\frac{\partial \phi}{\partial x} \qquad (3.45)$$

Along the edge of the cross section, the Prandtl stress function must be constant and is usually taken as zero for convenience. The search for an elastic solution to torsion problems therefore lies in finding a function of x and y, $\phi(x,y)$, that can satisfy the above conditions. Once that function is known, the corresponding torque applied to the member can be calculated because the two are related by the following relationship:

$$T = 2\iint \phi \, \partial x \, \partial y \qquad (3.46)$$

The structure of the differential equation of the stress function (Equation 3.44 above) is identical to the one describing the deflection of a membrane (or soap bubble) of the same shape as the cross section of interest, tied at its edges and subjected to a uniform pressure (Prandtl's membrane analogy). A comparison of the two expressions shows that the stress function, ϕ, corresponds to the membrane deflection, and the shear stresses in the torsion problem are analogous to the slope of the membrane, as illustrated in Figure 3.15. Therefore, the torsional moment is equal to twice the volume of the bubble. The reader is referred to Popov 1968, Timoshenko and Goodier 1970, Ugural and Fenster 1995 (among many) for a complete derivation of Equations 3.43 to 3.46.

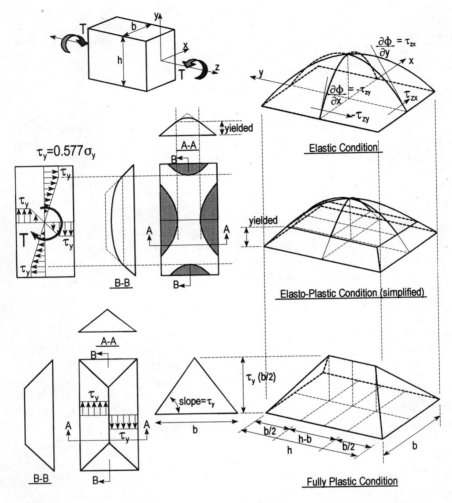

Figure 3.15 Rectangular cross section subjected to progressively increasing torque up to plastic torque.

3.5.2 Sand-heap analogy

Under progressively increasing applied torque, the cross section will begin to yield when the shear stress at any point within the cross section reaches $0.577\sigma_y$ (per the Von Mises yield criterion). Visibly, as predicted by the membrane analogy, this should occur somewhere along the edge of the cross section, at the edge point(s) closest to the center of twist (i.e., where the slope of the soap bubble would be the largest). It can also be visualized that, upon application of larger torques, yielding will spread inward (Figure 3.15), and the plastic torque, T_p, will be reached upon yielding of the entire cross section.

Plastic Behavior at the Cross-Section Level

In the fully yielded condition, because the resulting stress at any point cannot exceed the shear yield stress:

$$\tau = \sqrt{\tau_{zx}^2 + \tau_{zy}^2} = \sqrt{\left(\frac{\partial \phi}{\partial y}\right)^2 + \left(\frac{\partial \phi}{\partial x}\right)^2} = |\nabla \phi| = \tau_y = \frac{\sigma_y}{\sqrt{3}} \quad (3.47)$$

which demonstrates that the maximum slope at all points in the fully plastic condition must be equal to the shear yield stress value. From that knowledge, a plastic ϕ surface can be constructed rather easily, and it is still true that twice the area under this curve directly gives the magnitude of the corresponding plastic torque. The shape of that plastic surface can be compared with that of a sand heap. When one tries to pile up dry sand on a table having the same shape as the cross section of interest, the resulting shape will resemble the plastic surface for that shape because dry sand is stable only up to a specific slope. This slope is also constant over the entire cross section, making the sand-heap analogy obvious.

For example, for a rectangular cross section, the plastic surface becomes a rectangular pyramid of slope τ_y, as shown in Figure 3.15; the corresponding plastic torque is:

$$T_p = 2 \text{ (Volume under stress function } \phi)$$

$$= 2\left[\frac{b}{2}\left(\frac{b\,\tau_y}{2}\right)(h-b)\right] + \frac{1}{3}\left(\frac{b\,\tau_y}{2}\right)b^2 = 2\tau_y \frac{b^3}{6}\left[\frac{3}{2}\left(\frac{h}{b}-1\right)+1\right] \quad (3.48)$$

$$= \tau_y \frac{b^3}{3}\left[\frac{3}{2}\left(\frac{h}{b}-1\right)+1\right]$$

$$= Z_T \tau_y$$

where Z_T is defined as the torsional plastic section modulus. Results for a number of other cross sections are shown in Table 3.1, along with the corresponding ratio of fully plastic torque to maximum elastic torque.

3.6 Combined flexure and torsion

The interaction of flexure and torsion also requires consideration of the plastic relationship between axial and shear stresses, as expressed by the Von Mises yield criterion. Expanding on the previously established principles, the flexure torsion interaction equations can be established as follows.

First, it is assumed that, because of the flexure/torsional interaction, only a reduced axial stress capacity, $\sigma_R < \sigma_y$, is available to resist the applied moment, and only a reduced shear stress capacity, $\tau_R < \tau_y$, is available to resist the applied torque, as shown in Figure 3.16.

Table 3.1 Plastic Torques and Torsional Shape Factors for Some Structural Shapes

Cross section	T_P	T_P/T_E
I-shapes (W)	$\tau_y t^2 \left[\dfrac{d}{2} + b - \dfrac{7t}{6}\right]$	$\approx \dfrac{3}{2}\left[\dfrac{2\rho_f + \rho_t^3 \rho_w}{2(\rho_f - 0.63) + \rho_t^4(\rho_w - 0.63)}\right]$ where $\rho_f = \dfrac{b}{t}$; $\rho_t = \dfrac{w}{t}$ and $\rho_w = \dfrac{(d-2t)}{w} = \dfrac{h}{w}$
I-shapes (WWF)	$\tau_y\left[t^2\left(b - \dfrac{t}{3}\right) + \dfrac{w^2}{2}\left(d + \dfrac{w}{3}\right) - tw^2\right]$ $\approx \dfrac{1}{2}\left[2t^2 b + w^2(d - 2t)\right]$	≈ 1.80 for W shapes ≈ 1.67 for WWF shapes
T-shape (uniform)	$\dfrac{\tau_y t^2}{2}\left[b + d - \dfrac{4t}{3}\right]$	-
T-shape, $w < t$	$\dfrac{\tau_y}{2}\left[bt^2 + dw^2 - \dfrac{t^3}{3} - w^2 t\right]$	-
T-shape, $w > t$	$\dfrac{\tau_y}{2}\left[bt^2 + dw^2 - \dfrac{w^3}{3} - t^2 w\right]$	-
Rectangle $a \times b$	$\dfrac{1}{2}\tau_y a^3\left[\dfrac{b}{a} - \dfrac{1}{3}\right] \approx \dfrac{ba^2 \tau_y}{2}$ if $b \gg a$	1.605 if $b/a = 1.0$ 1.690 if $b/a = 2.0$ 1.590 if $b/a = 4.0$ 1.500 if $b/a = \infty$
Circular tube (2R, 2r)	$\dfrac{2\pi}{3}\tau_y(R^3 - r^3)$ $\approx 2\pi \tau_y\left(R - \dfrac{t}{2}\right)^2 t$ if $(R-r) = t$ = small	1.33 1.0
Triangle	$\dfrac{a^3}{12}\tau_y = 2\sqrt{3}\, r^3 \tau_y$	1.67
Ellipse $2(a+b) \times 2(a-b)$	$\dfrac{2\pi}{3}a^3\left[1 - 4.5\left(\dfrac{b}{a}\right)^2 + 4\left(\dfrac{b}{a}\right)^3\right]\tau_y$	$1.33\left[\dfrac{1 - 4.5\beta^2 + 4\beta^3}{1 - \beta - \beta^2 + \beta^3}\right]$ where $\beta = b/a$

Then, the Von Mises yield criterion is rewritten as:

$$\sigma^2 + 3\tau^2 = \left(\frac{M_{pr}^T}{Z}\right)^2 + 3\left(\frac{T_{pr}^M}{Z_T}\right)^2 = \sigma_y^2 \qquad (3.49)$$

Plastic Behavior at the Cross-Section Level

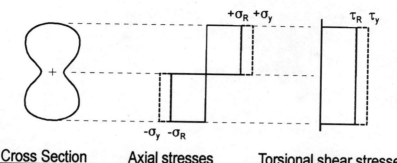

Figure 3.16 Reduced axial and shear stress conditions due to plastic flexure-torsion interaction.

where M_{pr}^T is the plastic moment reduced to account for the presence of torsion, and T_{pr}^M is the plastic torque reduced to account for the presence of flexure. Substituting into that equation the values of Z and Z_T, gives the following interaction equation:

$$\left(\frac{M_{pr}^T}{M_p}\right)^2 + \left(\frac{T_{pr}^M}{T_p}\right)^2 = 1 \tag{3.50}$$

Thus, for any given magnitude of applied moment, M, the corresponding maximum plastic torque that can be developed is:

$$T_{pr}^M = T_p \sqrt{1 - \left(\frac{M}{M_p}\right)^2} \tag{3.51}$$

Note that similar interaction equations could be developed following the same strategy to address the axial force and torque interaction, as well as flexure, axial force, and torque interaction. It is worthwhile to reemphasize that such interaction equations based on an assumed distribution of stresses satisfying equilibrium are usually conservative and safe, in accordance with the lower bound theorem presented in the next chapter.

References

1. AISC. 1992. *Manual of Steel Construction*. LRFD 2nd Edition. Chicago: American Institute for Steel Construction,.
2. ASCE. 1971. *Plastic Design in Steel—A Guide and Commentary*. ASCE manual and reports on engineering practice No. 41. New York: American Society of Civil Engineers.
3. CISC. 1995. *Handbook for Structural Steel Design*. Toronto, Ontario: Canadian Institute for Steel Construction.
4. Closkey, D.J. 1988. "Report of the Commissioner's Inquiry—Station Square Development, Burnaby, B.C." Inquiry Commissioner. Province of British Columbia.

5. Galambos, T.V., and Ravindra, M.K. 1976. "Tentative Load and Resistance Factor Design Criteria for Steel Beam Columns." Research Report No. 32. Civil Engineering Department, Washington University, St. Louis, MO.
6. Kasai, K., and Popov, E.P. 1986. "A Study of the Seismically Resistant Eccentrically Braced Steel Frame Systems." Report UCB/EERC-86/01. Earthquake Engineering Research Center. California: University of California at Berkeley.
7. Popov, E.P. 1968. *Introduction to Mechanics of Solids*. Englewood Cliffs, New Jersey: Prentice-Hall.
8. Timoshenko, S.P., and Goodier, J.N. 1970. *Theory of Elasticity*. 3rd Edition. New York: McGraw-Hill.
9. Ugural, A.C., and Fenster, S.K. 1995. *Advanced Strength and Applied Elasticity*. 3rd Edition. Englewood Cliffs, New Jersey: Prentice-Hall.

Chapter 4

Concepts of Plastic Analysis

4.1 Introduction to simple plastic analysis

Many of the cross-sectional plastic capacities developed in the previous chapter have already been integrated into ultimate or limit states design codes, alongside related equations that address member stability. Today, these equations are generally used in a design process in which all actions on the structural members (i.e., shear and axial forces, moments, and torques) have been obtained by elastic structural analyses. Although provisions exist in most steel design standards and specifications to allow plastic analysis as an alternative analysis procedure, the availability of matrix-based elastic analysis computer programs that effectively eliminate tedious calculations have made the elastic structural analysis the more popular procedure. As a result, most designs are now driven by member strength, not global structural strength. Although this is certainly expedient for standard designs, in many instances knowledge of the true ultimate resistance and strength is still necessary. This chapter presents simple plastic analysis methods suitable for hand calculations to determine such ultimate global structural capacities.

For simplicity's sake, throughout this chapter, unless stated otherwise, the moment capacities used for calculations will not be reduced to account for the effects of axial, shear, or torsion, as described in the previous chapter. These reductions can be considered when these effects are known (or suspected) to have a major impact on the results, but for many structures, this impact is negligible.

Various levels of modeling and analytical sophistication are possible in plastic analysis, with some approaches suitable only for computer nonlinear analyses. Some efforts have been invested recently to bring some of the more advanced analysis techniques within the reach of practicing engineers (SSRC 1993). This chapter, however, concentrates on simple but effective analysis methods that can be carried out by hand (i.e., simple plastic theory).

A number of assumptions are therefore necessary:

- Plasticity along a structural member can exist only at plastic hinges idealized as rigid-perfectly plastic hinges of zero-length and of capacity M_P.
- Small deformation theory is applicable and geometric nonlinearity is not considered.
- The effect of strain hardening is neglected.
- Structural members are properly braced to prevent instability due to either local buckling or lateral-torsional-buckling.
- Plastic hinges can undergo an infinite amount of plastic deformation (or at least, deformations sufficiently large to allow the structure to align its ultimate strength).
- Loads of constant relative magnitude are monotonically applied—that is, progressively increased without load reversal.

Of those assumptions, only the concept of the zero-length plastic hinge which is fundamental to simple plastic theory, deserves some additional explanations. It is best described through an example, such as the cantilever beam loaded up to its plastic moment, shown in Figure 4.1. In such a member, some level of cross-sectional plastification exists at all points where the moment exceeds M_Y. The top fiber of the cross-section has just reached its yield stress at M_Y. Plastification spreads down the flanges and eventually into the web as the moment exceeds M_Y, and eventually, a fully plastified cross-section is reached at M_P.

Thus, plasticity is spread over some length of the member, called the real plastic hinge length, L_P. In this case, because the moment diagram is linear, one can calculate the value of L_P by similar triangles, knowing the ratio of M_P over M_Y. In steel structures where wide flange members with shape factors of approximately 1.12 are typically used, this value is usually taken as approximately 10% of the distance from the point of maximum moment to the inflection point (and slightly more at the center of beams subjected to distributed loads).

Accurate calculation of the deflected shape of partly plastified members is possible, if one knows that the mechanics-of-material relationship between deflection, y, slope, θ, and curvature, ϕ:

$$\frac{d^2 y}{dx^2} = \frac{d\theta}{dx} = \phi \tag{4.1}$$

remains valid even in the inelastic range. To do this calculation, one must obtain the magnitude of this curvature at any point along the length of the beam directly from the actual nonlinear moment-curvature

Concepts of Plastic Analysis 105

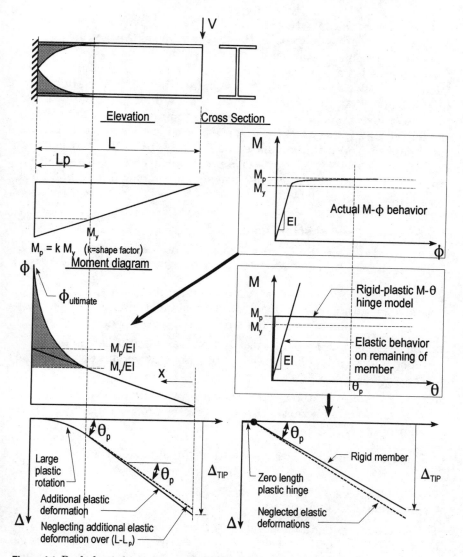

Figure 4.1 Real plastic hinge length and deflection model versus simplified zero-length plastic hinge model.

relationship for this structural shape, as shown in Figure 4.1 (methods to obtain moment-curvature relationships were presented in the previous chapter). Because maximum deformations are usually those of structural engineering interest, one can calculate the cantilever's tip deflection using the moment area method:

$$\Delta_{TIP} = \int \phi x dx = \int_0^{(L-L_p)} \frac{M_x}{EI} dx + \int_{(L-L_p)}^{L} x \phi dx \qquad (4.2)$$

The large curvatures that exist over the real plastic hinge length make a substantial contribution to the value of the cantilever's tip deflection. This is schematically illustrated in Figure 4.1.

Although such a procedure is theoretically exact, it is doubtful that such a level of accuracy is needed for engineering purposes. Therefore, a more expedient approach is to use a zero-length plastic hinge instead of the real plastic hinge length. Assuming a rigid-plastic hinge model, with a capacity M_P, the essence of the plastic structural behavior is captured, and the calculated deflections are sufficiently close to the more accurate values previously obtained. Hence, in this book, unless mentioned otherwise, the term *plastic hinge* will refer to a zero-length plastic hinge.

4.2 Simple plastic analysis methods

There are three ways to calculate the ultimate capacity of a structure using plastic analysis:

- A systematic event-to-event calculation (also known as the step-by-step method), taking into account structural changes when they occur as the magnitude of the loading is progressively increased
- The equilibrium method (also known as the statical method), in which a statically admissible equilibrium state is directly proposed as a potential solution
- The kinematic method (also known as the virtual-work method), wherein a collapse mechanism is directly proposed as a potential solution.

These three methods are reviewed in the following sections.

4.2.1 Event-to-event calculation (step-by-step method)

The step-by-step method, or event-to-event method, consists of simply following the structural behavior by a series of analyses, or steps, from the initial elastic behavior, through the formation of individual plastic hinges, and eventually to collapse. Although tedious, the method is straightforward, and best explained by an example.

As shown in Figure 4.2a, a W530x138 (W21x93 in U.S. units) wide-flange beam fully fixed at both ends and having a yield strength of 350 MPa (50 ksi) is loaded by a point load of progressively increasing magnitude until its plastic moment is reached. As a first step, elastic analysis of the structure is conducted, and the resulting moment diagram is computed (Figure 4.2b). This can be done with any standard methods of structural analysis or, alternatively for such a simple structure, with a standard solution available in design handbooks (e.g., AISC 1992, CISC 1995). The latter approach is more expedient and used here. Based on the results obtained, the plastic moment, M_p, will be reached first at point A under an applied load of 1238 kN (278 kips). The corresponding moment diagram normalized in terms of M_p is shown in Figure 4.2c. The incremental deflection under the applied load is calculated to be 7.0 mm (0.28 in).

According to simple plastic theory, M_p is the maximum moment that can be applied at point A. Therefore, the fixed-end condition that existed at point A cannot prevent the development of plastic rotations at that point (unless load reversal occurs, as described in Chapter 3). As a result, under increased applied loads, the support at point A now behaves as a simple support. This is consistent with the rigid-plastic moment-curvature hinge model described in Figure 4.1. Therefore, for the second step of analysis, a modified structure is analyzed, as shown in Figure 4.2e. In that modified structure, the fixity condition at point A has been changed to a simple support, with the implicit understanding that rotations that will develop there will actually be plastic rotations. The resulting moment diagram is shown in Figure 4.2f. These moments must be compared with the remaining capacities along the length of the structural member. In other words, the results at any given step are incremental results that must be added to those obtained during the previous analysis steps.

For the current example, the moments M_B' and M_C' are the incremental moments resulting from the load P' applied during step 2, and these need to reach only 0.43 M_p and 0.6 M_p, respectively, before another plastic hinge develops, as shown in Figure 4.2f. The smallest value of P' that will produce such a plastic hinge is 465 kN (105 kips). The resulting normalized incremental moment diagram is shown in Figure 4.2g, and the corresponding moment diagram at the end of the second analysis step (i.e., for an applied load of 1238+465=1703kN) is shown in Figure 4.2h. For a beam simply supported at one end and fixed at the other, the incremental deflection under the applied load of 465 kN is calculated to be 7.5 mm (0.30 in); the total deflection at the end of the second step is therefore 14.5 mm (0.57 in).

For the third step of analysis, following the same logic as above, the plastic hinges that now exist at points A and B are replaced by hinges in a new modified structure, as shown in Figure 4.2i. The incremental moment diagram resulting from this third analysis is shown in Figure 4.2j, and these moments must be compared with the capacities remaining at the end of step 2 along the structural member. In this case, the incremental moment M_C'' needs to reach only 0.26 M_p before a plastic hinge develops there. This corresponds to a value of P'' of 66 kN (14.8 kips). The resulting normalized incremental

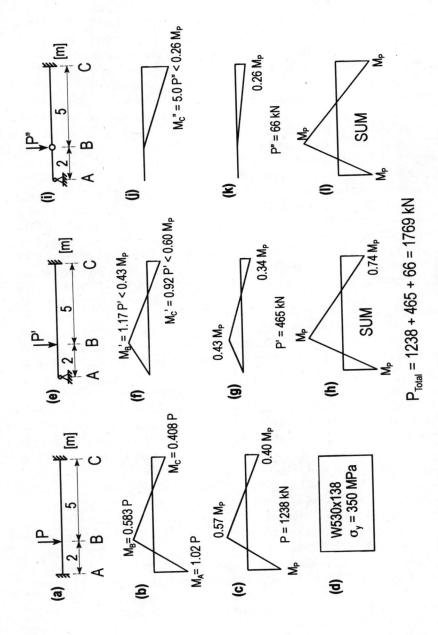

Figure 4.2 Example of step-by-step plastic analysis.

moment diagram is shown in Figure 4.2k, and the corresponding moment diagram at the end of the third analysis step (i.e., for an applied load of 1238+465+66=1769 kN) is shown in Figure 4.2l. The incremental deflection under the applied load of 66 kN is calculated to be 16.0 mm (0.63 in) for a total deflection of 30.5 mm (1.2 in) at the end of the third analysis step.

At this point, because three plastic hinges have developed in a structure having only two degrees of indeterminacy, a mechanism is formed; that is, the structure is unstable, and it will deform as a system of rigid-link members between the plastic hinges. Unrestricted plastic deformations will develop when this load is reached, so this condition is called the plastic collapse mechanism. For many applications, knowledge of this maximum capacity is sufficient, but for some others, it is also necessary to know that the structure can remain stable for some level of plastic deformation beyond formation of this plastic mechanism, as will be seen later.

It is also conceivable that, eventually, at very large deformations, catenary action may develop and that even larger loads could be resisted. Although this would be possible if members and their connections in structures could resist the large catenary-induced tension forces and allow such a large-displacement mechanism to develop in a stable manner (which is rarely the case), this type of behavior is beyond the scope of simple plastic theory defined by the assumption stated in the previous section.

Although the step-by-step method requires time-consuming calculations, it is the only suitable method if one wishes to know the load-deflection or moment-rotation relationships at some specific point along the structural member, over the entire loading history. For instance, as a result of the above example, a complete load-deflection curve can be drawn, as shown in Figure 4.3. This illustrates well the changes in structural stiffness that occur at the formation of plastic hinges. It is notable that, in this example, the load that produces the plastic collapse mechanism is 43% larger than the maximum load permitted by elastic structural analysis methods (i.e., step 1 results), which typically do not account for the redistribution of load possible after formation of the first plastic hinge. The ratio of the load at the formation of the plastic collapse mechanism to the load at the formation of the first plastic hinge is frequently called the redistribution factor (equal to 1.43 in this example).

Finally, note that, for all but the simplest structures, the step-by-step method does not easily lend itself to parametric representation of results. The calculations for the above fixed-ended beam loaded by a single point load would have to be repeated if some numerical value were changed (such as span length or point load location). This is not necessarily the case for the other two methods described below.

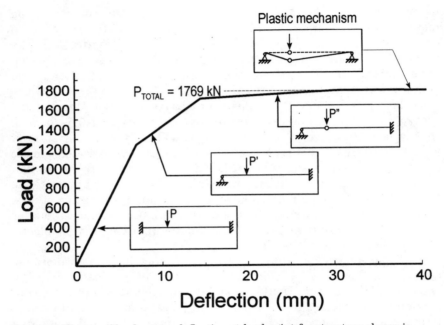

Figure 4.3 History of load versus deflection at load point for structure shown in Figure 4.2.

4.2.2 Equilibrium method (statical method)

The basic premise of the statical or equilibrium method is that any moment diagram in equilibrium with the externally applied loads, and for which the moment at any point along the structure does not exceed the specified member capacities, will provide an estimate of the plastic collapse load. The estimate will be the true collapse load if the moment diagram shows that a sufficient number of plastic hinges exist to form a plastic collapse mechanism; at worst, if not enough hinges are found, the calculated value of the collapse load is a conservative estimate.

A systematic description of the procedure applicable to any structure could be formulated as follows:

1. For any redundant structure, eliminate a number of internal redundancies (such as moment, shear, or axial points of resistance) to make the structure statically determinate.

2. Draw the moment diagram of the resulting statically determinate structure.

3. Draw the moment diagrams resulting from the application of each redundant action (i.e., those removed in step 1) onto the statically determinate structure.

4. Construct a composite moment diagram by combining the moment diagrams obtained in the two previous steps.

5. From the composite diagram, establish the equilibrium equations.

6. Establish at which points the moment diagram will reach the members' plastic capacities such that a sufficient number of plastic hinges will exist to form a plastic collapse mechanism and integrate this additional information into the equilibrium equations.

7. Solve for the plastic collapse load using the equilibrium equations.

As a good practice, it is worthwhile to check that the moments do not exceed M_p anywhere for the given resulting load, and that a collapse mechanism is indeed formed, although this is supposed to be ensured by step 6.

To demonstrate the effectiveness of this procedure, the same example problem presented in the previous section is reanalyzed using the statical method. Moreover, the solution is obtained in a parametric form before its is solved numerically.

The procedure is illustrated in Figure 4.4. In this example, as shown in Figure 4.4a to 4.4d, the two end moments are selected as the redundants, converting the fixed-ended beam into a statically determinate simply supported beam. This is an arbitrary choice because the moment and shear at midspan, for example, would have been an equally appropriate choice of redundants. The loading applied to this simply supported beam (Figure 4.4b) produces the moment diagram shown in Figure 4.4e; the redundant moments that load the same beam (Figures 4.4c and 4.4d) produce the moment diagrams shown in Figures 4.4f and 4.4g. The resulting composite moment diagram is shown in Figure 4.4h.

In this case, a simple equilibrium equation can be written, based on the graphical information presented in Figure 4.4 and adopting the arbitrary sign convention that positive moments produce tension at the bottom of the beam:

$$M_B - \left(\frac{b}{L}\right) M_A - \left(\frac{a}{L}\right) M_C = \frac{Pab}{L} \tag{4.3}$$

with the constraints that:

$$|M_A| \le M_P \quad ; \quad |M_B| \le M_P \quad ; \quad |M_C| \le M_P \tag{4.4}$$

Three hinges are necessary in a beam having two degrees of indeterminacy to form a collapse mechanism. Therefore, observation of the moment diagram in Figure 4.4.h reveals that a plastic collapse mech-

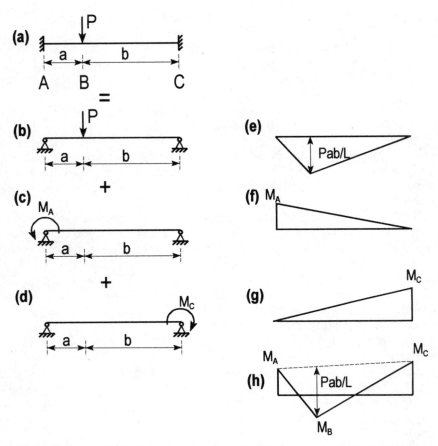

Figure 4.4 Example of plastic analysis by the equilibrium method: parametric representation.

anism would form if the values of M_A, M_B, and M_C all reach M_P. Hence, the equilibrium equation can be rewritten as:

$$M_P - \left(\frac{b}{L}\right)(-M_P) - \left(\frac{a}{L}\right)(-M_P) = M_P + \left(\frac{a+b}{L}\right)M_P = 2\,M_P = \frac{Pab}{L} \quad (4.5)$$

Solving for the applied load that would produce this collapse mechanism, and substituting numerical values from the previous example, gives:

$$P = \frac{2M_P L}{ab} = \frac{2(1263 \text{ kN-m})\,(7\text{m})}{(2\text{m})\,(5\text{m})} = 1769 \text{ kN (397.7 kips)} \quad (4.6)$$

which is the same result obtained previously, with the difference that a general solution has simultaneously been obtained for a beam having fixed ends and a point load applied anywhere along its length.

The difficulty of this method lies in step 6, which requires some judgment and experience, except for the simplest structures. The risk is to obtain a moment diagram that does not reach M_P at the number of locations necessary to produce a collapse mechanism. Such an oversight is unlikely to occur in simple structures; however, to illustrate its consequences without introducing an undue amount of calculations, the same simple example of Figure 4.4 is again used for this purpose. Although the incomplete plastic collapse mechanism is noticeable at first glance, it would not be so obvious in a more realistic and complex structure (such structures are presented in the next chapter).

Thus, assuming that all results of Figure 4.4 have again been obtained, but that, in step 6, the engineer erroneously assumes that plastic hinges will form only at points A and B, Equation 4.3 can be rewritten as:

$$M_P - \left(\frac{b}{L}\right)(-M_P) - \left(\frac{a}{L}\right)\left[-P\left(\frac{ba^2}{L^2}\right)\right] = \frac{Pab}{L}$$

$$M_P\left(1 + \frac{b}{L}\right) = P\left[\frac{ab}{L} - \frac{a}{L}\left(\frac{ba^2}{L^2}\right)\right]$$

(4.7)

where the elastic solution in a fixed-ended beam (CISC 1995) has been substituted for M_C. Solving this equation gives a maximum load, P, equal to 1651 kN (371 kips). This result gives a moment diagram that satisfies equilibrium, with moments equal to 1263 kN-m (932 k-ft) at points A and B, and 673 kN-m (497 k-ft) at point C, as shown in Figure 4.5a. This solution would therefore be an acceptable conservative estimate, per the equilibrium method; however, note that a complete collapse mechanism does not form, which suggests that a better solution exists.

If it had been assumed that a plastic hinge would form only at point B, and if the elastic solutions were used for moments M_A and M_C, a maximum load, P, of 2167 kN (487 kips) would have been obtained. However, that solution would not be acceptable according to the equilibrium method, because the resulting moment diagram would exceed M_P at point A, as shown in Figure 4.5b. This demonstrates the importance of step 6 in the above procedure.

The power of the statical method is that, for hand calculations, the above formal mathematical treatment can be greatly simplified through use of graphical solution methods. Two examples are provided in Figure 4.6, showing how a solution can rapidly be obtained when one graphically combines moment diagrams. In the first case, a uniformly distributed load is applied to a fixed-ended beam, whereas a two-span continuous beam is considered in the second case. In those cases, making all peaks of the composite moment diagram equal to M_p automatically ensures that equilibrium will be satisfied and that the moments will not exceed M_p anywhere along the span(s).

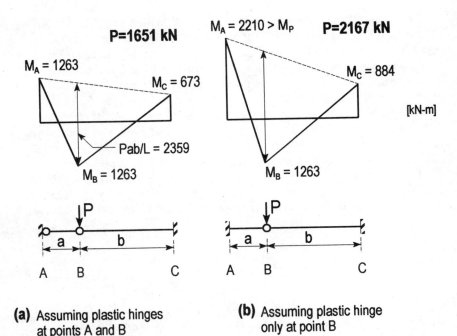

(a) Assuming plastic hinges at points A and B

(b) Assuming plastic hinge only at point B

Figure 4.5 Example of plastic analysis by the equilibrium method in which an insufficient number of plastic hinges is considered (incomplete plastic collapse mechanism).

For the first example, working with absolute values, the graphical combination of moment diagrams gives:

$$\left(\frac{M_P}{2}\right) + \left(\frac{M_P}{2}\right) + M_P = 2\,M_P = \frac{\omega L^2}{8} \quad \Rightarrow \quad \omega = \frac{16\,M_P}{L^2} \quad (4.8)$$

Similarly, the composite moment diagram or the second example gives, for each span:

$$\left(\frac{M_P}{2}\right) + M_P = \frac{3}{2} M_P = \frac{PL}{4} \quad \Rightarrow \quad P = \frac{6 M_P}{L} \quad (4.9)$$

These examples illustrate the effectiveness of the equilibrium method. However, for complex structures, a systematic approach suitable for computer implementation is needed. Such an approach is described in the next chapter.

4.2.3 Kinematic method (virtual-work method)

The basic premise of the kinematic, or virtual-work, method is that if the correct collapse mechanism is known (by either an educated

Concepts of Plastic Analysis

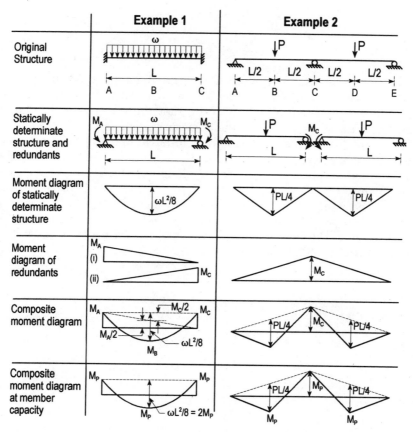

Figure 4.6 Example of graphical approach to solution of plastic analysis by equilibrium method.

guess or some of the techniques to be demonstrated later), the load that produces this plastic collapse mechanism is the exact value of the plastic collapse load. This collapse load is calculated by the virtual-work method considering only the plastic deformations from the rigid-link mechanism action. The drawback of this method is that incorrect estimates of the collapse mechanism will give unconservative results. Therefore, it is crucial to verify that the results obtained by the kinematic method produce moment diagrams that do not exceed at any point the plastic moment of the individual structural members.

In essence, once a plastic collapse mechanism has developed, the external work produced by the applied loads $W_{external}$ must be equal to the internal work produced at the plastic hinges $W_{internal}$. This can be generically expressed as follows:

$$W_{internal} = W_{external}$$

$$\sum_{i=1}^{N} M_P \theta_i = \sum_{j=1}^{N} P_j \delta_j + \int_{x=0}^{L} \omega(x)\, \delta(x)\, dx \qquad (4.10)$$

where θ_i is the plastic rotation at hinge i having a plastic moment M_{pi}, P_j and δ_j are applied load, and deflection at point j, respectively, and $w(x)$ and $\delta(x)$ are expressions for the distributed applied load and deflections along the length of the structure, respectively.

The simplicity of this method is best illustrated by an example. The fixed-ended beam previously solved by the step-by-step and equilibrium methods is shown in Figure 4.7 along with an assumed collapse mechanism.

Using the same arbitrary sign convention as before (i.e., positive moments and positive rotations are those that produce tension on the bottom fiber of the beam) and the parameters defined in Figure 4.7, one can write the following work equations:

$$W_{internal} = W_{external}$$

$$-M_P(-\theta) + M_P\left(\theta + \frac{a\theta}{b}\right) + (-M_P)\left(-\frac{a\theta}{b}\right) = P(a\theta)$$

$$2M_P\left(\frac{b+a}{b}\right)\theta = P(a\theta) \qquad (4.11)$$

$$\frac{2M_P L \theta}{b} = P(a\theta)$$

$$\frac{2M_P L}{ab} = P$$

which is the same result obtained previously by the equilibrium method. Equilibrium is also checked parametrically in Figure 4.7. Note that moments are always of the same sign as their corresponding rotations; this observation is used advantageously to further simplify calculations in the remainder of this chapter.

Because three plastic hinges are needed to create a mechanism for this beam, it is relatively easy to guess the correct collapse mechanism. However, to illustrate how an incorrect solution can be identified, another (and obviously incorrect) collapse mechanism is considered for this same problem, as shown in Figure 4.8. Solving the work equation gives:

$$W_{internal} = W_{external}$$

$$M_P(1 + 2 + 1)\theta = P(a\theta)$$

$$4M_P \theta = P(a\theta) \qquad (4.12)$$

$$4M_P/a = P$$

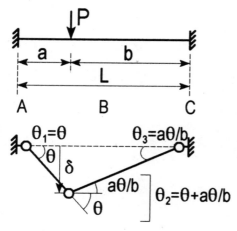

Checking statics:

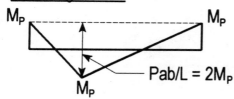

Figure 4.7 Example of plastic analysis by the kinematic method: parametric representation.

which is a larger collapse load than before because a is smaller than b. Two approaches are possible to verify whether equilibrium is satisfied. First, by graphical combination of individual bending moments, as shown in Figure 4.8c, it can be observed that the maximum positive moment exceeds M_p under the applied load of P equal to $4M_p/a$. Second, for more complex structures for which the graphical method may be difficult to apply, equilibrium can be checked through use of free-body diagrams.

Indeed, the knowledge that a moment equal to M_p exists at each plastic hinge provides the necessary additional information to check equilibrium as if each segment of the structure's collapse mechanism were statically determinate. This is demonstrated in Figure 4.8d, in a parametric manner for this example, although in most cases this would be best done numerically. For this problem, two segments must be checked. The right segment satisfies equilibrium as the moment diagram varies linearly between $+M_p$ and $-M_p$ and therefore does not exceed M_p anywhere. For the left segment, the applied load (from the solution presented in Equation 4.12) and the two end moments ($= M_p$) are known. Shears at each end of that segment must first be calculated; results are shown in Figure 4.8d. From those results, the moment diagram in that segment is drawn, as shown in Figure 4.8e,

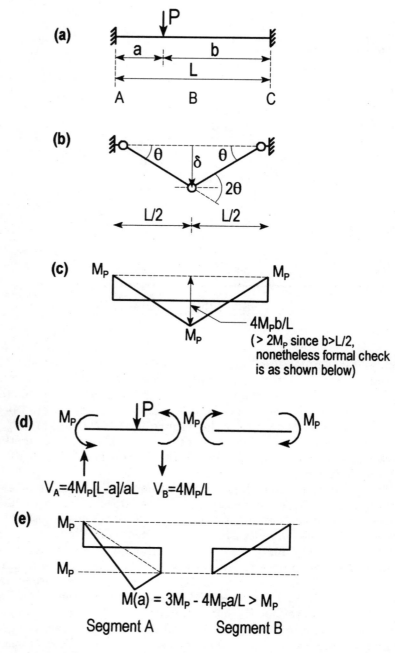

Figure 4.8 Example of plastic analysis by the kinematic method in which an incorrect plastic collapse mechanism is considered.

and observed to exceed M_P locally. In fact, the maximum moment under the applied load is:

$$M_a = -M_P + V_A a$$

$$= -M_P + \left[\frac{4M_P(L-a)}{aL}\right]a \qquad (4.13)$$

$$= 3M_P - 4M_P\left(\frac{a}{L}\right) > M_P \text{ since } a \leq \frac{L}{2}$$

Therefore, this assumed plastic collapse mechanism is not acceptable, and other mechanisms must be tried until a satisfactory solution is found. Unless the solution obtained by the kinematic method also satisfies equilibrium, the results obtained shall not be used because they are unconservative. The degree of unconservatism is quite variable, depending mostly on how far the trial plastic mechanism is from the correct solution. However, although the method is a trial-and-error procedure, observation of the moment diagram obtained in a given attempt usually provides guidance to better locate the plastic hinges in the subsequent trial. For example, the resulting moment diagram in Figure 4.8 suggests that moving the plastic hinge from midspan to under the applied load (a point of maximum moment) would provide a better solution. In most cases, with experience, the correct collapse mechanism for a given structure can be found in only a few trials, sometimes even at the first or second attempt.

To provide some additional examples, the same two problems analyzed previously by the statical method (Figure 4.6) are reanalyzed using the kinematic method. Results are shown in Figure 4.9. The same answers are obtained, thus showing the power of the kinematic method. A systematic description of the kinematic procedure, needed to handle more complex structures, is postponed to Section 4.4 because it relies on some additional concepts that must first be presented. Indeed, in the previous examples, a number of important observations have indirectly introduced some important theorems of plastic analysis that must now be presented more formally.

4.3 Theorems of simple plastic analysis

The previous examples have illustrated that, in any plastic analysis problem, three conditions must be satisfied:

- Equilibrium must exist between the externally applied loads and the internal actions that resist these loads.
- The calculated moment at any cross section must never exceed the plastic moment at that cross section.

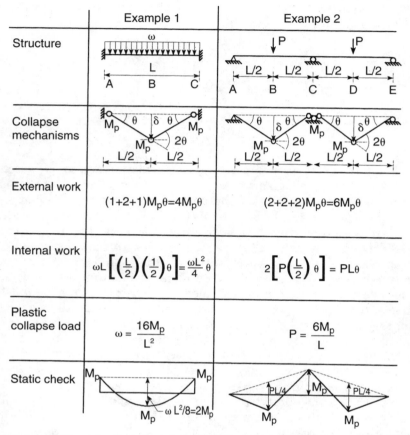

Figure 4.9 Example of solution of plastic analysis by the kinematic method.

- A valid plastic collapse mechanism must develop when the plastic collapse load is reached.

Except for the step-by-step analysis method, which painstakingly follows the progression of yielding under increasing load to ensure that no plastic or elastic condition is violated, the above examples demonstrate that an expedient plastic analysis procedure is possible when one operates on only one or two of the above three conditions, checking the other(s) only after a result is obtained. The implications from this approach are described in the following sections.

4.3.1 Upper bound theorem

When the kinematic method is used, a collapse mechanism is assumed and the collapse load is computed based on a plastic energy balance principle, as expressed by the virtual-work equation. In the computation of that collapse load, no attention is paid to whether the moment diagram exceeds the plastic moment resistance, M_p, at any cross section (other than as a checking step after a solution is obtained), so the estimated collapse load is greater than, or at best equal to, the true solution. Therefore, in the search for the correct answer, any value calculated during a given trial is an upper bound to the true solution.

From this observation, the upper bound theorem can be formulated as follows:

> *A collapse load computed on the basis of an assumed mechanism will always be greater than or equal to the true collapse load.*

4.3.2 Lower bound theorem

When the statical method is used, a moment diagram is drawn to satisfy equilibrium. Because no attention is paid to whether a valid plastic collapse mechanism develops (again, other than by checking after a solution has been reached), the estimated collapse load is less than, or at best equal to, the true solution. Therefore, any value calculated during a given trial is a lower bound to the true solution, and a lower bound theorem can be stated as follows:

> *A collapse load computed on the basis of an assumed moment diagram in which the moments are nowhere greater than M_p is less than or equal to the true collapse load.*

4.3.3 Uniqueness theorem

When all three plastic conditions are satisfied, both methods will give the correct and unique answer. Hence, the lower bound and upper bound meet at the exact solution, which allows the formulation of the following uniqueness theorem:

> *The true collapse load is the one that has the same upper and lower bound solution.*

Although the above three theorems may at first look trivial, they are most useful for plastic analysis, as well as for general structural engineering applications. In fact, many engineers have unknowingly used lower bound approaches in structural design by always giving preference to solutions that satisfy equilibrium rather than to solutions that satisfy compatibility of deformations.

4.4 Application of the kinematic method

The above examples demonstrate that although the kinematic method is a powerful tool to determine the plastic collapse mechanism, the risk of finding an unconservative solution exists, particularly for more complex structures in which the number of possible mechanisms can be quite large. Therefore, to eliminate this risk, the following procedure must be systematically followed for the kinematic method:

1. For any structure, clearly identify where plastic hinges can potentially develop to produce possible plastic collapse mechanisms. Hinges usually form at load points, supports, corners of frames, etc.
2. Define the basic independent mechanisms and possible combined collapse mechanisms.
3. Solve the work equations for all plastic collapse mechanisms. The best estimate of the true solution is provided by the mechanism that gives the lowest value for the collapse load.
4. Check whether the moment diagram for that best estimate exceeds the plastic moments of individual numbers at any point. If it does not, the true solution has been found; otherwise, return to step 2 because a potential plastic collapse mechanism has been overlooked.

A key step in the above procedure is the definition of basic independent and combined mechanisms; this deserves additional explanation, as provided in the following sections.

4.4.1 Basic mechanism types

For beam and column frame type structures, four basic independent mechanisms constitute the building blocks from which the most complex plastic collapse mechanisms can be constructed and understood. They are the beam, panel, joint, and gable mechanisms. Some tentative definitions are provided hereunder, with the understanding that these are purposely broad and intended solely to help the reader develop an appreciation of the concepts. In some of the more complex examples presented in this book, "gray areas" of interpretation exist because some structures, or parts of structures, can be broken down into basic independent mechanisms in more than one way.

4.4.1.1 Beam (member) mechanism A beam mechanism (sometimes called member mechanism) can be defined as any mechanism produced by loads applied between the ends of a member (point loads or distributed loads) that does not require displacement of the ends of that member (Figure 4.10a). Note that all examples presented in this chapter so far have involved only beam mechanisms.

4.4.1.2 Panel (frame or sidesway) mechanism A panel mechanism (also known as a frame mechanism or sidesway mechanism) is produced by the racking action of a square panel; members displace while remaining parallel, and plastic hinges form only at the ends of members (Figure 4.10b).

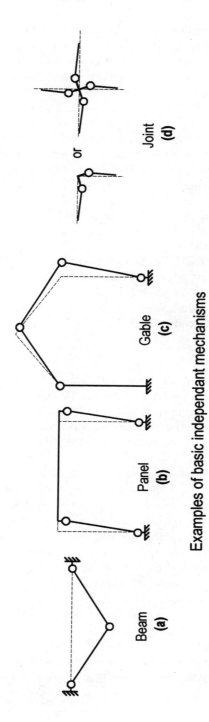

Figure 4.10 Examples of basic independent mechanisms.

4.4.1.3 Gable mechanism
Gables (sometimes called portal frames) are special frames having pitched roofs that have been widely used in industrial construction. As shown in Figure 4.10c, the gable mechanism is defined by the downward movement of the apex of the roof, accompanied by a sideways displacement of only one of the vertical supporting members. If one of the supporting members is conveniently kept vertical throughout the development of the plastic mechanism, the various angles of rotation involved in this mechanism can be more easily related to each other. However, in essence, the gable mechanism requires only that all members of the gable change angles with respect to each other.

4.4.1.4 Joint mechanism
A joint mechanism can be defined by the rotation of a joint, with plastic hinges being developed in all members framing into that joint (Figure 4.10d). Thus, a joint mechanism involves rotation only locally at the joint and does not otherwise affect the geometry of the structure. Although this mechanism does not produce any external work, it is frequently a convenient mechanism to reduce internal work when one is trying to find the plastic collapse mechanism of a complex structure.

4.4.2 Combined mechanism

The objective of the kinematic method is to find the lowest possible collapse load, so it could be advantageous for some structures to combine the basic mechanisms in ways that would reduce the amount of internal work, increase the amount of external work, or both. The examples in the following sections illustrate this concept.

4.4.2.1 Simple frame example
A frame is loaded by both a lateral load, H, and a gravity load, P, as shown in Figure 4.11a. For the purpose of this example, all frame members have the same plastic moment capacity, M_p. Two solution paths are possible, depending on whether one wishes to consider the joint mechanisms. When only two members frame into a joint, the basic joint mechanism does not serve a useful purpose and can usually be neglected (or implicitly considered, depending on the perspective). The simplest solution in this case is to neglect the joint mechanism. As a result, in this example, two basic independent mechanisms are possible: a beam and a panel, as shown in Figures 4.11b and 4.11c.

Although it happens at a floor level, the beam mechanism is essentially the same one considered in previous examples, and equating internal work to external work would give results identical to those

Concepts of Plastic Analysis

already presented in Equation 4.11. For the panel mechanism of Figure 4.11c, the virtual-work equation gives:

$$W_{internal} = W_{external}$$

$$M_P(1 + 1 + 1 + 1)\beta = H(h\beta)$$

$$4M_P\beta = H(h\beta) \tag{4.14}$$

$$\frac{4M_P}{h} = H$$

Although it is customary to use θ for all mechanisms, as is done throughout in this book, for the present discussion only, different parameters are used to identify the plastic rotations of the two different basic mechanisms.

In this simple example, there is only one possible way to combine basic independent mechanisms. The objective is to search for the lowest collapse load (by either increasing the external work or decreasing the internal work) and, in this example, combining the two basic mechanisms increases the amount of external work. Moreover, if β is made equal to θ, the plastic hinge rotations at the left corner of that frame cancel each other, but this is somewhat offset by a greater plastic rotation at the right corner of that frame. The resulting combined mechanism is shown in Figure 4.11d. The virtual-work equation becomes:

$$W_{internal} = W_{external}$$

$$M_P\left(1 + 1 + \frac{a}{b} + 1 + \frac{a}{b} + 1\right)\theta = H(h\theta) + Pa\theta \tag{4.15}$$

$$\left(4 + 2\frac{a}{b}\right)M_P\,\theta = (Hh + Pa)\,\theta$$

$$\left(4 + 2\frac{a}{b}\right)M_P = (Hh + Pa)$$

No conclusions can be reached with results expressed in this parametric form. However, if a and b are equal, h is equal to L, and H is equal to half of P, the above expression can be simplified to:

$$6\,M_P = \frac{PL}{2} + \frac{PL}{2} = PL \tag{4.16}$$

$$\frac{6M_P}{L} = P$$

The corresponding results for the basic independent mechanisms are:

$$\frac{2 M_P L}{ab} = P$$

$$\frac{2 M_P L}{\left(\frac{L}{2}\right)\left(\frac{L}{2}\right)} = \frac{8 M_P}{L} = P \qquad (4.17)$$

for the beam mechanism, and

$$\frac{4 M_P}{h} = H$$

$$\frac{4 M_P}{L} = P/2 \qquad (4.18)$$

$$\frac{8 M_P}{L} = P$$

for the panel mechanism. These two values of P are both higher than the plastic collapse load for the combined mechanism. By virtue of the upper bound theorem, the combined mechanism is closer to the true solution. At this point, it is worthwhile to draw the moment diagram and check whether the plastic moments are exceeded anywhere. The results are shown in Figure 4.11i. Figure 4.11j shows how this result is obtained when the structure is broken into free body diagrams between the plastic hinges. Numbers in circles indicate the sequence in which the calculated values were obtained to complete the moment diagram. As shown in Figure 4.11i, the plastic moment is not exceeded, and the plastic collapse load calculated for the combined mechanism is therefore the true solution.

As mentioned earlier, for joints connecting only two framing members, the joint basic mechanisms can be neglected. However, until some experience is gained with the kinematic method, there is a temptation to conduct plastic analysis as systematically as possible and therefore to include the joint basic mechanisms in the solution process even when this is not necessary. Therefore, Figures 4.11e to 4.11h illustrate, as an alternative solution, how joint mechanisms can be included as part of the solution process. In that solution, the beam and panel basic independent mechanisms remain the same as before, while two joint mechanisms are possible (as shown in Figure 4.11g). Again, different parameters are used to identify the plastic rotations of the two different basic mechanisms. A combined beam and panel mechanism is shown in Figure 4.11h. The virtual-work equation for that combined mechanism would give:

$$W_{internal} = W_{external}$$

$$\frac{2 M_P L}{b} \theta + 4 M_P \beta = Hh\beta + Pa\theta \tag{4.19}$$

which is unsolvable because β is unrelated to θ. Although both angles could be arbitrarily assumed to be equal, this decision could not be justified solely based on the shape of the combined mechanism. However, Figure 4.11h clearly illustrates that with the joint mechanism, one could eliminate three plastic hinges by rotating both joints by θ. At that point, it becomes justifiable to make β, θ, γ, and α equal because doing so reduces the amount of internal work by eliminating plastic hinges. The resulting combined mechanism, shown in Figure 4.11d, is obviously the same.

Finally, depending on the relative values of a, b, h, P, and H, the collapse mechanism giving the lowest plastic collapse load may change. For example, if b is equal to a, h is equal to L/4, and H to $P/4$, the beam collapse mechanism will give the lowest value, namely, 9.8 M_p/L compared with 64 M_p/L and 10.7 M_p/L for the panel and combined mechanisms, respectively. Hence, engineers who frequently work with the same frames would benefit from conducting parametric studies to investigate which conditions provide the most cost-effective design. Incidentally, structural optimization using plastic design has received a considerable attention in the past, but is beyond the scope of this book.

4.4.2.2 Two-span continuous beam example Frequently, more than one basic independent mechanism of a given type is possible in a structure or structural member. For example, for the two-span continuous beam of Figure 4.12, there are three possible beam mechanisms: one acting on the left span and two on the right span. For the left span, the plastic collapse load is given by Figure 4.12b:

$$W_{internal} = W_{external}$$

$$M_P (0 + 2 + 1) \theta = P\left(\frac{L}{3}\right) \theta$$

$$3 M_P \theta = \frac{PL\theta}{3} \tag{4.20}$$

$$\frac{9 M_P}{L} = P$$

Notice that the real hinge at the left end of that span rotates without producing plastic work. It would therefore be a mistake to include it in the internal work formulation.

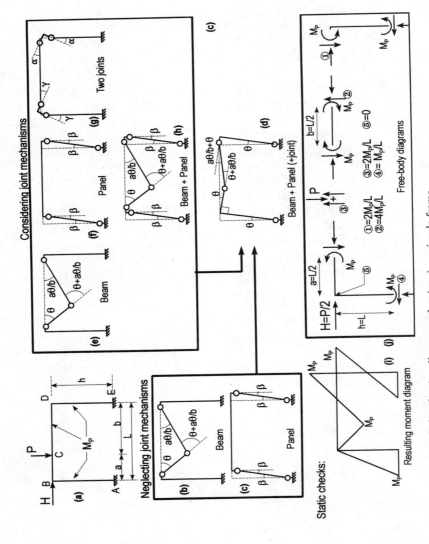

Figure 4.11 Search for plastic collapse mechanism in a simple frame.

The two-beam mechanisms possible on the right span are shown in Figures 4.12c and 4.12d. For the first of these two mechanisms, the work equation becomes:

$$W_{internal} = W_{external}$$

$$M_P\, \theta + 2M_P\,(1.5 + 0.5)\,\theta = 2P\left(\frac{L}{3}\right)\theta + \left(\frac{P}{2}\right)\left(\frac{L}{3}\right)\left(\frac{\theta}{2}\right)$$

$$5\,M_P\,\theta = \frac{3PL\theta}{4} \qquad (4.21)$$

$$\frac{20 M_P}{3L} = \frac{6.67\,M_P}{L} = P$$

Here, one must be careful to correctly locate the left hinge of that plastic mechanism immediately to the left of the middle support, that is, in the weaker member having a resistance of M_p instead of in the stronger member having a resistance of $2M_p$. Indeed, moment equilibrium at that interior support would reveal that it is impossible to develop $2M_p$ there. This indirectly accounts for the possible joint mechanism, as described earlier. Furthermore, as shown in the formulation of that equation, one must also include the external work contribution produced by all loads that displace in the collapse mechanism.

Following a similar approach (Figure 4.12d), a plastic collapse load of 11 M_p/L can be obtained for the second beam mechanism of that right span. Based on the above results, the lowest plastic collapse load is the one that produces the collapse mechanism shown in Figure 4.12c.

It would not be appropriate to combine two beam mechanisms acting on the same span as shown in Figure 4.12e, for demonstration purposes. This erroneous combination results in a collapse mechanism having too many hinges; only three hinges are needed to produce a beam mechanism. As a result, the virtual-work equation cannot be solved, unless some arbitrary decision is made to relate β to θ, a decision for which no rational basis can be found. The only instances in which more plastic hinges than necessary to produce collapse should be acceptable is when two independent plastic collapse mechanisms are found to have the same plastic collapse load. Such an example was presented in Figure 4.9 (Example 2), and although the plastic deformations of the left and right spans cannot be related in that example either, this was not needed because the simultaneous formation of a mechanism in both spans was merely coincidental and did not result from a deliberate attempt to combine mechanisms in ways they should not be combined. Incidentally, collapse mechanisms that develop more plastic hinges than otherwise necessary (as in

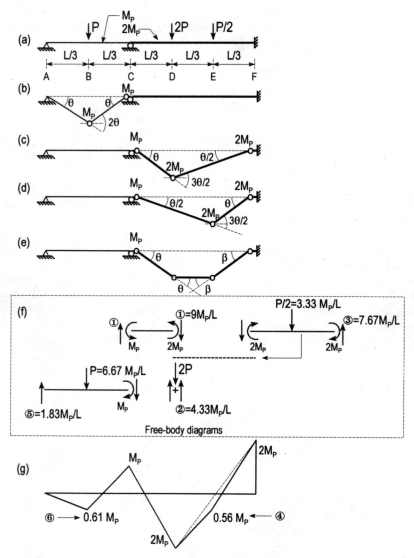

Figure 4.12 Search for plastic collapse mechanism in a two-span continuous beam.

Example 2 of Figure 4.9) are called "overly complete plastic mechanisms."

Finally, to verify that the above result is the true solution, the structure is broken into free body diagrams between the plastic hinges (Figure 4.12f), and statics is used to find the unknown actions and draw the moment diagram (Figure 4.12g). It is verified that this moment diagram nowhere exceeds M_p, which confirms that the true solution has been found.

4.4.2.3 Example of an overhanging propped cantilever beam

For part of the structure shown in Figure 4.13 adjacent to the fixed support, reinforcing cover plates are welded to the wide-flange beam. These locally increase the capacity of this beam from M_p to M_{pr}. The objective of this problem is to determine the length of the cover plate reinforcement (i.e., minimum value of x) needed to develop this value of M_{pr} and to calculate the resulting collapse load, P, when Q is equal to $0.25P$.

Using the kinematic method, three possible beam collapse mechanisms should be considered, as shown in Figure 4.13. In two cases, the cantilever tip load produces negative external work, thus increasing the collapse load compared with the case for which there is no tip cantilever load. If the parametric expressions for the first two mechanisms are equated, the required length of the cover plate reinforcement is equal to $0.12L$. The resulting plastic collapse load is $10.37\, M_p/L$. It can be checked that the third mechanism gives a plas-

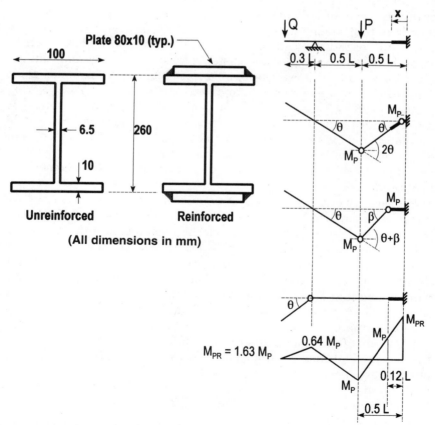

Figure 4.13 Search for plastic collapse mechanism in an overhanging propped cantilever beam.

tic collapse load of 13.33 M_p/L, and consequently does not govern (although if Q is progressively increased, it would eventually become the governing failure more).

This problem is actually easier to solve with the equilibrium method, when one realizes that the effect of the cantilever is to provide a moment equal to $0.3QL$ over the simple support and that the desired failure mode should develop a plastic capacity of M_{pr} at the fixed end. The desired resulting moment diagram can directly be drawn, and by similar triangles, the point nearest the fixed support where M_p is reached can be determined; cover plates would therefore need to extend at least up to that point.

4.4.3 Mechanism analysis by center of rotation

For some types of mechanisms, it can be demonstrated that the articulated links of a resulting plastic collapse mechanism actually rotate around a point called the instantaneous center of rotation. This is generally the case for gable frames and other structures having inclined members that displace as part of the resulting plastic mechanisms. The center of rotation can be particularly useful to establish relationships between the angles of plastic rotation at the various hinges involved in the collapse mechanism. The following example gable frame demonstrates how this can be accomplished.

Here, loading and geometry of the gable frame are arbitrarily selected. Although the examples are treated parametrically, as would normally be done to allow further parametric sensitivity analyses or design optimization, it should be clear that other equations would need to be derived to study gable frames having different geometry or load conditions.

The key to a successful analysis using the center-of-rotation method lies in the ability to visually identify how the various rigid-link members involved in a given plastic mechanism rotate around various points. For example, for the gable frame and collapse mechanism shown in Figure 4.14, it is relatively easy to visualize that, during development of the plastic mechanism, member BC rotates around point B and that member DE rotates around point E, and most will be able to draw Figure 4.14b without difficulty.

To determine the center of rotation of a member having two displacing ends (such as member CD) is a bit more challenging. First, the small displacement theory still holds, so each displacing joint must move perpendicularly to a line connecting the centers of rotation of the two members framing into that joint. Thus, joint C must move perpendicularly to lines BC and CF, which implies that point F is located somewhere on an extension of line BC. Likewise, extending this reasoning to joint D, point F must also be located on an extension of line DE. One can therefore schematically locate point F, in this case, at the intersection of the extension of lines BC and DE. To quantitatively find the location of that point, as well as the relative plastic rotation angles at each hinge, one must develop Figure 4.14c, starting from Figure 4.14b and take the following steps (numbered in Figure 4.14c):

1. Quantitatively locate point F: First, the known inclined distance from B to C is arbitrarily defined as s (step 1), and then the inclined (step 2) and vertical (step 3) distances from C to F are calculated from the properties of similar triangles.
2. Arbitrarily define one of the mechanism rotations as the reference angle θ (step 4), preferably where a joint does not displace (point E is chosen here), and calculate the angle at the other center of rotation that shares the same hinge. In this case, knowing that joint D displaces by Δ_H perpendicularly to member DE and line DF enables one to obtain the rotation angle ϕ at point F by the following (step 5):

$$\Delta_H = \theta h_1 = \phi\left(\frac{L}{a}\right)h_2 \quad \Rightarrow \quad \phi = \left(\frac{h_1}{h_2}\frac{a}{L}\right)\theta \qquad (4.22)$$

3. Realizing that all points displacing around this center of rotation will undergo the same rotation ϕ (step 6) and knowing that the joint C displaces by Δ_C perpendicularly to member BC and line CF (step 7), calculate the rotation angle β at point B by the following (step 8):

$$\Delta_C = \beta s = \phi\left(\frac{b}{a}\right)s \quad \Rightarrow \quad \beta = \left(\frac{b}{a}\right)\phi = \left(\frac{h_1}{h_2}\frac{b}{L}\right)\theta \qquad (4.23)$$

4. Calculate the displacements in the directions parallel to the applied loads to be able to calculate the external work. In this example, Δ_{CV} is needed and it is equal to βa (step 9).
5. Calculate all remaining unknown plastic rotation angles. By simple geometry, this angle at point G is the sum of β and ϕ (step 10), and at point H, the sum of ϕ and θ (step 11).
6. Calculate the plastic collapse mechanism by the usual work equation. For this problem, this gives:

$$W_{internal} = W_{external}$$

$$M_P[\beta + (\beta + \phi) + (\phi + \theta) + \theta] = Pa\beta$$

$$2M_P(\beta + \phi + \theta) = Pa\left(\frac{h_1}{h_2}\frac{b}{L}\right)\theta \qquad (4.24)$$

$$2 M_P \left[\frac{h_1}{h_2} + 1\right] \theta = \frac{Pab}{L}\left(\frac{h_1}{h_2}\right)\theta$$

$$\frac{2 M_P (h_1 + h_2)L}{abh_1} = P$$

Further simplifications of such a result are usually possible once all geometric dimensions are related to a common parameter. For example, if $L = 2h_1$, $a = b = h_1$, and $h_1 = 4\, h_2$, the plastic collapse load becomes $10\, M_p/L$.

Another example of calculation of the instantaneous center of rotation is provided in Figure 4.15, this time for a combined beam-panel-

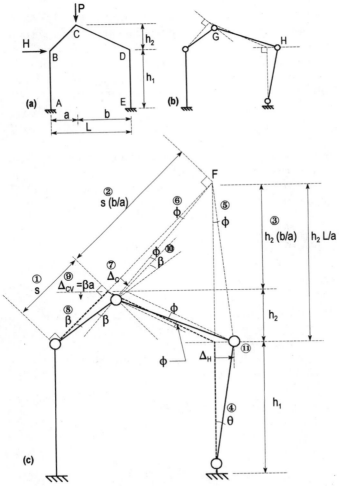

Figure 4.14 Basic plastic collapse mechanism in a gable using the center of rotation method.

gable mechanism, which is frequently the governing failure mode of gable frames. The same procedure described above can be followed, with the difference that the centers of rotation are now A, G, and F for the respective rigid-links ABC, CDE, and EF.

The key step of the entire procedure lies in the schematic determination of point G, located at the intersection of lines AC and EF (a

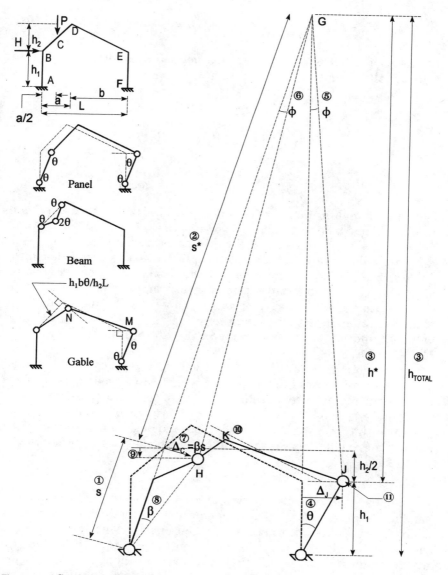

Figure 4.15 Combined beam-panel gable plastic collapse mechanism in a gable using the center of rotation method.

common mistake made in solving such a problem is to extend line BC instead of line AC, which is incorrect because B is not a center of rotation in this case). Again, solving this problem parametrically following the same procedure, gives the following equations:

$$\Delta_J = \theta h_1 = \phi\left[\left(\frac{2L}{a}-1\right)h_1 + \frac{h_2 L}{a}\right] = \phi h^* \quad \Rightarrow \quad \phi = \left(\frac{h_1}{h^*}\right)\theta \quad (4.25)$$

$$\Delta_C = \beta s = \phi\left(\frac{2L}{a}-1\right)s = \phi s^* \quad \Rightarrow \quad \beta = \left(\frac{s^*}{s}\right)\phi = \left(\frac{s^*}{s}\frac{h_1}{h^*}\right)\theta \quad (4.26)$$

where all geometric parameters are defined graphically in Figure 4.15. Finally, from the work equation:

$$W_{internal} = W_{external}$$

$$M_P[\beta + (\beta + \phi) + (\phi + \theta) + \theta] = P\left(\frac{a}{2}\right)\beta + Hh_1\beta$$

$$2M_P\left[\left(\frac{s^*}{s}\frac{h_1}{h^*}\right) + \left(\frac{h_1}{h^*}\right) + 1\right]\theta = \frac{2Pa + Hh_1}{2}\left(\frac{s^*}{s}\frac{h_1}{h^*}\right)\theta$$

(4.27)

Further simplifications are possible, if a relationship between P and H is established. For example, if $L = 3l$, $a = 2l$, $b = l$, $h_1 = 2l/3$, $h_2 = l$, and $H = P$, the plastic collapse load for this structure becomes $87 M_p / 20l$.

A comprehensive treatment of the plastic analysis and design of gable frames is beyond the scope of this book. However, the plastic analysis, design, and optimization of these frames, even including the effect of beam haunches at the eaves and apex, has been the subject of extensive study in the past, and the interested reader will easily find valuable information on that topic in the literature (e.g., Horne and Morris 1981).

Finally, note that the center of rotation method is not limited exclusively to gable frames or structures having inclined members. In fact, the combined plastic collapse mechanism of the regular frame shown in Figure 4.11 can also be analyzed by the center of rotation method. However, that method is obviously more tedious, so it is advisable to avoid it for regular frames that can be easily analyzed by other ways.

4.4.4 Distributed loads

Distributed loads can be easily and directly considered when one is conducting plastic analysis by the equilibrium method, but a special treatment is sometimes necessary when one is using the kinematic method. The added difficulty arises from the fact that the kinematic method requires the assumption of a plastic mechanism and, therefore, identification of the possible plastic hinge locations. This is relatively easy in structures loaded by point loads, but sometimes far less obvious in members subjected to distributed loading. In accordance with the upper bound theorem, the true solution can be obtained only if the correct plastic mechanism is found, so some strategies must be adopted to expediently overcome this added complication.

Three approaches are commonly used to handle the effects of distributed loads:

- Exact analytical solution
- Iterative solution
- Equivalent load-set (or replacement loads) solution

4.4.4.1 Exact analytical solution As the name implies, the exact analytical method requires direct calculation of the exact plastic hinge locations. Therefore, the virtual-work equation is first formulated parametrically, as a function of the unknown plastic hinge locations. Then, from the knowledge that the true solution is the one that gives the lowest plastic collapse load (i.e., upper bound theorem), one obtains the corresponding plastic hinge locations by setting the first derivative of this parametric equation to zero.

Some classic solutions are shown in Figure 4.16. For example, for the propped cantilever of Figure 4.16a, assuming that a plastic hinge will form at a distance x from the fixed support (as shown in Figure 4.16d), the work equation could be written as follows:

$$W_{internal} = W_{external}$$

$$2 M_P \theta \left(\frac{L}{L-x}\right) + 2 M_P \theta \left(\frac{x}{L-x}\right) = \omega L \theta \left(\frac{x}{2}\right)$$

$$2 M_P \theta \left(\frac{L+x}{L-x}\right) = \omega L \theta \left(\frac{x}{2}\right) \quad (4.28)$$

$$\frac{4 M_P}{L^2} \left[\frac{1}{k}\left(\frac{1+k}{1-k}\right)\right] = \omega \quad \text{if} \quad x = kL$$

The first derivative of the work equation, with respect to x (or k according to the above substitution of variables), becomes:

$$\frac{\partial \omega}{\partial x} = \frac{\partial \omega}{\partial k} = 0 = \frac{(1+k)(-1)}{(1-k)(k)^2} + \frac{-(1+k)(-1)}{(1-k)^2(k)} + \frac{1}{(1-k)(k)}$$

$$0 = \frac{k^2 + 2k - 1}{(1-k)^2(k)^2} \quad (4.29)$$

$$0 = k^2 + 2k - 1 \quad \text{if} \quad (1-k)^2 (k)^2 \neq 0$$

From the roots of the above equation, $k = 0.4142$, the plastic hinge location is determined to be located at $x = 0.4142L$. The corresponding plastic collapse load is $\omega = 23.31 \, M_p/L^2$.

In Figure 4.16b, the same propped cantilever is considered, with the difference that the positive plastic moment is larger than the negative plastic moment, as would typically be the case in a composite floor beam. Following the same procedure as above:

$$W_{internal} = W_{external}$$

$$3M_P\,\theta\left(\frac{L}{L-x}\right) + 2M_P\,\theta\left(\frac{x}{L-x}\right) = \omega L\theta\left(\frac{x}{2}\right)$$

$$\left(\frac{6M_PL + 4M_Px}{L-x}\right)\frac{1}{Lx} = \omega$$

$$\left(\frac{10M_PL - 4M_PL + 4M_Px}{L-x}\right)\frac{1}{Lx} = \omega \qquad (4.30)$$

$$\left(\frac{10M_PL}{Lx(L-x)} - \frac{4M_P}{Lx}\right) = \omega$$

$$\frac{2M_P}{L^2}\left(\frac{5}{k(1-k)} - \frac{2}{k}\right) = \omega \quad \text{if} \quad x = kL$$

Then, taking the first derivative of this parametric solution gives:

$$\frac{\partial\omega}{\partial x} = \frac{\partial\omega}{\partial k} = 0 = \frac{5(-1)}{(1-k)(k)^2} + \frac{5(-1)(-1)}{(1-k)^2(k)} - \frac{2(-1)}{(k)^2}$$

$$0 = \frac{2k^2 + 6k - 3}{(1-k)^2(k)^2} \qquad (4.31)$$

$$0 = 2k^2 + 6k - 3 \quad \text{if} \quad (1-k)^2(k)^2 \neq 0$$

From the roots of the above equation, $k = 0.4365$, the plastic hinge location is determined to be exactly located at $x = 0.4365L$. The corresponding plastic collapse load is $\omega = 31.49 \, M_p/L^2$.

Clearly, the parametric derivative can rapidly become complex to solve, even for the simple cases shown above, and it is sometimes more expedient to numerically search for the minimum solution. As a result, curves like those shown in Figure 4.16e can be generated, in which the collapse load mechanism values are plotted as a function of the plastic hinge position. Nonetheless, even when the minima search is performed numerically and graphically, obtaining a parametric work equation of the collapse mechanism sometimes remains the most tedious aspect of the exact solution method.

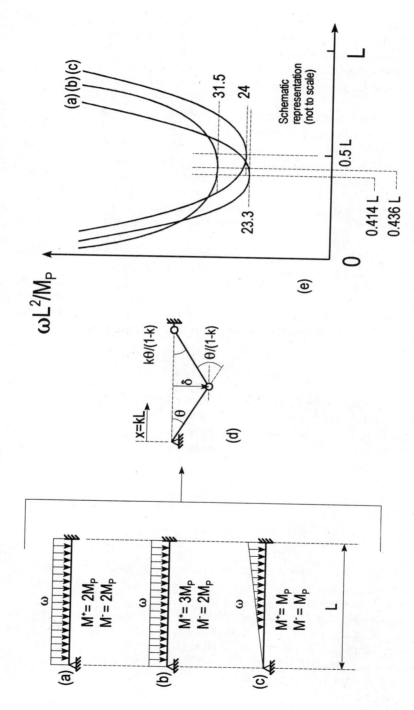

Figure 4.16 Plastic collapse mechanism of some beams subjected to distributed loads using exact analytical solution.

Fortunately, as can be observed from the curves plotted in Figure 4.16e, the curves are rather flat near the true solution. In fact, the error in plastic collapse load introduced by assuming the plastic hinge to be at midspan is small, on the order of approximately 2%. This provides an additional incentive for a more expedient method.

4.4.4.2 Iterative solution The iterative procedure requires that plastic hinge locations be first assumed and relocated through subsequent iterations until the incremental accuracy gained in the plastic collapse load is deemed to be sufficiently small. This effectively eliminates the need to account parametrically for the possible plastic hinge locations and greatly simplifies calculations. However, to provide a more effective iteration strategy, it is sometimes worthwhile (but not necessary) to develop a standard expression for the plastic hinge location as a function of member loading and end-moments. For example, for a beam subjected to a uniformly distributed load, a general expression for the moment by superposition of the bending moment diagrams can be derived from the moment diagrams shown in Figure 4.17. Because plastic hinges frequently occur at (or near) the center span, the origin of the x-axis variable is arbitrarily located at the middle of the span. The resulting expression for the bending moment diagram is:

$$M = -M_L\left[\frac{1}{2} - \frac{x}{L}\right] - M_R\left[\frac{1}{2} + \frac{x}{L}\right] + \frac{\omega}{2}\left[\frac{L^2}{4} - x^2\right] \qquad (4.32)$$

One can obtain the third term of that equation using a free-body diagram of half the simply supported beam under a uniformly distributed loading, noting that the shear force is equal to zero at midspan in that case. The point of maximum moment can be again obtained by taking the first derivative:

$$\frac{\partial M}{\partial x} = \frac{M_L}{L} - \frac{M_R}{L} - \omega x = 0 \qquad (4.33)$$

$$\frac{M_L - M_R}{\omega L} = x$$

The above equation is valid only for beams subjected to uniformly distributed loads, but equivalent relationships can easily be derived for other load patterns.

Conceptually, for the multistory frame example shown in Figure 4.17, solution by the iterative method requires that a first set of plastic hinge locations be assumed and that the corresponding plastic collapse load be calculated. From the resulting moment diagram, one can find the new plastic hinge locations in each beam, either by using the above equation or by moving the plastic hinges to points where the moments are found to exceed M_p. The plastic collapse load corre-

Concepts of Plastic Analysis 141

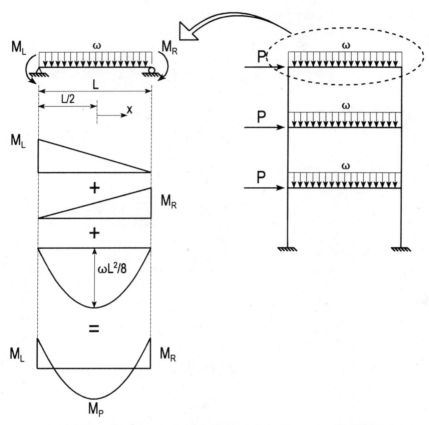

Figure 4.17 Moment-diagram constructions needed for determination of plastic collapse mechanism of beams subjected to uniformly distributed loads with arbitrary end-moments using iterative solution method.

sponding to these new hinge locations would then be computed, and the process would be repeated until a satisfactory solution is found.

This method is numerically illustrated when one uses the propped cantilever of Figure 4.16a. In the first step of this procedure, plastic hinges are assumed to be located at midspan (this is usually a simple and effective starting assumption) and at the final support. The resulting work equation gives:

$$W_{internal} = W_{external}$$

$$2M_P\theta\,(0 + 2 + 1) = \omega L\left[\theta\left(\frac{L}{2}\right)\right]\frac{L}{2} \qquad (4.34)$$

$$\frac{24M_P}{L^2} = \omega$$

Iterating to find the new plastic hinge location, knowing that a plastic hinge must also exist at the right support, gives the following for Equation 4.33:

$$\frac{\partial M}{\partial x} = \frac{M_L}{L} - \frac{M_R}{L} - \omega x = 0$$

$$\frac{M_L - M_R}{\omega L} = x \quad (4.35)$$

$$\frac{0 - 2M_P}{\left(\frac{24M_P}{L^2}\right)L} = \frac{-L}{12} = -0.0833L = x$$

Based on previous observations, the engineer may wish to stop here, knowing that the above estimate of plastic collapse load will be sufficiently accurate and that the newly obtained plastic hinge position will not change substantially in subsequent iterations. However, if it is deemed necessary, one can continue the iteration process by calculating the updated plastic collapse load corresponding to the new plastic hinge locations:

$$W_{internal} = W_{external}$$

$$2M_P\theta\left(\frac{L}{0.583\,L}\right) + 2M_P\theta\left(\frac{0.417\,L}{0.583\,L}\right) = \omega L\theta\left(\frac{0.4167\,L}{2}\right) \quad (4.36)$$

$$\frac{23.36\,M_P}{L^2} = \omega$$

This equation would give a new plastic hinge location of -0.0856L, clearly indicating that a satisfactory solution has been reached. Checking that the moment diagram does not exceed the specified capacities (within some tolerance) would validate this conclusion.

4.4.4.3 Equivalent load-set solution The two solution methods presented above are manageable only for the simplest structures or to verify results obtained by other means. Direct solutions can be difficult to formulate, and the iteration process can be slow and tedious for large structures. The equivalent load-set (or replacement loads) method is proposed as a more direct procedure, also more suitable for large structures.

The equivalent load-set solution consists of replacing the distributed load applied on a member by the smallest possible number of equivalent point loads that would produce the same effect on both the

Concepts of Plastic Analysis   143

member and the structure under consideration. Ideally, in order of priority, the replacement loads must exert the same gravity loading on the structure and produce the same maximum moments, at the same locations, on the member, as the original distributed load. Practically, however, it is not always possible to satisfy all of these conditions.

Some examples of equivalent load-sets are shown in Figure 4.18. These examples show that some trial and error (and at times some ingenuity) is initially needed to establish an acceptable set of replacement loads. For a uniformly distributed load, as shown in the first example of Figure 4.18, a resultant placed at midspan, although producing the same gravity load reactions and location of maximum

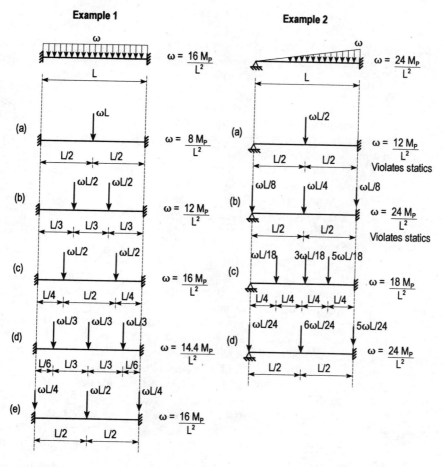

Figure 4.18 Examples of equivalent load-sets for beams subjected to distributed loads.

moments, does not give an adequate plastic collapse load value. Some improvement occurs when the total load is split into two equidistant point loads (Case b).

There is no compelling reason to try only equally spaced point loads, and an even better solution results from locating the two point loads at the quarter points of the span from the ends (Case c). However, if the beam subjected to this equivalent load-set is part of a frame also subjected to lateral loads, the resulting combined gravity and sway mechanism would develop an in-span plastic hinge under one of the point loads and thus quite a distance from where it would occur under the uniformly distributed load (i.e., closer to the middle of the span).

Thus, under some loading conditions, this choice of equivalent load may not provide appropriate results. Dividing the total load into three equivalent loads makes it possible to keep a load at the middle of the span. By trial and error, a good solution for equal magnitude point loads can be found (Case d), but there is again no particular reason to require that the loads be equal, and the best solution is found when the loads are divided into a $\frac{1}{4}:\frac{1}{2}:\frac{1}{4}$ ratio (Case e). In all cases, as long as the equivalent loads are acting in the same direction and produce a collapse mechanism similar to what would have occurred under the distributed loads, the plastic collapse load of the equivalent load set will never exceed that for the true loading.

A triangular loading for a propped cantilever is considered in the second example of Figure 4.18. Again, following closely the approach adopted for the previous example, one first tries a single equivalent load (Case a), but this choice gives poor results. To force the development of a plastic hinge at midspan, the best solution obtained for the uniformly distributed load (Example 1) is then tried, and the correct plastic collapse load is obtained (Case b). However, both this and the previous solutions violate the static condition because the resulting support reactions do not equal those of the original loading. Given that it is more important to match the support reactions (static compatibility) than the maximum moments (mechanism condition), a better solution must be sought. Further, to some other trials (Cases c and d), a satisfactory solution is obtained (Case e).

Although sets of equivalent loads can be used directly in the analysis of complex frame structures, it must be remembered that the existence of a statically admissible moment diagram must always be checked once a tentative solution has been reached under the kinematic method. In that final check, it is worthwhile to consider the true loading conditions and assess the significance of the error introduced by the equivalent load-set.

4.4.4.4 Additional comments Knowledge of the exact plastic hinge location was shown to have a limited impact on the calculated plastic collapse load, but other reasons may require this location to be accurately known. For example, as will be shown later, lateral bracing must be provided at plastic hinges to prevent instability and ensure effective development of the plastic rotations needed to reach the calculated collapse load.

Fortunately, in many instances, point loads and distributed loads will be present simultaneously and the point loads will dominate the response. This would be the case, for example, when girders support beams or joists in addition to their own tributary distributed load (and self-weight). In those cases, the distributed loads could be lumped with the point loads, when one uses basic tributary area principles, without the introduction of a significant error.

4.5 Shakedown theorem (deflection stability)

Up to now in this chapter, only monotonically proportional loading has been considered (i.e., loads that progressively increase while remaining proportional to each other). In most practical situations, loading is nonproportional, and one could conclude that the above plastic theory is not applicable to structures subjected to nonproportional cyclic loading. Fortunately, this is not the case. To demonstrate this, the shakedown phenomenon (also known as deflection stability) must be explained.

Unfortunately, the shakedown theorem was christened much before the advent of earthquake engineering, and in the contemporary engineering context, the term *shakedown* may erroneously conjure images of falling debris and of steel structures "shaken downward," thus collapsing. Nothing could be further from the truth. Instead, in the context of alternating cyclic-loading, the shakedown theorem can be formulated as follows:

> *If there exists an extreme state, beyond the threshold of first yielding, at which point the structure can behave in a purely elastic manner in spite of a previous plastic deformation history, the structure will be said to have shaken down to that complete elastic behavior if the applied cyclic loads do not exceed those that produced this extreme state.*

The following classic example should help clarify this abstract statement of the shakedown concept.

First consider the case of monotonically applied loading. The two-span continuous beam shown in Figure 4.9 is used for this example (Figure 4.19a). One conducts step-by-step analysis using a bilinear elasto-perfectly plastic moment-curvature model (Figure 4.19m). The first plastic hinge is found to occur at point C under an applied load of $16\,M_p/3L$ (Figure 4.19b). The corresponding elastic deflection at both points B and D is $(7/144)M_pL^2/EI$ (Figure 4.19c).

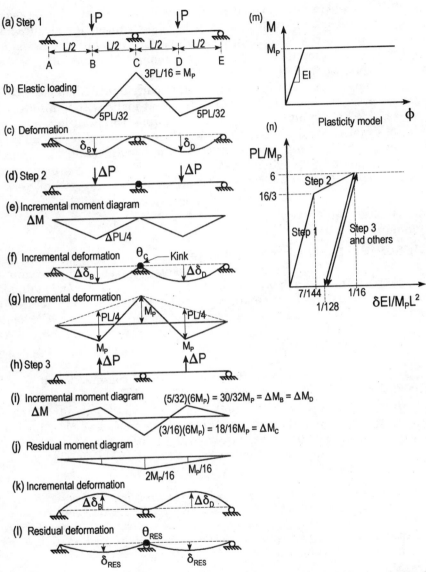

Figure 4.19 Two-span continuous example of shakedown for proportionally applied loads.

At the second step of the calculation, an incremental load of $2M_p/3L$ can be added on the resulting simply supported beams before plastic hinges develop simultaneously at points B and D (Figures 4.19d and 4.19e), with a corresponding additional deflection of $(2/144)M_pL^2/EI$ (Figure 4.19f), for a total deflection of $M_pL^2/16EI$ at those locations. A plastic hinge rotation of $M_pL/12EI$ also develops at point C (being twice the elastic support rotation of a simply supported beam of span L under the incremental loading of $2M_p/3L$). The resulting moment diagram corresponding to the plastic collapse load of $6M_p/L$ (= $16M_p/3L + 2M_p/3L$) is shown in Figure 4.19g.

In the third step of analysis, both loads are removed simultaneously (Figure 4.19h). Interestingly, the entire load can be removed in an elastic manner (Figure 4.19i), leaving a residual moment diagram (Figure 4.19j). Moreover, a residual rotation remains at point C because the plastic rotations there are not recovered, and residual deflections of $M_pL^2/128EI$ remain at points B and D (Figure 4.19k) because the removed loads generate an elastic deflection recovery of smaller magnitude than the deflections reached at attainment of the plastic collapse load. Thus, in its new unloaded condition, the two-span continuous beam is no longer straight, and it has a kink at point C (Figure 4.19l). The resulting load-deflection diagram for this loading history is shown in Figure 4.19n. It shows that the structure can now be loaded and unloaded elastically up to its shakedown load of $6M_p/L$. In other words, the structure has shaken down to a complete elastic behavior for positive cyclic proportional loading.

Hence, if there is a residual bending moment diagram from which the structure will behave elastically under a given extreme state of loading, the structure will find its way to that moment diagram after only a finite amount of plastic flow. However, that residual bending moment diagram is not unique and depends on the structure's particular loading history.

Now, to illustrate how nonproportional positive cyclic loading can affect the value of the shakedown load, the same two-span continuous beam considered is reanalyzed. The difference is that the maximum possible load at point B (i.e., P_B) is first applied and sustained while the maximum possible load that can be applied at point D (i.e. P_D) is successively applied and removed.

Using a step-by-step analysis and the same bilinear elasto-perfectly plastic moment-curvature model of Figure 4.19m shows the first plastic hinge to occur at point B under an applied load of $P_B = 64\ M_p/13L$ (Figure 4.20b). The corresponding elastic deflection at point B is $(23/312)M_pL^2/EI$ (Figure 4.20c). A second step of calculation reveals that an incremental load of $14M_p/13L$ can be added at point B before a plastic hinge develops at point C (Figures 4.20d and 4.20e), with a corresponding additional deflection at point B of $(7/52)M_pL^2/EI$ (Figure 4.20f), for a total deflection of $(5/24)M_pL^2/EI$ there. At the end of that step, a plastic hinge rotation of $7M_pL/12EI$ has developed at point B. Interestingly, the maximum load P_B that can be applied remains $6M_p/L$, as obtained in the previous example, but the moment diagram corresponding to this plastic collapse load is different, as shown in Figure 4.20g. As shown in Figure 4.20t, the calculated deflections at point B at the development of the first plastic hinge and at the plastic collapse mechanism are respectively 1.52 and 3.33 times larger than for the case in which both loads are applied simultaneously (Figure 4.19n).

The second point load, P_D, is then applied (Figure 4.20h). Note that this load cannot be applied in an elastic manner to the beam model of Figure 4.20a because the negative plastic moment at point C was

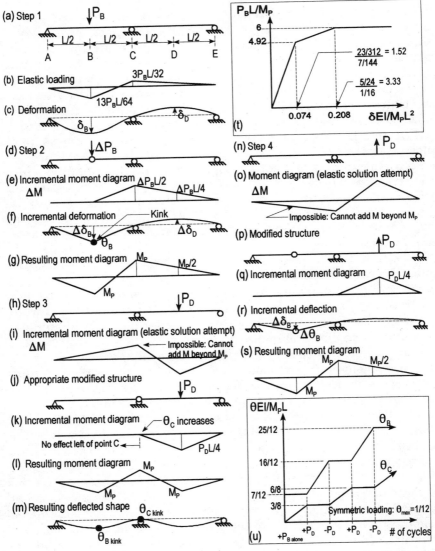

Figure 4.20 Two-span continuous example of shakedown for nonproportionally applied loads.

already reached during application of P_B (Figure 4.20i). Therefore, a modified beam having a hinge at point C is analyzed (Figure 4.20j), and the resulting incremental moment diagram shown in Figure 4.20k is obtained when P_D equals $6M_p/L$. The resulting moment diagram (obtained by superposition of the moment diagrams) and deflected shape are shown in Figures 4.20l and 4.20m, respectively. As expected, because the plastic collapse load does not depend on the loading history, the ultimate plastic capacity of this continuous beam is reached when P_B and P_D both equal $6M_p/L$, as in the previous example. However, the final state of deformation is quite different, with plastic kinks at points B and C. As a rule, plastic deformations are loading-history dependent and greater in magnitude when nonproportional loading is applied.

Next, the entire load P_D (= $6M_p/L$) is removed (Figure 4.20n). As shown in Figure 4.20o, this load cannot be removed in an elastic manner considering the beam model of Figure 4.20a because doing so would require increasing positive moment at point B, which is already stressed to its positive plastic moment capacity. However, elastic removal of P_D is possible considering a modified structure having a hinge at point B (Figure 4.20p). The resulting incremental moment diagram and deflection diagrams are shown in Figures 4.20q and 4.20r, respectively. Clearly, in the process of removing P_D, additional plastic deflections and rotations are introduced at point B.

The moment diagram after P_D is removed (Figure 4.20s) is identical to that obtained when P_B was first applied. Therefore, each subsequent cycle of application and removal of P_D will increase plastic rotations at point C (application of P_D) and plastic rotations and deflections at point B (removal of P_D). The plastic deformations will continue to grow without bounds until progressive collapse, as shown in Figure 4.20u. This implies that the above applied loads of $6M_p/L$ are in excess of the shakedown load for nonproportional loading on this structure; the shakedown load is the maximum load for which no unbounded plastic incremental deformations would occur.

The above calculations suggest that the value of the shakedown load for this beam is between the elastic and plastic load limits of $4.92M_p/L$ and $6M_p/L$. To systematically determine the shakedown load, one must formulate equations of equilibrium at each point of possible plastic hinge in terms of the incremental loads at that point (for complex structures, this undertaking can be formidable).

In this example, the equilibrium equation is established from the moment diagram obtained after a plastic hinge has formed at point B (Figures 4.21a and 4.21b). The moment at point C can be calculated as:

$$M_C = \frac{3}{32} P_{BY} L + \Delta P_B \frac{L}{2} \tag{4.37}$$

where P_{BY} is the load applied in step 1 (Figure 4.20a) necessary to form a plastic hinge at point B, and ΔP_B is the incremental load applied in step 2 at point B (Figure 4.20d). The two terms in Equation 4.37 contributing to the moment at point C result from the sequential application of these loads. Given that the solution sought must permit application of loads P_B and P_D in any sequence, and assuming that both loads are constrained to have the same maximum value, then, $P_{D\text{-}MAX}$ must also be equal to $(P_{BY} + \Delta P_{BY})$. The incremental moment diagram that results from a load applied at point D (Figure 4.21c) until a plastic hinge forms at point C is shown in Figure 4.21d. The resulting expression for that incremental moment at point C is:

$$M'_C = \frac{3}{32} P_{D\text{-}MAX} L = \frac{3}{32} (P_{BY} + \Delta P_B) L \qquad (4.38)$$

As long as no collapse mechanism is allowed to form, no incremental plastic deformations can be introduced into the structure by nonproportional loading. Therefore, to prevent formation of a plastic hinge at point C, the following equation must be satisfied:

$$M_C + M'_C = \left[\frac{3}{32} P_{BY} L + \Delta P_B \frac{L}{2}\right] + \frac{3}{32} (P_{BY} + \Delta P_B) L < M_P \qquad (4.39)$$

Because M_P and P_{BY} are known, this equation can be solved, and a value of ΔP_B equal to $32 M_p / 247 L$ is found. The corresponding shakedown load limit is therefore the sum of P_{BY} and ΔP_B, which is $96 M_p / 19 L$ in this case, or $5.05\ M_p/L$. This load is only 2.6% more than the elastic limit of $4.92 M_p/L$. Given this limited gain and the fact that nearly all live loads are generally nonproportional in nature, the reader may wonder at this point whether plastic design is still worth considering. The answer is yes.

The above example is an extreme case because all the loads are applied and removed. This is not the case in real structures because the dead load is permanent. In fact, in this example, if the loads P_B and P_D included a dead load equal to $2M_p/L$, the likelihood of shakedown would be zero because the first plastic hinge would always form at point C, regardless of the nonproportional sequence of application and removal of the live load component of P_B and P_D. Many researchers have investigated whether the shakedown load limit could invalidate plastic theory, and all have concluded that normal wind and gravity loads are highly unlikely to cause such deflection instability problems in plastically designed structures because a rather large ratio of live load to dead load is necessary to trigger this process (typically the live loads would need to exceed two-thirds to three-fourths of the total load, depending on the type of structure). However, engineers using plastic design concepts should be familiar with the shakedown phenomenon and its consequences, to properly identify those instances when it may need to be considered.

Finally, although rare and unusually intense earthquakes can induce rather large loads in structures, the shakedown problem is typically not addressed during analysis for seismic response. Rather, dynamic nonlinear inelastic step-by-step analysis using earthquake ground motion records is preferred. Cumulative plastic deformations and rotations obtained from such analyses are useful to develop or validate simpler design rules more suitable for hand calculation.

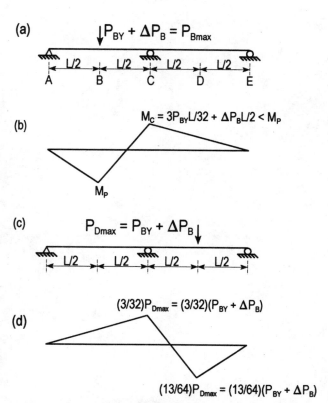

Figure 4.21 Systematic determination of shakedown load for nonproportionally applied loads using equation of equilibrium at points of possible hinge formation in terms of incremental loads.

References

1. AISC. 1992. *Manual of Steel Construction — LRFD 2nd Edition*. Chicago, Illinois: American Institute for Steel Construction.
2. CISC. 1995. *Handbook for Structural Steel Design*. Toronto, Ontario: Canadian Institute for Steel Construction.
3. Horne, M.R., and Morris, L.J. 1981. *Plastic Design of Low-Rise Frames*. Cambridge, Massachusetts: MIT Press.
4. SSRC. 1993. *Plastic Hinge Based Methods for Advanced Analysis and Design of Steel Frames—An Assessment of the State-of-the-Art*. D. W. White and W. F. Chen, editors. Bethlehem, Pennsylvania: Structural Stability Research Council.

Chapter 5
Systematic Methods of Plastic Analysis

The examples of combined plastic mechanisms presented in Chapter 4 show that the challenge in analyzing complex structures lies in finding the collapse mechanism that will give the true (unique) plastic collapse load. Systematic methods of plastic analysis are therefore needed to efficiently identify this correct mechanism. A number of such methods have been developed in the past, and two are reviewed in this chapter: direct combination of mechanisms, which is more suitable for hand calculation, and the method of inequalities, more suitable for linear programming. However, in both these methods, knowledge of the number of basic mechanisms present in a given structure is imperative for a successful solution search, and this topic is therefore presented first.

5.1 Number of basic mechanisms

Four basic mechanism types were identified in the previous chapter:

- Beam
- Panel
- Gable
- Joint

In any given structure, the first task in systematic plastic analysis is to locate and identify those basic mechanisms. To provide some assistance in that endeavor, it is worthwhile to formalize the relationship between the degree of indeterminacy of a structure, X, the number of potential plastic hinge locations, N, and the number of possible basic mechanisms, n. This is accomplished by the following equation:

$$n = N - X \tag{5.1}$$

Figure 5.1 illustrates how this equation can be used. Note that the same single-bay moment-resisting frame is used throughout the example shown in that figure. Simple observation reveals this frame to be indeterminant to the first degree, $X=1$. Using the knowledge developed in Chapter 4, one can identify the number of potential plastic hinge locations, N. Cases 1 and 2 in Figure 5.1 illustrate that it doesn't matter whether one considers only one plastic hinge at each corner of that frame or two (i.e., one at the end of each member framing into the joint). Case 1 is a simplified version of Case 2 in which the joint mechanisms have been implicitly considered. In Case 1, the engineer must determine whether the flexural capacity at the corner joint is equal to the plastic moment of the column or the beam, whereas the more explicit approach used in Case 2 systematically manages this information.

For hand calculations, the approach of Case 1 is recommended when only two members frame into a joint, and the approach of Case 2 must be used at all joints in which more than two members meet (as shown in Figures 5.2 and 5.3). Real hinges (such as at the base of the frames in Figure 5.1) are never included in the value of N because they cannot do plastic work. Note that the number of potential plastic hinge locations is also a function of loading, as shown by a comparison of Cases 2 and 3 in Figure 5.1.

Once the number of basic mechanisms is known, judgment is often necessary to identify these possible mechanisms. Although it is straightforward to find the 16 basic mechanisms in Figure 5.2, more complex structures, such as the one shown in Figure 5.3, can be more challenging.

In some cases, the mechanism identification procedure can be subjective because two (or more) different sets of basic mechanisms can be obtained for the same structure. This is illustrated for the Veerendel truss in Figure 5.4. The basic mechanisms constitute only the building blocks for systematic plastic analysis, so any appropriate set of basic mechanisms will eventually lead to a correct solution (although the length of the "path" toward that goal may vary).

5.2 Direct combination of mechanisms

The direct-combination-of-mechanisms method is a systematic extension of the upper bound method presented in the previous chapter. In the upper bound method, various basic and combined mechanisms are added or subtracted by trial and error until the mechanism having the lowest plastic collapse load is found. By virtue of the uniqueness the-

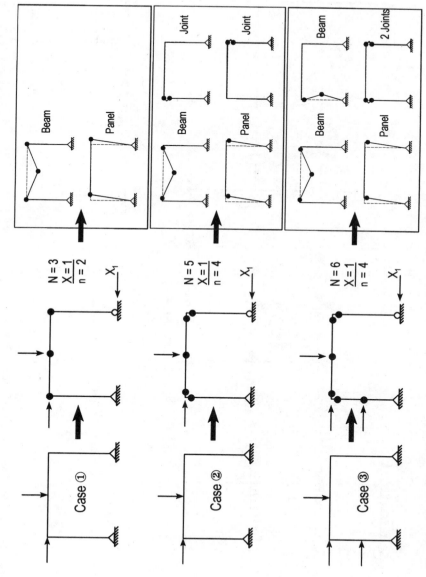

Figure 5.1 Examples of identification of basic mechanisms for a simple frame indeterminate to the first degree.

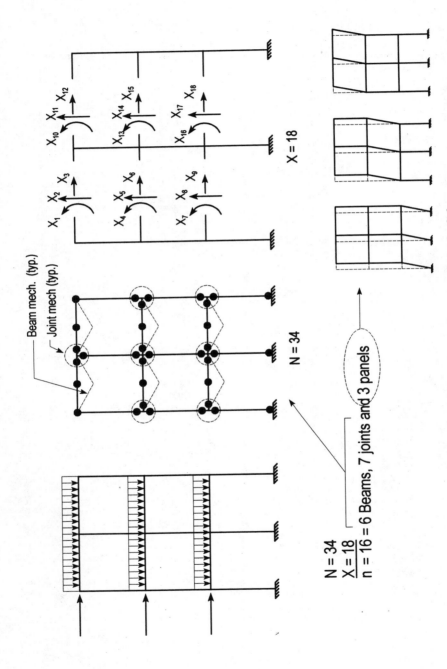

Figure 5.2 Examples of identification of basic mechanisms for a three-story frame loaded as shown.

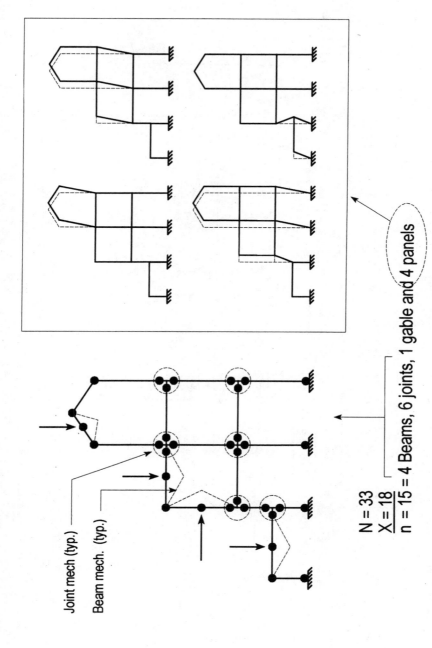

Figure 5.3 Examples of identification of basic mechanisms for a three-story frame loaded as shown.

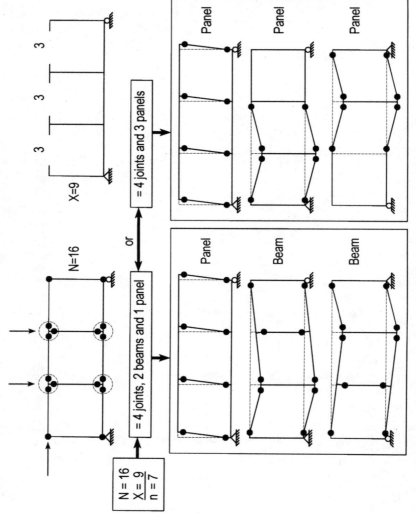

Figure 5.4 Examples of basic mechanisms identification for a Veerendel truss.

orem, this calculated collapse load is the true solution if it also produces a statically admissible moment diagram.

Therefore, in this method, the search for the plastic collapse load proceeds through systematic combinations of mechanisms to either increase the external work or decrease the internal work. A tabular procedure is adopted to keep a manageable record of basic mechanisms and previously attempted combinations as well as to facilitate the identification of those combinations that can be best combined. This systematic tabular approach is best described through the use of examples.

Example 5.2.1: One-bay, one-story frame

The one-bay, one-story frame previously analyzed in Chapter 4 with a trial-and-error approach (see Figure 4.11) is considered here for the case $a = L/2$. Obviously, a systematic method of plastic analysis is not necessary for such a simple example, but the objective at this point is to illustrate how to use the tabular procedure; additional complexities at this early stage would only detract from this goal.

First, in this systematic method, an arbitrary sign convention must be adopted. It is expedient to draw a dotted line along the frame members to visually define the adopted sign convention, as shown in Figure 5.5; moments and rotations are positive when they produce tension on the side of structural members adjacent to the dotted lines. Then, one establishes the number of basic mechanisms for the structure using the procedure described in Section 5.1.

With this information, the general layout of the table needed for the structure under consideration can be drawn, as shown in Table 5.1. In this example, there are five potential plastic hinge locations and the structure is indeterminant to the third degree, producing two basic mechanisms: one beam and one panel, as shown in Figure 5.5. Note that the sign of the plastic rotations is retained with the plastic hinge data in Table 5.1. Also, for the sake of clarity, basic and combined mechanisms are separated in this table.

The characteristics of the previously sketched basic mechanisms are first entered in this table. Normalized coefficients are used as much as possible for clarity and to expedite the mechanisms combination process. For each mechanism, the internal work that develops at each plastic hinge location is inventoried under Plastic Hinge Data. Thus, one can directly obtain the total internal work corresponding to any given mechanism by adding the absolute values of the plastic hinge data across a row of the table. For structures composed of more than one section shape, each member would have a different plastic

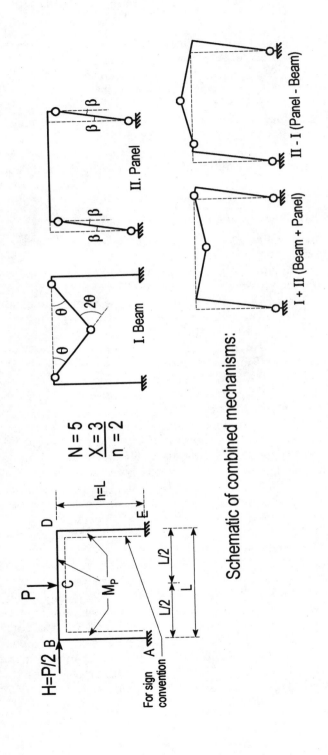

Figure 5.5 One-bay, one-story frame example.

TABLE 5.1 Systematic Direct Combination of Mechanisms for One-Bay, One-Story Example (Figure 5.5)

Mechanism		Plastic Hinge Data $(\alpha M_p)(\beta\theta)/(M_p\theta)$					Internal Work	External Work		P	$P/(M_p/L)$
Identifier	Title	A	B	C	D	E	$W_i/M_p\theta$	$W_e/PL\theta$			
				Basic Mechanisms							
I	Beam	0	−1	+2	−1	0	4	0.5		$8\,M_p/L$	8
II	Panel	−1	+1	0	−1	+1	4	0.5		$8\,M_p/L$	8
				Combined Mechanisms							
I + II	Beam + panel	−1	0	+2	−2	+1	6	1.0		$6\,M_p/L$	6 ✓
II − I	Panel - beam	−1	+2	−2	0	+1	6	0		∞	∞

moment. The plastic hinge data therefore simultaneously accounts for the magnitude of the plastic rotation ($\beta\,\theta$) and plastic moment ($\alpha\,M_p$), as shown in the header of Table 5.1.

One should always draw rough sketches of all mechanisms (using the tabulated plastic hinge data) to keep track of the physical meaning of each combination. The total external work should always be computed directly from these sketches. For small structures, the rough sketches can be inserted directly into the table.

Once the basic mechanisms have been logged, one must search for combinations that minimize the internal work, maximize the external work, or do both. The tabular format presented here facilitates the identification of combinations that will result in the cancellation of plastic hinges (i.e., reduction of internal work). In this simple example, the number of possibilities is limited. The two basic mechanisms are first directly combined by simple addition, column by column, of the normalized plastic rotations listed in the respective rows of Table 5.1 corresponding to these two basic mechanisms. Then, these are subtracted to demonstrate that some combinations do not progress the solution. Inspection of the sketches of the mechanism reveals that subtracting the beam mechanism from the panel mechanism reduces the amount of external energy.

For each combination, the corresponding plastic collapse load must be calculated from the ratio of the internal work to the external work because these results from separate rows in Table 5.1 are not additive. Whenever one believes that the lowest plastic collapse load has been found, the resulting moment diagram should be drawn. A statically admissible moment diagram will reveal that the true solution has been found, whereas, in the alternative, the resulting diagram will indicate where plastic hinges should develop and thus provide guidance for the most promising subsequent combination.

Example 5.2.2: Two-story frame with overhanging bay

The two-story frame with an overhanging bay, shown in Figure 5.6, is analyzed to illustrate how to combine mechanisms in a more general situation. Per the systematic methodology presented in the previous example, a sign convention is selected, the potential plastic hinge locations are determined, and the basic plastic mechanisms are identified. The layout of Table 5.2 is then established, and the plastic collapse loads for each basic mechanism are calculated. Note that joint mechanisms are used only to eliminate internal work: therefore, no external work or plastic collapse load is calculated for those mechanisms.

TABLE 5.2 Systematic Direct Combination of Mechanisms for Two-Story Frame with an Overhanging Bay (Figure 5.6)

Mechanism		Plastic Hinge Data $(\alpha M_p)(\beta\theta)/(M_p\theta)$																			Internal Work	External Work		
Identifier	Name/sketch	1	2	3	4	5	6	7	8	9	10	11	12	13	14	15	16	17	18	19	$W_i/M_p\theta$	$W_e/PL\theta$	P	$P/(M_p/L)$
											Basic Mechanisms													
I	Beam 1					−1	2	−1													4	0.5	$8\,M_p/L$	8
II	Beam 2										−1	2	−1								4	0.5	$8\,M_p/L$	8
III	Beam 3													−1	2	−1					4	0.5	$8\,M_p/L$	8
IV	Beam 4				−1													−1	2	−1	4	0.5	$8\,M_p/L$	8
V	Panel 1				−1					1			1	1			−1				6	1.0	$6\,M_p/L$	6
VI	Panel 2	−1	+1	−1					−1				1								4	2.0	$2\,M_p/L$	2
VII	Panel 3										−1		−1				−1			1	4	1.0	$4\,M_p/L$	8
VIII	Joint 1			−1	1	1															3	—	—	—
IX	Joint 2							−1	1	−1	1										4	—	—	—
X	Joint 3															−1	1	1			3	—	—	—
								Combined Mechanisms																
XI	V+VI	−1	+1	1	−1					1			1	1							10	3	$3.33\,M_p/L$	3.33
XII	XI+VIII+IX	−1	+1			1		−1			1		1	1			−1			−1	9	3	$3\,M_p/L$	3.0
XIII	XII+VII	−1	+1			1		−1					0	1			−1	−1			7	4	$1.75\,M_p/L$	1.75
XIV	XIII+X	−1	+1			1		−1						1		−1					6	4	$1.5\,M_p/L$	1.5 ✓
XV	XIV+I+III	−1	+1				2	−2							2	−2					10	6	$1.66\,M_p/L$	1.66

163

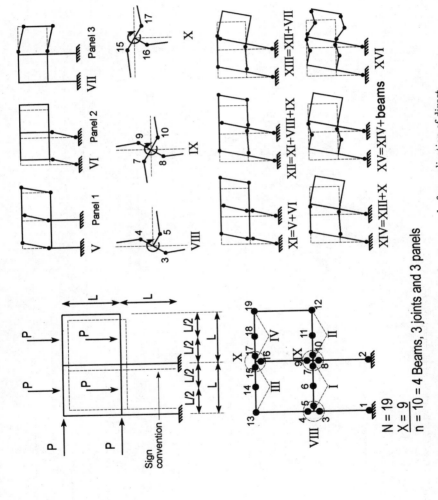

Figure 5.6 Two-story frame with an overhang example for application of direct combination of mechanism method.

There is no single strategy to find the combined mechanism that will produce the lowest plastic collapse load because the particulars of a given problem will suggest different approaches. (In fact, even for a given structure, different engineers may adopt different search sequences.) Here, results for the basic mechanisms indicate that the panel mechanisms produce the largest amount of external work, the lowest plastic collapse load being that of mechanism VI. Therefore, as an arbitrary starting point, one could first try to combine some panel mechanisms to increase the amount of external work. However, as shown by the combination of mechanisms V and VI (noted XI in the table), this also results in significantly more internal work because the plastic hinges of the two mechanisms are additive. This, as well as the sketch of this first combination, suggests that joint mechanisms must be introduced to reduce internal work. A combination of mechanisms V, VI, VIII, and IX is attempted (mechanism XII), but only one plastic hinge is eliminated in the process.

Examination of the results so far reveals that adding panel 3 would reduce the amount of internal work (trading hinges 10, 17, and 12 for 19) and increase the amount of external work. This is tried with mechanism XIII, which gives the lowest value of the plastic collapse load at this point. As examination of that mechanism in Figure 5.6 suggests, adding mechanism X removes one further plastic hinge (mechanism XIV) and gives a lower value for the collapse load. Mechanism XV investigates if the additional external work produced by the beam mechanisms exceeds the extra internal work thus introduced; the resulting plastic collapse load is found to be higher than for mechanism XIV. Although one could attempt some additional trial combinations if deemed necessary (such as mechanism XVI in Figure 5.6), it appears that no other combination of mechanisms would provide a lower value for the plastic collapse load. The moment diagram for mechanism XIV (not shown here) is found to be statically admissible confirming that the true solution has been found.

As demonstrated by the above examples, without engineering judgment, the direct combination method, even using a systematic procedure, can rapidly become excruciatingly laborious as structural complexity grows. Partly for that reason, the method presented in the following section, although nearly suitable only for implementation in computer programs, is preferable when one is dealing with extremely large problems (or for small problems whenever one is ready to trade control against expediency, reserving engineering judgment for the task of checking the validity of the generated computer results).

5.3 Method of inequalities

Although the systematic method of plastic analysis formulated in the previous section can also be computerized, the most computationally efficient solution procedure for plastic analysis is the method of inequalities, based on a matrix formulation of the lower bound method. However, because formal presentation of this mathematical procedure would be abstract, an example is presented instead to illustrate how a systematic lower bound method can proceed from the method of inequalities.

Considering the small size of this example problem, the corresponding amount of computations necessary for the solution presented hereafter may appear excessive, particularly when one can find the correct answer within minutes and with a minimum of calculations by using the graphical approach presented earlier. Nonetheless, the principal objective here is to illustrate how a systematic computerized procedure can be designed to automatically converge to the correct solution. In the process, for further clarity, expanded equations instead of matrix notation have been used (the matrix methods implemented in computer programs are too numerically intensive for hand calculations).

The frame considered in this example is shown in Figure 5.7, along with the four potential plastic hinge locations corresponding to the loading condition. Given that this frame is indeterminate to the second degree ($x = 2$), it is possible to express the moments at each of the potential plastic hinge locations as a function any two arbitrarily selected redundant forces. The chosen sign convention, indicated by the dotted line, is drawn along the frame members in Figure 5.7. In this problem, the two reactions at the right support are selected as the redundant forces, and a set of equations can be written for the moment at the potential plastic hinge locations as a function of these forces.

$$M_4 = -2RL \tag{5.2a}$$

$$M_3 = -2RL + VL \tag{5.2b}$$

$$M_2 = -2RL + 2VL - 2FL \tag{5.2c}$$

$$M_1 = 2VL - 2FL - 6FL \tag{5.2d}$$

This set of four equations cannot be solved because it contains six unknowns (four moments and two redundant forces). However, one can eliminate the two redundant forces using algebraic manipulations and substitutions. As a result, the above system of equations is reduced to the following two equations and four unknowns:

$$-M_2 + 2M_3 - M_4 = 2FL \tag{5.3a}$$

$$-M_1 + M_2 - M_4 = 6FL \tag{5.3b}$$

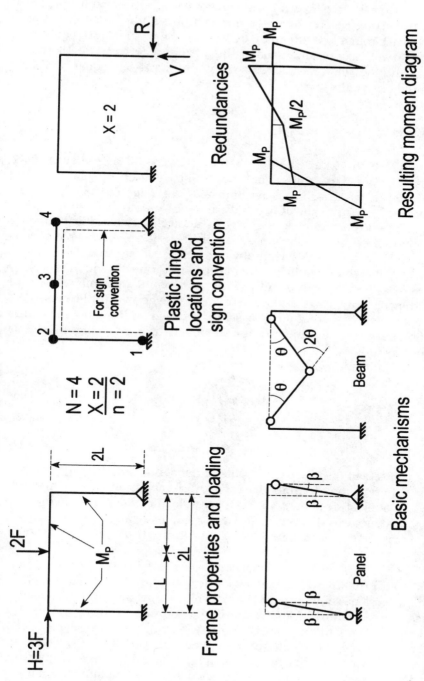

Figure 5.7 Frame example of the method of inequalities.

Note that these two equations represent two equilibrium equations for this structure. Incidentally, one can obtain the same two equations, with much less difficulty, by writing the virtual-work equations for the two basic mechanisms of this structure. Hence, for the basic beam and the panel mechanisms shown in Figure 5.7, equating internal and external work (Chapter 4) gives:

$$-M_2\theta + 2M_3\theta - M_4\theta = 2FL\theta \quad (5.4a)$$

$$M_2\theta - M_1\theta - M_4\theta = 6FL\theta \quad (5.4b)$$

from which the θs can be eliminated. This should not be surprising because, in essence, work equations are also equilibrium equations.

However, for these basic mechanisms, the hinge moments, M_i, are expressed not in terms of M_p, but rather as unknowns whose values remain to be determined. For large structural systems, this approach to obtain a set of equilibrium equations is preferred because it is considerably easier to write virtual-work equations for a number of basic mechanisms than equilibrium equations in terms of redundants.

Nonetheless, at this point, the above remains a system with more unknowns (four) than equations (two). However, within the context of plastic design, a solution is possible because additional constraints exist that limit the magnitude of moment at each potential plastic hinge as follows:

$$-M_p \leq M_1 \leq M_p \quad (5.5a)$$

$$-M_p \leq M_2 \leq M_p \quad (5.5b)$$

$$-M_p \leq M_3 \leq M_p \quad (5.5c)$$

$$-M_p \leq M_4 \leq M_p \quad (5.5d)$$

In practical situations, the plastic moments would likely differ at the different plastic hinge locations; different positive and negative values of plastic moment are even possible at a given location for composite structures. However, for this example, a single value of plastic moment is assumed throughout the entire frame for the sake of simplicity.

The above equilibrium equations and inequality relationships, coupled with the knowledge that a lower bound approach is a search for the largest possible applied loads (while respecting all above equalities and inequalities), make possible a solution to this problem. The algorithm to achieve such a solution typically uses the equilibrium equations and the inequality conditions to systematically eliminate the unknowns, one by one, performing all possible cross-comparisons of inequality equations in the process, until all remaining equations

Systematic Methods of Plastic Analysis 169

are expressed in terms of known quantities. One can then find the largest possible value for the applied loads (often expressed as a function of a common load, F) simply by scanning all resulting inequalities. The largest value of F that satisfies all resulting constraints gives the collapse load.

For the example at hand, the equilibrium equations are first used to express two of the remaining unknowns in terms of the others:

$$M_3 = FL + \frac{1}{2}(M_2 + M_4) \tag{5.6a}$$

$$M_1 = -6FL + M_2 - M_4 \tag{5.6b}$$

These results from the equilibrium equations are then substituted in the appropriate inequalities:

$$-M_p \leq -6FL + M_2 - M_4 \leq M_p \tag{5.7a}$$

$$-M_p \leq M_2 \leq M_p \tag{5.7b}$$

$$-M_p \leq FL + \frac{1}{2}(M_2 + M_4) \leq M_p \tag{5.7c}$$

$$-M_p \leq M_4 \leq M_p \tag{5.7d}$$

From this point on, further elimination of unknowns through the inequality relationships proceeds by construction of all possible inferences from the inequality set. For example, to eliminate M_2 from the inequality set, all inequality equations that contain M_2 must be rearranged to isolate that term within the inequalities. This gives:

$$-M_p \leq M_2 \leq M_p \tag{5.8a}$$

$$-2M_p - 2FL - M_4 \leq M_2 \leq 2M_p - 2FL - M_4 \tag{5.8b}$$

$$-M_p + 6FL + M_4 \leq M_2 \leq M_p + 6FL + M_4 \tag{5.8c}$$

Then the left side of the inequalities given in Equation 5.8 are systematically compared with the right side of the same inequalities. More explicitly, the left side inequality of Equation 5.8a is compared with the right side inequalities of Equations 5.8a, 5.8b, and 5.8c; the left side inequality of Equation 5.8b is compared with the same three right side inequalities; and finally, the left side of Equation 5.8c is similarly compared. This gives:

$$-M_p \leq M_p \tag{5.9a}$$

$$-M_p \leq 2M_p - 2FL - M_4 \tag{5.9b}$$

$$-M_p \leq M_p + 6FL + M_4 \tag{5.9c}$$

$$-2M_p - 2FL - M_4 \leq M_p \tag{5.9d}$$

$$-2M_p - 2FL - M_4 \le 2M_p - 2FL - M_4 \qquad (5.9e)$$

$$-2M_p - 2FL - M_4 \le M_p + 6FL + M_4 \qquad (5.9f)$$

$$-M_p + 6FL + M_4 \le M_p \qquad (5.9g)$$

$$-M_p + 6FL + M_4 \le 2M_p - 2FL - M_4 \qquad (5.9h)$$

$$-M_p + 6FL + M_4 \le M_p + 6FL + M_4 \qquad (5.9i)$$

Equations 5.9a, 5.9e, and 5.9i are automatically satisfied. Then, by finding matching pairs, one can write equation 5.10a (from 5.9c and 5.9g), equation 5.10b (from 5.9b and 5.9d), and 5.10c (from 5.9 f and 5.9h) as follows:

$$-2M_p - 6FL \le M_4 \le 2M_p - 6FL \qquad (5.10a)$$

$$-3M_p - 2FL \le M_4 \le 3M_p - 2FL \qquad (5.10b)$$

$$-\frac{3}{2}M_p - 4FL \le M_4 \le \frac{3}{2}M_p - 4FL \qquad (5.10c)$$

$$-M_p \le M_4 \le M_p \qquad (5.10d)$$

By repeating the above process to eliminate M_4, that is, by systematically comparing each left side of the inequalities given in Equation 5.10 with the corresponding four right sides of the same set of inequality equations, one obtains the following set of 16 inequalities:

$$-2M_p - 6FL \le 2M_p - 6FL \qquad (5.11a)$$

$$-2M_p - 6FL \le 3M_p - 2FL \qquad (5.11b)$$

$$-2M_p - 6FL \le \frac{3}{2}M_p - 4FL \qquad (5.11c)$$

$$-2M_p - 6FL \le M_p \qquad (5.11d)$$

$$-3M_p - 2FL \le 2M_p - 6FL \qquad (5.11e)$$

$$-3M_p - 2FL \le 3M_p - 2FL \qquad (5.11f)$$

$$-3M_p - 2FL \le \frac{3M_p}{2} - 4FL \qquad (5.11g)$$

$$-3M_p - 2FL \le M_p \qquad (5.11h)$$

$$-\frac{3M_p}{2} - 4FL \le 2M_p - 6FL \qquad (5.11i)$$

$$-\frac{3M_p}{2} - 4FL \le 3M_p - 2FL \qquad (5.11j)$$

$$-\frac{3M_p}{2} - 4FL \le \frac{3M_p}{2} - 4FL \qquad (5.11k)$$

$$-\frac{3M_p}{2} - 4FL \le M_p \qquad (5.11l)$$

$$-M_p \leq 2M_p - 6FL \tag{5.11m}$$

$$-M_p \leq 3M_p - 2FL \tag{5.11n}$$

$$-M_p \leq \frac{3M_p}{2} - 4FL \tag{5.11o}$$

$$-M_p \leq M_p \tag{5.11p}$$

By eliminating equations 5.11a, 5.11f, 5.11k, and 5.11p, which provide no useful information, one obtains the following 12 inequalities for F, respectively:

$$-\frac{5M_p}{4L} \leq F \tag{5.12a}$$

$$-\frac{7M_p}{4L} \leq F \tag{5.12b}$$

$$-\frac{M_p}{2} \leq F \tag{5.12c}$$

$$F \leq \frac{5M_p}{4L} \tag{5.12d}$$

$$F \leq \frac{9M_p}{4L} \tag{5.12e}$$

$$-2M_p \leq F \tag{5.12f}$$

$$F \leq \frac{7M_p}{4L} \tag{5.12g}$$

$$-\frac{9M_p}{4L} \leq F \tag{5.12h}$$

$$-\frac{5M_p}{8L} \leq F \tag{5.12i}$$

$$F \leq \frac{M_p}{2} \tag{5.12j}$$

$$F \leq \frac{2M_p}{L} \tag{5.12k}$$

$$F \leq \frac{5M_p}{8L} \tag{5.12l}$$

The range of values of F that can satisfy all the above inequalities is $-M_p/2 \leq F \leq M_p/2$, or in absolute value, $F = M_p/2$. From this result, one can calculate the actual value of each previously eliminated unknown by going in reverse through the elimination order. Hence, one calculates M_4 first by substituting the value for F into equation

5.10. This gives, for the positive value of F (negative value would simply give a reversed moment diagram):

$$(-2M_p - 3M_p = -5M_p) \leq M_4 \leq (-M_p = 2M_p - 3M_p) \quad (5.13a)$$

$$(-3M_p - M_p = -4M_p) \leq M_4 \leq (2M_p = 3M_p - M_p) \quad (5.13b)$$

$$(-1.5M_p - 2M_p = -3.5M_p) \leq M_4 \leq (-0.5M_p = 1.5M_p - 2M_p) \quad (5.13c)$$

$$-M_p \leq M_4 \leq M_p \quad (5.13d)$$

These inequalities (in particular Equations 5.13a and 5.13d) constrain M_4 to be equal to $-M_p$, which indicates presence of a plastic hinge at that location. Then, substituting this result together with the value of F into equation 5.8 gives:

$$-M_p \leq M_2 \leq M_p \quad (5.14a)$$

$$(-2M_p - 2(\tfrac{1}{2})M_p - (-M_p) = -2M_p) \leq M_2 \leq (2M_p = 2M_p - 2(\tfrac{1}{2})M_p - (-M_p)) \quad (5.14b)$$

$$(-M_p + 6(\tfrac{1}{2})M_p + (-M_p) = M_p) \leq M_2 \leq (3M_p = M_p + 6(\tfrac{1}{2})M_p + (-M_p)) \quad (5.14c)$$

From those inequalities it follows that M_2 must be equal to M_p. Finally, when one substitutes all previously obtained results into the equilibrium equations (Equation 5.6), the remaining two unknowns can be calculated:

$$M_3 = \tfrac{1}{2}M_p + \tfrac{1}{2}(M_p - M_p) = \tfrac{1}{2}M_p \quad (5.15a)$$

$$M_1 = -6\left(\tfrac{1}{2}\right)M_p + M_p - (-M_p) = -M_p \quad (5.15b)$$

These results indicate, according to the previously defined sign convention, that the plastic collapse mechanism for this structure for the given loading condition is the basic panel mechanism. The resulting statically admissible moment diagram is shown in Figure 5.7.

Formally, for any given structure, a computerized solution is possible provided that the following two mathematical constructions can be formulated:

- A set of $n = N\text{-}X$ equilibrium equations, obtained directly from the basic independent mechanisms expressed in the following matrix format:

$$\begin{bmatrix} c_{11} & c_{12} & c_{13} & c_{14} & \cdots & c_{1N} \\ c_{21} & c_{22} & c_{23} & c_{24} & \cdots & c_{2N} \\ c_{31} & c_{32} & c_{33} & c_{34} & \cdots & c_{3N} \\ \cdots \\ c_{n1} & c_{n2} & c_{n3} & c_{n4} & \cdots & c_{nN} \end{bmatrix} \begin{bmatrix} M_1 \\ M_2 \\ M_3 \\ M_4 \\ \cdots \\ M_N \end{bmatrix} = \begin{bmatrix} k_1 \\ k_2 \\ k_3 \\ k_4 \\ \cdots \\ k_N \end{bmatrix} P \qquad (5.16)$$

- A series of N inequalities expressing the plastic moment constraints at each potential plastic hinge location, such that:

$$-M_{p1} \leq M_1 \leq M_{p1}$$
$$-M_{p2} \leq M_2 \leq M_{p2}$$
$$\cdots \qquad (5.17)$$
$$-M_{pN} \leq M_N \leq M_{pN}$$

The matrix solution of this problem per the philosophy expressed in the above example can be accomplished through use of the standard simplex method algorithm developed for multidimensional optimization in linear programming (Livesley 1975, Cambridge 1992, Dantzig 1963).

References

1. Cambridge University Press. 1992. *Numerical Recipes in Fortran*. Cambridge, Massachusetts.
2. Dantzig, G.B. 1963. *Linear Programming and Extensions*. Princeton, New Jersey: Princeton University Press.
3. Livesley, R.K. 1975. *Matrix Methods of Structural Analysis, 2nd Edition*. Permagon Press.

Chapter

6

Applications of Plastic Analysis

Previous chapters have demonstrated how the plastic collapse mechanism for a given structure and set of loads can be determined. Although the techniques described in those chapters are conceptually simple, computations can rapidly become tedious when one is analyzing complex structures. Fortunately, from a design perspective, engineers familiar with plastic-analysis concepts can effectively couple this knowledge with the creative freedom of the design process to eliminate many of these analysis complications. As a result, many plastic-*design* tools can be, and have been, created. These vary greatly from one application to the other because different tools are needed for different tasks. In some, numerous plastic-based design solutions can be rapidly developed and assessed. In other cases, the engineer can decide a priori which specific plastic collapse mechanisms are desirable and design the structure such that no other failure mechanism can develop.

The number of applications using the plastic-design concepts has grown considerably in recent years; only a sample of these can be presented in this chapter wherein the concepts are stressed over the details. More comprehensive examples are provided in the following chapters in which emphasis is on the earthquake-resistant design of ductile steel structures, which implicitly considers many of the concepts presented in this chapter.

Incidentally, for design, specified applied loads are multiplied by load factors, and various load combinations must be considered. Currently, slight variations exist in the magnitude of those factors and the required load combinations among the various international codes and standards based on Limit States Design (LSD) or Load and Resistance Factor Design (LRFD). Although load factors have not been used so far in this book to provide a broader presentation of concepts untied to any particular code, it must be understood that they should always be considered. Likewise, in a design process, the plastic moments provided should always be slightly larger than the required values calculated by analysis. In compliance with the LRFD (and LSD) philosophy, the required plastic moments should be reduced by the appropriate resistance factor (typically expressed as ϕ). This will be illustrated in the design examples provided in later chapters.

6.1 Moment redistribution design methods

6.1.1 Statical method of design

Using the statical method to design continuous beams constitutes one of the simplest possible applications of plastic analysis principles. In the example shown in Figure 6.1, a 4-span continuous beam must be designed to resist a set of factored point loads identified to constitute the critical load-combination. As for most design processes, many satisfactory solutions can be found.

If the beam is constrained to be one continuous member of constant cross-section over its entire length, the statical method (i.e., superposition of the moment diagrams for span loading [Figure 6.1b] and redundants [Figure 6.1c] is schematically shown in Figure 6.1d) indicates that the left span will reach its beam plastic collapse mechanism first. This solution is shown in Figure 6.1e. Systematically, one could find this by expressing mathematically the statical moment combination for each span, although spans visibly not critical need not be checked. For example, for span AC, for the beam plastic collapse mechanism to form, considering all moments in absolute value, the following expression can be written:

$$\frac{M_A + M_C}{2} + M_B = \frac{0 + M_p}{2} + M_p = 250 \text{ kN-m } (184 \text{ k.ft}) \tag{6.1}$$

which gives $M_p = 166$ kN-m (123 k-ft), or the minimum strength that must be provided to resist the specified loads. Following the same logic, $M_p = 160$ kN-m (118 k-ft) would be obtained for span C E. Spans E G and G I would not need to be checked because it can be deducted visually from Figure 6.1e that they would not govern. Therefore, a beam having a plastic moment of 166 kN-m would be required as a minimum satisfactory design. This approach makes for an expedient design process.

However, as a result of the above design, all spans except span AC would be overdesigned, and it may be of interest to investigate whether a minimum weight solution would provide much savings in material. Although systematic optimum design techniques have been developed for plastic analysis, these are beyond the scope of this book. Nonetheless, using simple reasoning, one can find a more economical solution. Observation of the result previously obtained reveals that span GI would be the most severely overdesigned because its statical moment is only 187.5 kN-m (138 k.ft). Therefore, one could focus initially on that span. Writing an expression similar to Equation 6.1, one finds a required plastic moment of 93.75 kN-m (69.1 k-ft) (Figure 6.1f). Note that because of the change in cross-section over span GI, the maximum value of M_G can now be only 93.75 kN-m. As a result, the required plastic moment capacity of span EG is given by:

$$\frac{M_E + M_G}{2} + M_F = \frac{M_p + 93.75}{2} + M_p = 240 \text{ kN-m (177 k-ft)} \qquad (6.2)$$

which translates into a required plastic moment of 128.75 kN-m (94.9 k-ft). Repeating the process for span CE would give:

$$\frac{M_C + M_E}{2} + M_D = \frac{166 + 128.75}{2} + M_p = 320 \text{ kN-m (236 k-ft)} \qquad (6.3)$$

and a required moment of 172.65 kN-m (127 k.ft). Although this required plastic moment exceeds the previous value of 166 kN-m, an overall saving is achieved when the new weight of the entire structure is compared with the previous solution (Figure 6.1g). This optimization is somewhat academic because many other factors would need to be addressed in a practical design. First, no beam would be spliced immediately over the support, and it may be more economical to extend the heavier beams a short distance beyond the supports, allowing for more efficient use of material at minimal extra cost. Moreover, consideration would be given to fabrication costs, maximum member length (for transportation logistic), and erection costs. For those reasons, the optimization process illustrated is not pursued further. However, the above concepts can be easily adapted to address whatever practical considerations are encountered.

Note that for moment redistribution to occur, significant plastic rotations need to develop at the plastic hinges. Therefore, it is important that only properly laterally supported compact sections be used in all plastic design applications (this also requires braces at all potential plastic hinge locations).

6.1.2 Autostress design method

The Autostress Design Method (ADM), which is an integral part of the Load and Resistance Factor Design (LRFD) edition of the AASHTO Bridge Design Code, is a relatively new design procedure that recognizes

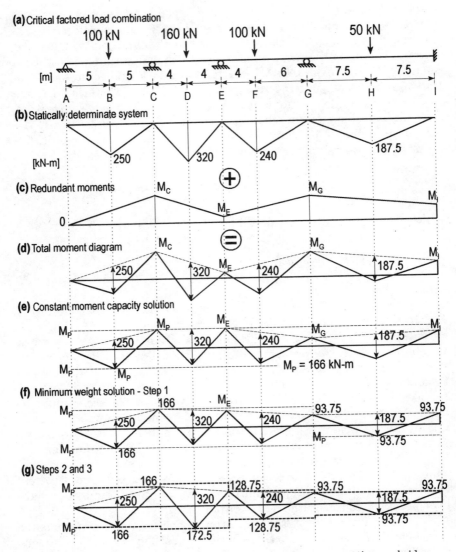

Figure 6.1 Example of statical method of design using four-span continuous bridge.

the ability of continuous steel members to redistribute moments plastically. It takes advantage of the fact that a steel structure will shakedown under a truck overload and thereafter behave elastically over an extended range (the shakedown range, as described in Chapter 4). In a less abstract way, the fundamental concept underlying the Autostress Design Method is usually explained by the following analogy (e.g., Haaijer et al. 1983).

If a two-span continuous steel bridge girder could be lifted in one piece and placed over three non-level supports, as shown in Figure 6.2a, with the middle support higher than the two exterior supports, the girder would bend under its own weight until it came in contact with the three supports (Figure 6.2b). In the process, given a sufficient difference in the height of the supports, the plastic moment of the girder would be reached at the middle support. The resulting moment diagram is shown in Figure 6.2c.

This process is equivalent to introducing into the girder a residual moment diagram varying linearly from 0 at the exterior supports to $(M_p - \omega L^2/12)$, if the dead load could be removed, as shown in Figure 6.2d (see Section 4.5). Upon first crossing of the maximum overload across that bridge, additional plastic rotation will develop over support C, but unloading and reloading will be elastic over an extended range without further plastic rotations (Figure 6.2f). At that point, the girder has shaken down according to the definition presented in Chapter 4.

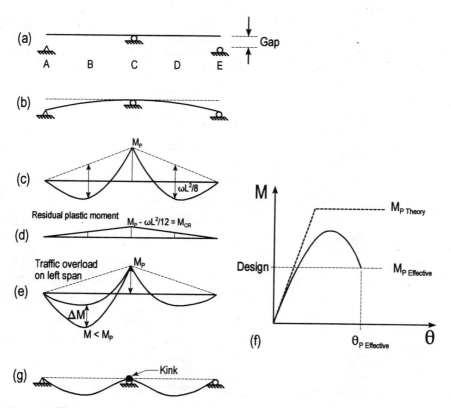

Figure 6.2 Illustration of the Autostress design method.

However, it is necessary neither to place the girders on unequal supports nor to jack up the middle support after construction (as some astute engineers might propose) to reach the above shakedown condition. In fact, no physical intervention is needed other than a large traffic overload. Indeed, the plastic hinge at support C and an equivalent residual moment diagram will automatically develop once a live load sufficiently large to produce yielding at that location occurs. The structure will then behave elastically until a larger load occurs. That larger load will introduce some additional plastic rotations at point C and extend further the elastic range. Then all subsequent loads of equal or lesser magnitude will be resisted elastically. The maximum load resistance will obviously be dictated by the plastic collapse mechanism that could be calculated by the methods shown in Chapter 4. The term Autostress emphasizes that this plastic moment redistribution occurs automatically.

Special design requirements must be satisfied to ensure that the steel girders designed under the Autostress Design Method will have sufficient plastic rotation capacity. Because most steel bridge girders are generally noncompact shapes that do not meet the width-to-thickness limits generally prescribed for plastic design, effective slenderness ratio and effective plastic moment concepts have been derived and verified experimentally. A detailed review of these concepts is not presented in this book, but the underlying general ideas are as follows:

- Noncompact steel beams will typically reach a maximum strength smaller than their theoretical M_p because of local buckling.
- The effective stresses at which local buckling of the flanges and web initiates can be obtained by algebraic manipulations of the design equations limiting width-to-thickness ratios.
- The ratio of the effective stress to the yield stress can be calculated.
- This ratio multiplied by the theoretical plastic moment of the cross-section defines an effective plastic moment used for design.

These concepts have been proven to provide a satisfactory design basis. Indeed, when a noncompact section is subjected to progressively increasing rotations, after reaching its maximum moment (which is larger than the calculated effective plastic moment because of strain hardening and other factors), it slowly loses some flexural capacity. An effective plastic rotation capacity for a given section can therefore be defined as the point at which the descending branch of its moment–curvature relationship crosses the effective plastic moment defined by the above procedure (see Figure 6.2f); values obtained are typically greater than those required to achieve the shakedown condition described earlier.

Finally, as shown in Figure 6.2g, a kink in the girders will obviously be introduced as a result of the overload and plastic hinge formation at point C. However, calculations and experimental results confirm that this kink is small and unlikely to be visually perceptible or felt by vehicles driving over the bridge.

6.2 Capacity design

6.2.1 Concepts

The concept of capacity design is very important in earthquake engineering practice, and although a pure capacity design approach has not been adopted in North America at this time, aspects of this philosophy are implicitly embedded in many code-detailing requirements (for both reinforced concrete and steel structures).

Capacity design was developed in the late 1960s in New Zealand as an approach to resist the effects of severe earthquakes. In capacity design, acknowledging that inelastic action is unavoidable during severe earthquakes, the designer dictates where inelastic response should occur. Such zones of possible inelastic action are selected to be regions where large plastic deformations can develop without significant loss of strength; these regions are detailed to suppress premature undesirable failure modes, such as local buckling or member instability in the case of steel structures. Then, one eliminates the likelihood of inelastic action or failure elsewhere in the structure by making the capacities of the surrounding structural members greater than that needed to reach the maximum capacity of the so-called plastic zone.

The classical example to explain this concept is the capacity-designed chain (Figure 6.3). In this chain, one link is designed to absorb a large amount of plastic energy in a stable manner prior to failure (e.g., link 4). Therefore, the other links (e.g., 1, 2, 3, 5, 6, and 7) can be designed without concern for plastic deformations, provided their capacities exceed the maximum capacity of the plastic link, thus avoiding the need for special detailing in all but one link.

Many other examples can be created based on the same philosophy. One such illustration of capacity design is shown in Figure 6.4. There, a cantilever beam of total length L consists of a brittle segment (such as a fiber composite material) of length a at the fixed end and a ductile steel segment of length b. A traditional design approach would require the use of large safety factors to provide protection against failure of the brittle material. Alternatively, a capacity design approach would aim at making the brittle material stronger than

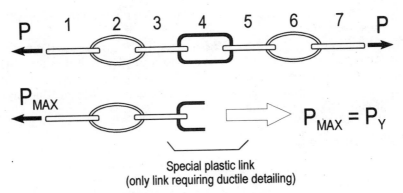

Figure 6.3 Illustration of capacity design principle.

needed to ensure that plastic hinging occurs first in the steel segment of the cantilever. Therefore, the moment resistance of the brittle segment would only need to exceed:

$$M_{BRITTLE} \geq \alpha \left(\frac{L}{b}\right) M_{P\text{-}STEEL} \qquad (6.4)$$

where all parameters are defined in Figure 6.4, and α is a number greater than 1.0 to account for the possible reserve strength of the steel cantilever beyond its nominal yield strength.

Clearly, capacity design is deeply rooted in plastic analysis and design. In theory, once a fully plastic state (also known as a plastic collapse mechanism) has been reached, no additional force can be imparted to the structure, and as a result, regions outside the critical plastic locations are protected against the effects of additional loading. For the small one-bay moment frame shown in Figure 6.5, if the plastic moment capacity of the beam is less than that of the columns, yielding will occur only at the base of the columns and at the ends of the beam. The rest of the structure is therefore certain to remain elastic (Figure 6.5) and requires no special ductile detailing. Practically, this remains true, although some allowance (such as the α factor in the previous example) must be made for the statistical variability of material properties (particularly the yield stress), the possible development of strain hardening in the critical plastic locations, dynamic-loading effects (i.e., strain-rate effects), and a few other case-dependent factors.

6.2.2 Shear failure protection

Capacity design can be used to check the potential adverse impact of nonstructural elements on key structural members. For example, in a frame subjected to lateral loads, the shear force in columns can be

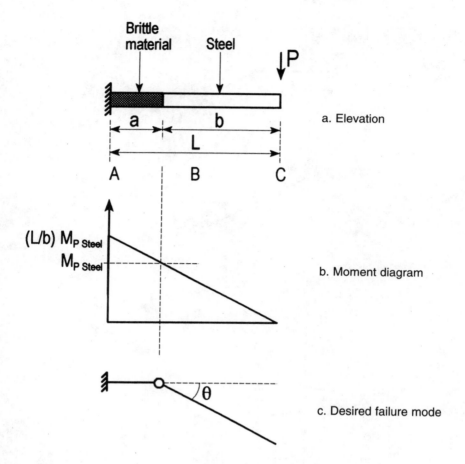

Figure 6.4 Illustration of an application of capacity design.

considerably larger than expected because of the presence of rigid non-structural elements not considered during the design process. This is clearly illustrated in Figure 6.6, in which a rigid partial-infill masonry wall restrains the elastic deformations of the steel columns. As a result, the plastic hinges required in the columns to produce a plastic collapse mechanism must relocate from the base of the columns to just above the infill where frame-action is unrestrained. A higher lateral force, H, is required to develop the collapse mechanism, and higher shear strength is required of the structural members and the connections to ensure development of this ductile mechanism. Mathematically, using simple free-body diagrams, the column shear strength required to form plastic hinges in this frame with masonry infill is:

$$V = \frac{2 M_P}{h^*} \tag{6.5}$$

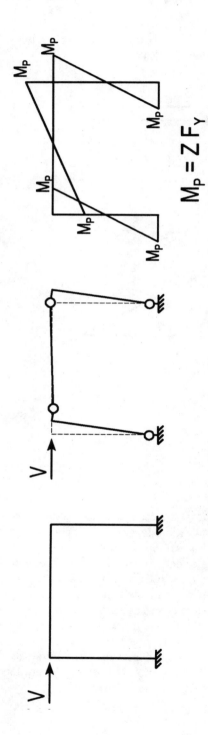

Figure 6.5 Illustration of a capacity design application.

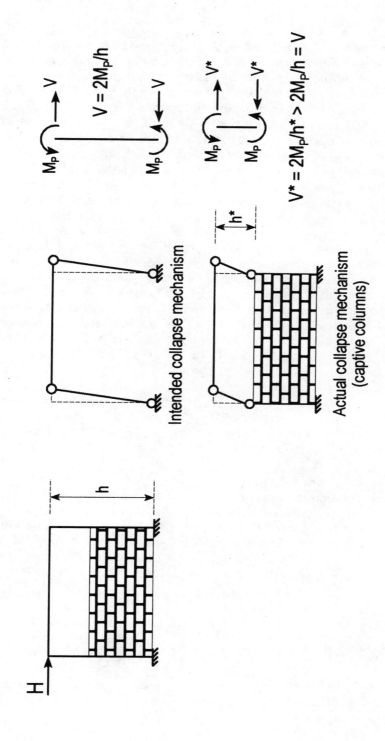

Figure 6.6 Impact of rigid nonstructural elements on shear force in columns.

where h is the unrestrained column height, as shown in Figure 6.6. This shear strength is h/h^* more than the shear strength that would have been sufficient to permit plastic hinges to form in the bare frame without any infills, where h is the full column height.

Fortunately, contrary to reinforced concrete columns for which this phenomenon has created a number of disastrous failures in past earthquakes (referred to as "short-column" or "captive column" failures in the literature), steel columns usually have a constant shear strength that is in excess of that required to form plastic hinges, and to date, steel columns have not suffered the same fate as some reinforced concrete columns. However, designers should be aware of this phenomenon and recognize instances in which it could lead to problems. For example, column splices located in such captive columns could be damaged if they are designed without consideration of the nonstructural walls.

A similar strategy can be used to protect against shear failures in gravity-resisting members. In this case, the effect of gravity loads must be considered, as must the fact that positive and negative moment capacities may differ (e.g., in composite constructions). This is illustrated in Figure 6.7. For example, for a segment of beam between two plastic hinges and subjected to a uniformly distributed load, the gravity shear force diagram must be added to the shear force diagram corresponding to the plastic moments, with the following result:

$$V_{left} = V_{gL} + V_{MP} = \frac{\omega L}{2} - \left[\frac{M_{PR} + M_{PL}}{L}\right]$$
$$V_{right} = V_{gR} + V_{MP} = \frac{\omega L}{2} + \left[\frac{M_{PR} + M_{PL}}{L}\right]$$
(6.6)

where all terms are defined in Figure 6.7.

Similar relationships could be derived for other loading distributions between the plastic hinges as shown in Figure 6.7.

Likewise, using these principles and the free-body diagrams presented in Figure 6.8, the maximum axial load that can be applied to columns at story i of a multistory frame as a result of a sway-type plastic collapse mechanism can be calculated as:

$$C_{max\text{-}i} = \sum_{i}^{n} \left[V_{gR\text{-}max\text{-}i} + V_{MP\text{-}i}\right] = \sum_{i}^{n} \left[\frac{\omega_{max\text{-}i} L}{2} + \left(\frac{M_{PR\text{-}i} + M_{PL\text{-}i}}{L}\right)\right]$$
$$T_{max\text{-}i} = \sum_{i}^{n} \left[V_{gL\text{-}min\text{-}i} - V_{MP\text{-}i}\right] = \sum_{i}^{n} \left[\frac{\omega_{min\text{-}i} L}{2} - \left(\frac{M_{PR\text{-}i} + M_{PL\text{-}i}}{L}\right)\right]$$
(6.7)

Compression is arbitrarily taken to be positive in that equation. An understanding of the concept used to derive this equation matters

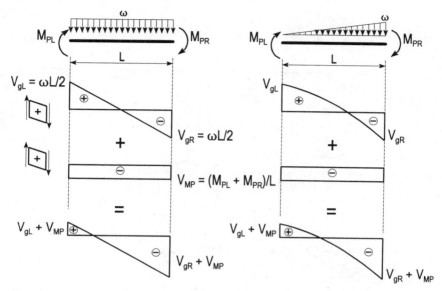

Figure 6.7 Examples of maximum shear force calculation in beams using capacity design principles.

more than the equation itself. For example, the same approach could be used to assess the impact of extreme load conditions, such as loss of a column due to an explosion or other causes, as shown in Figure 6.9.

6.2.3 Protection against column hinging

For many reasons, beam yielding is generally preferable to column yielding, particularly in multistory frames (see Chapter 8). Beam yielding greatly enhances the energy absorption capability of a structure because more plastic hinges are involved in the development of the plastic collapse mechanism. This is illustrated in Figure 6.10. In that example, for the same total roof displacement, the column plastic rotation demand for the column-sway mechanism is approximately eight times larger than the beam plastic rotation demand for the beam-sway mechanism, resulting in a greater risk of collapse because of limits in the plastic rotation capacity of structural members (see Chapters 8 and 10).

Although this philosophy, also known as "strong-column/weak-beam" design, has been widely accepted as desirable in reinforced concrete structures, its implementation in structural steel design code has met considerable resistance. In low-rise steel buildings, beams are generally considerably deeper than columns, and the adoption of such a philosophy may affect the economical balance between competing proposals in steel and other materials. However, many other capacity

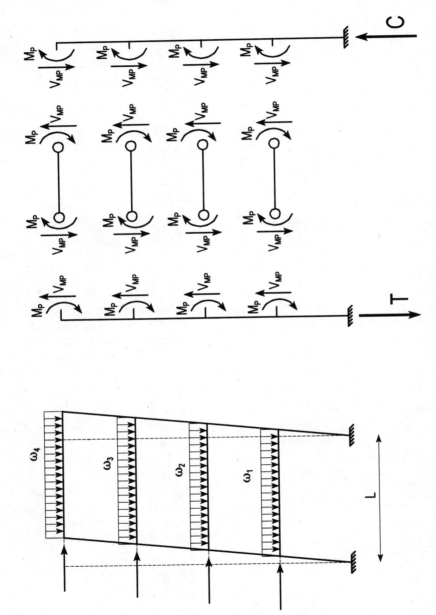

Figure 6.8 Calculation of maximum axial force in columns using capacity design.

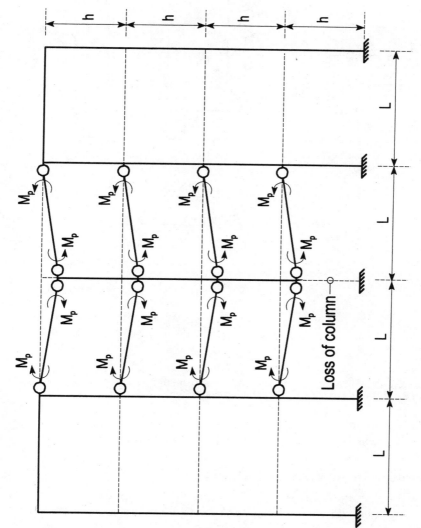

Figure 6.9 Plastic collapse mechanism due to loss of a column in a structural frame.

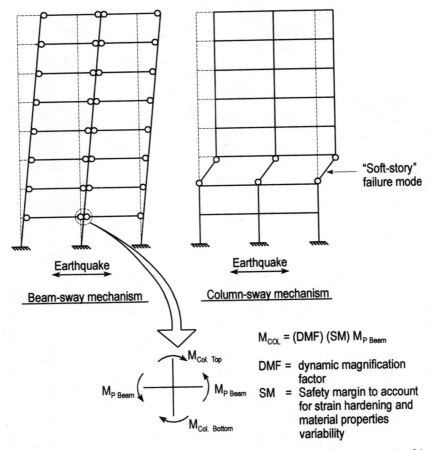

Figure 6.10 Comparison of plastic collapse mechanism in presence (beam sway) and in absence (column sway) of "strong-column/weak-girder" design philosophy.

design principles have found their way into steel design codes and standards, as will be seen in subsequent chapters.

6.3 Push-over analysis

A push-over analysis is basically a step-by-step plastic analysis for which the lateral loads of constant relative magnitude are applied to a given structure and progressively increased until a target displacement is reached, while gravity loads are kept constant. Thus, as the name implies, the structure is truly pushed sideways (or pushed over) to determine its ultimate lateral-load resistance as well as the sequence of yielding events needed to reach that goal, or the magnitude of plastic deformations at the target displacement. Typically, many engineers have accomplished this by repeatedly running linear elastic structural analysis computer programs, modifying the model of a structure as necessary to account for the progressive appearance of plastification at finite locations throughout the structure. As a result, the push-over analysis method is relatively accessible and has been used in addition to conventional analyses to determine the ultimate capacity of important existing structures, to validate proposed retrofit or design solutions, and to compare the ultimate capacity and, to some extent, the ductility of various design alternatives.

Although, in principle, nothing precludes the extension of this concept to conduct cyclic push-over analysis for a limited number of cycles, this rapidly becomes excessively arduous without the help of special purpose computer programs. Consequently, nearly all practical applications of the push-over analysis so far have considered monotonically increasing lateral loads. However, only a limited amount of information can be extracted from noncyclic push-over analyses, and extrapolating the findings from those analyses may lead to erroneous conclusions. The examples presented in the subsections below have been constructed to illustrate some of these limitations and risks of misinterpretation.

Finally, it must be recognized that the information acquired from a push-over analysis is highly dependent on the lateral load distribution adopted (Lawson et al. 1994). Therefore, whenever the chosen lateral load distribution is intended to capture the possible effects of dynamic excitation, it may be wise to consider multiple lateral-load distribution patterns.

6.3.1 Monotonic push-over analysis

The three-story braced frame shown in Figure 6.11a was designed in four different ways to resist a set of statically applied lateral loads. First, a tension-only design was considered (Case I). In a tension-only design approach, the braces in compression are ignored and the tension braces are designed to resist all the applied loads. Such braced frame designs have been popular and are still used in nonseismic regions to provide wind resistance.

In Case I, the brace slenderness ratio, KL/r, was limited to 200 and 300, as suggested by some buildings codes, for members in compression and tension, respectively. Therefore, double angles back to back were chosen for the brace members.

In Case II, both the compression and tension members were designed to resist loads. Design was governed by the compression capacity of the brace members. As a result, bigger double-angle braces were necessary to provide a satisfactory design. In Case III, to reflect that some earthquake-resistant design requirements restrict the maximum brace slenderness ratio to less than for nonseismic applications, the frame was redesigned as done for Case II

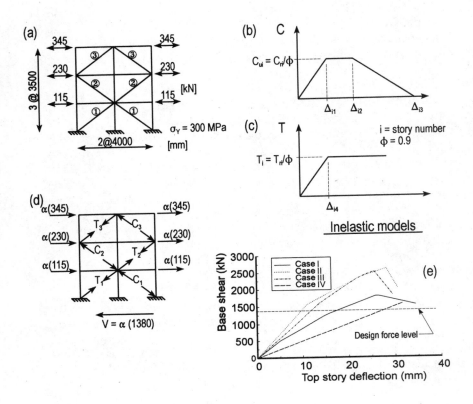

Figure 6.11 Monotonic push-over analysis examples on three-story braced frames with braces having various slenderness ratios.

but considering a maximum brace slenderness ratio of 110 which is the limit permitted for seismic design by some codes for steels having a specified yield strength of 300 MPa (43.5 ksi). As a result, W-shapes were chosen for the braces of that frame.

Finally, in Case IV, a tension-only design without any brace slenderness restrictions was undertaken, leading to braces made of steel plate. Information on the four resulting designs is presented in Table 6.1. In all cases here, the floor beams are joined to the columns using only shear connections (i.e., simply supported beams) because nothing prevents this practice in some parts of North America.

To simplify calculations, the inelastic models shown in Figure 6.11b and 6.11c were adopted for the compression (C) and tension (T) braces, respectively. Numerical values for the parameters of these models are listed in Table 6.1. The trilinear compression brace model represents the following states of behavior:

- Buckling occurs at a load of C_{ui}, which corresponds to a maximum elastic axial compression shortening of Δ_{i1}.
- When axial shortening reaches a value of Δ_{i2}, a flexural plastic hinge forms at the midlength of the brace as a result of the compression load acting eccentrically on the member that has buckled sideways.
- The axial strength progressively reduces to zero as a result of large plastic deformations that develop at the brace's midlength, reaching zero when the corresponding axial shortening reaches Δ_{i3}.

Values for Δ_{i2} were calculated by consideration of a simplified brace model, and the values of Δ_{i3} were arbitrarily set equal to twice Δ_{i2}. Likewise, only the contribution of brace elongation/shortening to the story drift is considered.

Push-over analysis was conducted to determine the base shear strength, V, of each proposed design and to obtain some knowledge of its ductile behaviors. Throughout this example, a factor α is used to express the ratio of the calculated capacity over the specified design loads. As shown in Figure 6.11d, all applied loads are scaled by this same value of α. Note that although factored member strengths ($\phi=0.9$) have been used for design purposes, the ultimate resistances ($\phi=1.0$) are used for all subsequent calculations in this section. Therefore, $\alpha = 1.11$ (i.e., 1/0.9) would correspond to a maximum base shear force equal to that required by design. Numerical results for the push-over analyses are presented in Table 6.2.

For Case I, the application of monotonically increasing lateral loads reveals that buckling of braces at the third, second, and first stories occurs sequentially and at loads much below those specified (i.e., at values of α equal to 0.364, 0.883, and 0.939, respectively, all less than 1.11). This is expected because the behavior of compression members was disregarded during the tension-only design. Tension yielding of a brace first occurs at the third story when α reaches 1.294. At that point, the axial displacement in the compression and tension braces (assuming the floor beams are axially rigid) is Δ_{i4}, and the axial force in the buckled compression member equals C_{ui} per Figure 6.11b because Δ_{i2} is larger than Δ_{i4}. However, as soon as Δ_{i2} is exceeded, both the force in the compression member and the shear strength of the third story drop. A stable story mechanism develops when $\alpha=1.11$—that is, exactly at the design level.

For Case II, buckling of braces simultaneously starts at the first and second stories ($\alpha=1.124$) and rapidly spreads to the third story ($\alpha=1.176$). These events occur at loads slightly above the design level. However, a significant strength capacity is available beyond that point as a result of the design philosophy that considers the resistance of both the tension and compression members. Loss of brace compression

Table 6.1 Characteristics of Braced Frames Design for Push-Over Analysis Example

Story	Analysis Results (Factored Loads) (kN)		Designed Members	Factored Resistance (kN)		Ultimate Capacities (kN)		KL/r	Inelastic Model Additional Parameters (mm)			
	C_f	T_f		C_r	T_r	C_u	T_u		Δ_{i1}	Δ_{i2}	Δ_{i3}	Δ_{i4}
Case I: Tension-only design with (KL/r) limits of 200 and 300 for compression and tension members respectively												
3	0	917	2L90x90x10	150	918	167	1020	193	1.3	8.6	17.2	∞
2	0	1528	2L150x150x10	607	1566	674	1740	112	3.1	6.4	12.8	∞
1	0	1833	2L150x150x13	774	2014	860	2238	114	3.1	6.4	12.8	∞
Case II: Elastic design considering tension and compression member capacities												
3	458	458	2L125x125x13	485	1663	539	1848	138	2.3	6.4	12.8	∞
2	764	764	2L150x150x13	774	2014	860	2238	114	3	6.4	12.8	∞
1	916	916	2L150x150x16	932	2454	1036	2727	115	3	6.4	12.8	∞
Case III: Elastic design considering tension and compression member capacities and (KL/r) limit of 110												
3	458	458	W200x46	693	1583	770	1759	104	3.5	7.2	14.4	∞
2	764	764	W200x52	800	1798	889	1998	102	3.5	7.2	14.4	∞
1	916	916	W310x67	958	2298	1064	2553	107	3.5	7.2	14.4	∞
Case IV: Tension-only design without (KL/r) limits												
3	0	917	Pl. 175x20	0	945	0	1050	921	-	-	-	∞
2	0	1528	Pl. 300x20	0	1620	0	1800	921	-	-	-	∞
1	0	1833	Pl. 350x20	0	1890	0	2100	921	-	-	-	∞

Note: 1 kN = 0.2248 kips and 1 mm = 0.0394 inch

Table 6.2 Push-Over Analysis Results for the Four Frames Considered (* = critical event at that step)

Ratio of Applied load over design load	Member Forces (kN)							Member Deformation (mm)			Top Story Lateral Deformation (mm)	Description of Event
α	C3	C2	C1	T3	T2	T1		Δ3	Δ2	Δ1	ΔTOT	
Case I: Tension-only design with (KL/r) limits of 200 and 300 for compression and tension members respectively												
0.364	167*	278	333	167	278	333		1.3	1.3	1.2	5.0	Buckling at C3
0.883	167	674*	809	642	674	809		5.0	3.1	2.9	14.6	Buckling at C2
0.939	167	674	860*	693	759	860		5.4	3.5	3.1	15.9	Buckling at C1
1.294	167	674	860	1020*	1302	1512		8.0	6.0	5.4	25.8 to 26.6	Yield at T3 Onset of C3's loss of compression strength
1.11	0*	674	860	1020	1022	1177		17.2	4.7	4.2	34.5	Top story mechanism
Case II: Elastic design considering tension and compression member capacities												
1.124	516	860*	1036*	515	859	1036		2.3	3.0	3.0	10.98	Buckling at C2 and C1
1.176	539*	860	1036	539	938	1120		2.3	3.3	3.4	11.9	Buckling at C3
1.733	539	860	1036	1051	1790	2144		4.6	6.4*	6.4*	23.1	Onset of C1 and C2's loss of compression strength
1.89	539	646	777	1190	2238*	2683		5.2	8.0	8.0	28.2	Yield at T2
1.47	539	0*	429	814	2238	2289		3.5	12.8	6.7	30.6	Mid-story mechanism
Case III: Elastic design considering tension and compression member capacities and (KL/r) limit of 110												
1.16	532	89*	1064*	532	887	1064		4.2	3.1	3.1	13.8	Buckling at C2 and C1
1.68	770*	889	1064	770	1676	2013		3.5	6.7	6.3	21.9	Buckling at C3
1.76	770	889	1064	843	1798	2162		3.8	7.2*	6.8	23.6	Onset of C2 loss of compression strength
1.83	770	790	1064	902	1998*	2281		4.1	8.0	7.1	25.6	Yielding at T2
1.31	770	0*	1064	809	1998	1338		3.7	14.4	4.2	29.6	Mid-story mechanism
Case IV: Tension-only design without (KL/r) limits												
1.14	0	0	0	1050*	1800	2100*		8.0	7.8	8.0	31.6	Top and bottom story mechanisms (simultaneous)

strength starts to develop ($\alpha=1.733$) at the first and second stories when the axial displacement of these compression braces exceeds Δ_{i2}. Because the tension brace in each story is elastic at that displacement, with an axial stiffness greater than the negative stiffness of the corresponding compression brace = $C_{ui}/(\Delta_{i3} - \Delta_{i2})$, it can carry the loss in brace compression strength for a given incremental lateral displacement. As a result, the lateral loads can be increased until the second-story tension brace yields ($\alpha=1.89$). For displacements beyond this point, lateral load resistance drops, and a plastic collapse mechanism develops at the second story.

Structural behavior in Case III closely parallels that of Case II, with the only notable difference being that buckling in the third story is delayed with respect to that at the other levels. In Case IV, as a result of the extreme brace slenderness ratios, buckling is immediate, and tension yielding and plastic collapse mechanisms occur in the first and third stories simultaneously ($\alpha=1.14$).

The baseshear versus lateral displacement diagrams for each of the four frames is shown in Figure 6.11e. These push-over analyses reveal that, for Cases II and III respectively, lateral forces 70% and 65% greater than those considered during design are necessary to trigger their collapse failure mechanisms, whereas Cases I and IV have little (16.5%) or no reserve strength. Figure 6.11e also provides some evidence that Cases II and III have a relatively more ductile behavior, with more plastic energy dissipated for a given frame displacement. The push-over analyses also exposed the lack of force redistribution in these frames; story plastic collapse mechanisms always developed following tension yielding of the brace at a given story, with Case IV being the only exception.

Attempts to extrapolate these results beyond the context of monotonic loading may lead one to erroneously conclude that the more stringent member slenderness limits imposed for seismic-design provide no tangible benefit because Case III exhibits less ductile behavior than does Case II. One could also argue that Case IV is superior by virtue of it being the only case for which a plastic collapse mechanism develops in two stories, with energy dissipation better distributed along the height. Such conclusions are faulty because they address behavioral aspects that are beyond the scope of the push-over analysis. Evaluation of the hysteretic energy dissipation merits of various designs can be reliably supported only by results from cyclic inelastic analyses, not by monotonic push-over analyses. Cyclic push-over analysis, as a minimum, is needed to reveal the impact of brace slenderness on hysteretic energy dissipation, as demonstrated in the next section.

6.3.2 Cyclic push-over analysis

Cyclic push-over analyses can be useful to investigate the cyclic inelastic behavior of structures or subassemblies, particularly when the intent is to investigate the impact of unidirectional energy dissipation mechanisms on structural response. For hand calculations, simple element models and small structures are preferable.

Here, a few cycles of push-over analysis are used to illustrate the detrimental impact of brace slenderness on seismic response. In this example, a one-bay frame having X-braces with a large slenderness ratio, KL/r, is analyzed (Figure 6.12a). Such braced-frames are typical of tension-only designs, as described in the previous example.

A reasonably accurate force-elongation model of very slender braces can be constructed by assuming that:

- Elastic buckling of a brace occurs as soon as compression forces are introduced in the brace.
- All buckling deformations are elastically recovered upon unloading.
- Each brace behaves as an elasto-perfectly plastic material in tension.

This example uses generic members that yield at forces and axial deformations of 100 and 10, respectively (units are not needed). The history of the cyclic displacement applied at the top of the braced frame is shown in Figure 6.12b, along with the resulting forces and deformations in braces A and B respectively (Figure 6.12c and 6.12d). One obtains the resulting force-displacement diagram of that braced frame by combining the force-elongation contributions of each brace as shown in Figure 6.12e. For the sake of clarity in parts c, d, and e of Figure 6.12, many lines that should actually be superimposed have been separated.

Walking step by step through the applied cyclic displacement history, one observes that after the first yielding excursion and first unloading, the brace member that yielded is now longer in its stress-free condition and must buckle when the frame returns to its initial zero-deflection position. The second brace undergoes a similar process upon its first yielding and unloading. Hence, both braces become buckled when the frame returns to its initial position. As very slender compression braces provide little or no lateral resistance, the frame must drift until one member recovers all of its elastic buckling deformation before the structure can resist loads anew, and it must reach its previous maximum drift before any new plastic energy dissipation can take place. Therefore, when subjected to severe cyclic loading or dynamic excitations such as those produced by large earthquakes, this

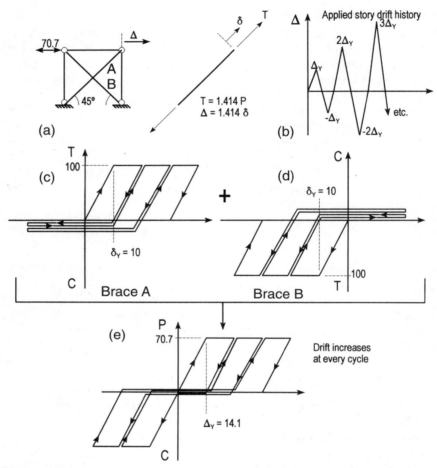

Figure 6.12 Cyclic push-over analyses examples on a single-story braced frame having slender braces.

frame will progressively drift to very large deformations if a given amount of plastic energy must be dissipated during each cycle. This partly explains why brace slenderness ratios are limited to relatively low values in seismic applications.

6.4 Seismic design using plastic analysis

Current design procedures recognize that structures cannot economically be designed to elastically resist the effects of earthquakes. Therefore, inelastic response (i.e., plastification of structural members) will develop if an earthquake as large as anticipated by the design procedures occurs. The role of the designer is to ensure that this plastification can develop in a stable manner without the risk of structural collapse. Engineers may elect to forgo this requirement, but as a trade-off, they will have to absorb a code-specified penalty of much larger required design forces. However, if a rare and unusually intense earthquake occurs, one that is greater than expected by the design code, structures that have been designed for ductile response could have a significant advantage over those that have not, despite the higher design forces considered in the latter case.

A more comprehensive assessment of the philosophy of earthquake-resistant design is presented in Chapter 9. However, with an appreciation that ductile structural systems are by far preferable to nonductile systems in seismic regions, the following two chapters address the behavior, design, and detailing of ductile braced frames and ductile moment-resisting frames. It should transpire from the following that, to a large extent, satisfactory seismic performance can be achieved through use of capacity, design principles and good ductile detailing practice.

References

1. Haaijer, G. Carskaddan, P.S. and Grubb, M.A. 1983. "Autostress Design of Steel Bridges." *ASCE Structural Journal*, Vol. 109, No. 1: 188–199.
2. Lawson, R.S. Vance, V. and Krawinkler, H. 1994. "Nonlinear Static Push-Over Analysis: Why, When, and How?" Proc. Fifth US National Conference on Earthquake Engineering, Chicago, Illinois. Ed.: Earthquake Engineering Research Institute. Vol. 1: 283–292.

Chapter 7

Design of Ductile Braced Frames

7.1 Introduction

The use of braced frames in buildings to resist wind loads dates back to the early part of this century, although in early applications braced frames were typically used in conjunction with masonry-infilled frames and moment frames to resist lateral loads. Earlier examples of braced frames appear in the nineteenth century in bridges and industrial buildings.

Braced framing systems resist lateral loads primarily by developing high axial forces in selected framing members. Only a small (or zero) percentage of the imposed lateral load on a braced frame is resisted by flexural or bending actions in moment-resisting connections. In the early applications of braced frames, the frame configurations were typically knee-braced or X-braced, utilizing tension-only braces. The bracing components were often encased in fireproofing concrete.

More complete bracing systems were developed in the 1960s and 1970s, along with the promulgation of more detailed seismic regulations. Braced framing systems proved popular in regions of high seismicity because materials savings could be achieved with respect to moment-resisting frames and control of frame drift due to high earthquake-induced inertial forces could be efficiently realized. The latter advantage of braced framing systems was realized in the United States following the 1971 San Fernando earthquake.

Two types of seismic braced framing systems are discussed in this chapter: concentrically braced frames (CBFs) and eccentrically braced frames (EBFs). Attention is focused on these types of braced frames because of their widespread use worldwide. Design examples are presented for a CBF (Section 7.2.6) and an EBF (Section 7.3.10).

7.2 Concentrically braced frames

7.2.1 General

Unlike the moment-resisting frame, the concentrically braced frame (CBF) is a lateral force-resisting system that is characterized by high elastic stiffness. High stiffness is achieved by the introduction of diagonal bracing members that resist lateral forces on the structural frame by developing internal axial actions and relatively small flexural actions. Diagonal bracing members and their connections to the framing system form the core units of a CBF. Braces can take the form of I-shaped sections, circular or rectangular tubes, double angles stitched together to form a T-shaped section, solid T-shaped sections, single angles, channels, and tension-only rods and angles. Brace connections to the framing system are commonly composed of gusset plates with bolted or welded connections to the braces. Common CBF configurations are presented in Figure 7.1. The V-braced frames of Figures 7.1b and 7.1c are also known as chevron-braced frames.

In the 1960s and 1970s, before the development of the eccentrically braced frame (see Section 7.3), the CBF was commonly used to resist earthquake-induced forces. Prior to this time, CBFs in one form or another had been used to resist wind-induced forces on buildings. In the early 1960s, the rules developed for designing steel-braced frames to resist wind effects were employed for seismic design.

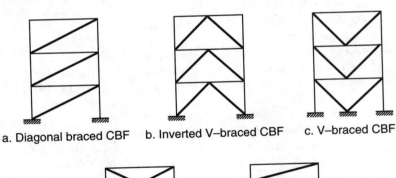

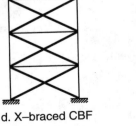
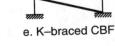

Figure 7.1 Sample CBF configurations.

Prior to the 1970s, many studies on the seismic response of braced frames focused on characterizing the inelastic cyclic response of braced tubular steel structures for the offshore oil industry. Significant research on the nonlinear behavior of bracing components was conducted in the 1970s and 1980s. Many current code provisions, details, and limitations are based on this research.

Seismic provisions for the analysis, design, and detailing of CBFs were gradually introduced into seismic regulations and guidelines in the United States in the early 1970s (ICBO 1970, 1976; SEAOC 1974). At the time of this writing, regulations and guidelines for the seismic design of CBFs can be found in the Structural Engineers Association of California (SEAOC) Recommended Lateral Force Requirements (SEAOC 1996), the Uniform Building Code (ICBO 1994), the NEHRP Recommended Provisions for the Development of Seismic Regulations for New Buildings (BSSC 1995), and the AISC LRFD Specification (AISC 1995). The rules presented in these current codes and resource documents are similar, with the exception that nonlinear analysis procedures are implemented in the NEHRP Guidelines for the Seismic Rehabilitation of Buildings (FEMA 1997).

7.2.2 Development of CBFs

Most of the CBF configurations presented in Figure 7.1 were initially developed to resist wind-induced actions in the linearly elastic range. Some of these configurations should not be used for seismic resistance because either the configuration may result in undesirable response in other structural components and elements or the configuration exhibits poor cyclic inelastic response.

One example of a CBF configuration that should not be used for seismic applications is the K-braced frame of Figure 7.1e. If one of the diagonal braces were to buckle, the force in the adjacent tension brace could become larger than the force in the buckled brace if the inertial forces increased in magnitude. The horizontal resultant of these two brace forces could then impose a large horizontal force at the mid-height of the column that may produce a plastic hinge in the column adjacent to the brace-to-column intersection point and result in column failure. Given that plastic hinge formation in columns is generally undesirable, K-braced frames cannot be used in regions of high seismicity and should not be used in other seismic regions unless other configurations are impractical.

So-called *tension-only* braced frames were commonly used to resist wind-induced lateral forces in buildings. Such frames are generally configured as X-braced frames (see Figure 7.1d) wherein the bracing elements exhibit high slenderness ($kL/r > 300$). The braces in tension-only frames were typically angles, rods, or flat sections. Although tension-only braced frames have been used to resist earthquake-induced lateral forces, the inelastic cyclic response of these frames is generally poor. The hysteresis for a tension-only braced frame is similar to that shown in Figure 7.2.

The cyclic inelastic behavior of a tension-only braced frame is characterized by yielding and elongation of the tension braces. The high slenderness of the braces results in buckling of the compression braces at low levels of axial load—the compression braces contribute little to the lateral strength of the frame. Upon repeated cyclic loading, each brace accumulates residual axial displacements, and the X-braced frame loses its lateral stiffness in the vicinity of zero frame displacement, defeating the intent of adding the braces to the frame. Tension-only braced frames are generally prohibited by many seismic regulations in North America, and limits are set on the relative tensile and compressive strengths of braces in braced frames. Where permitted, tension-only braced frames are not considered ductile. They are assigned a small response modification factor and are required to be designed for a 10 percent larger force than the nominal design force to account for the impact-type loading associated with the cyclic response of slender braces (CISC 1995).

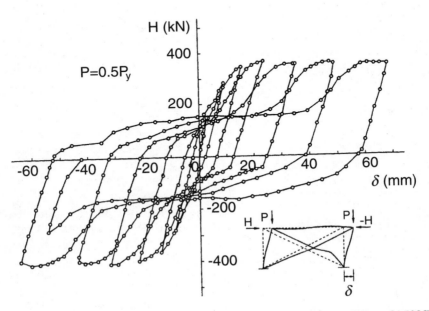

Figure 7.2 Hysteresis of an X-braced frame with slender braces. (*Wakabayashi 1986*)

7.2.3 Cyclic axial load response

A physical understanding of the inelastic response of a brace subjected to reversed cycles of axial load is necessary to best appreciate some of the concepts presented in the following sections.

The behavior of axially loaded members is commonly expressed in terms of the axial load (P) axial deformation (δ), and transverse displacement at midlength (Δ). According to convention, tension forces and deformations are positive, and compression forces and deformations are negative. A sample hysteresis curve for a brace component is presented in Figure 7.3.

Starting from the unloaded condition (point O in Figure 7.3), the brace is compressed in the linearly elastic range. Buckling is assumed to occur at point A. If the brace is sufficiently slender, the brace may buckle elastically ($P = C_r$) in which the applied axial load may be sustained as the brace deflects laterally (the plateau AB in Figure 7.3). Up to this point, the brace behavior has been elastic, and unloading would occur along BAO if the axial compression load was removed.

During buckling, the brace deflects transversely. A free-body diagram of one-half of the brace, from its pinned-end to its midlength, shows that the brace is subjected to varying moments along its length in addition to constant axial load. The largest value of moment in the brace is realized at the point of maximum transverse displacement. At a critical value of the transverse displacement of the brace, the moment in the brace will equal the plastic moment of the brace, and a plastic hinge will form at the midlength (point B in Figure 7.3). The value of the transverse displacement (Δ) corresponding to the formation of a flexural plastic hinge will depend on the degree of flexure-axial load interaction in the brace (see Chapter 3). Further increases in the axial displacement produce corresponding increases in Δ because of plastic hinge rotations at the midlength (segment BC)—producing the plastic kink evident in

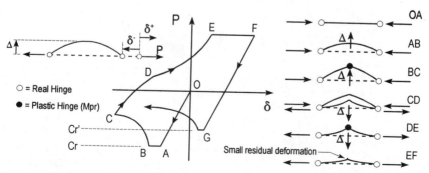

Figure 7.3 Sample hysteresis of a brace under cyclic axial loading.

the figure. The axial resistance of the brace drops in segment BC. Because the moment at midlength ($M = P\Delta$) cannot increase beyond the plastic moment, an increase in Δ is accompanied by a decrease in P. The transition from point B to point C is complex (and nonlinear) because of flexure-axial load interaction. A decrease in axial load produces an increase in moment capacity.

Upon unloading (from point C in Figure 7.3) to $P = 0$, the brace retains a residual axial deflection (δ) and a transverse deflection (Δ), with the residual transverse deflection being a visible kink in the brace.

When the brace is loaded in tension from $P = 0$ to point D, the behavior is elastic. At point D, the product of the axial load and the transverse displacement equals the plastic moment of the brace (similar to point B described earlier), and a plastic hinge forms at the midlength of the brace. However, along segment DE, the plastic hinge rotations act in the reverse direction of that along segment BC and effectively reduce the magnitude of the transverse deflection (Δ). As a result, axial forces larger than that at point D ($= P_D$) can be applied.

It is not possible to completely remove the transverse deflection. The theoretical axial force required to produce additional plastic hinge rotations tends to infinity as the transverse deflection approaches zero, but the axial force in the brace cannot exceed its tensile yielding resistance ($= AF_y$), and residual transverse deflections cannot be avoided. Upon reloading in compression, the brace behaves as a component with an initial midlength deflection, and its buckling capacity (C_r') is typically lower than its first buckling load ($=C_r$ at point A). The ratio C_r'/C_r depends primarily on the slenderness ratio (kL/r), and the following expressions have been used to capture this relationship:

[SEAOC, 1990]:
$$C_r' = \frac{C_r}{1 + 0.50\left(\frac{kL}{r\pi}\sqrt{\frac{0.5F_y}{E}}\right)} = \frac{C_r}{1 + 0.5\left(\frac{kL/r}{C_c}\right)} \quad (7.1a)$$

[CSA, 1994]:
$$C_r' = \frac{C_r}{1 + 0.35\left(\frac{kL}{r\pi}\sqrt{\frac{F_y}{E}}\right)} = \frac{C_r}{1 + 0.35\lambda^*} \quad (7.1b)$$

where λ^* is a slenderness coefficient. For an A36 steel brace with a slenderness ratio equal to 0, $C_r' = C_r$. If the slenderness ratio is increased to 130, $C_r' = 0.67C_r$.

The length of the elastic buckling plateau (segment AB) also reduces upon each subsequent inelastic cycle as a result of the residual initial deflection. The shape of the hysteresis curves (OABCDEF) in subsequent inelastic cycles, subject to the two changes described above, remains basically unchanged. Typically, a quantitative assessment of the hysteretic energy dissipation capacity of a brace can be obtained from testing using repeated cyclic inelastic loading. Traditionally, researchers have presented either all hysteresis curves for a given experiment's loading history or simply the envelope of all hysteresis curves (see Black et al. (1980)). Both approaches are used in the following sections.

Analytical expressions to express all phases of the hysteresis described above have been developed. For the derivation of these expressions, the reader is referred to Nonaka (1987, 1989) and Ikeda and Mahin (1984).

From the above discussion, it is evident that slenderness ratio has a dominant impact on the shape of the hysteresis curve. For a slender brace (large kL/r), segment OA will be rather small, whereas the plateau segment AB could be rather long, resulting in relatively small hysteretic energy dissipation capacity in compression. For stocky braces (small kL/r), the reverse is true, and segment AB may not exist.

7.2.4 Nonlinear response of concentric braces

7.2.4.1 General
Seismic design practice for standard occupancy buildings in North America is based on the use of forces substantially smaller than those expected in a fully elastic building during earthquake shaking. The reduction in forces is justified on the basis that the seismic response of the framing system is ductile. For a concentrically braced frame to demonstrate ductile response, the bracing members must be capable of sustaining large inelastic displacement reversals without significant loss of strength and stiffness.

As indicated in Section 7.2.1, the inelastic behavior of bracing components subjected to axial cyclic loading has been investigated by numerous researchers in the last 20 years. These investigations have included both experimental and analytical studies. These studies have identified three key parameters that affect the hysteretic behavior of a bracing component:

- Slenderness ratio
- End conditions
- Section shape

These three key parameters are discussed below. Much additional information can be found in the cited references.

7.2.4.2 Brace slenderness
The cyclic response of a component loaded axially in compression depends principally on its slenderness. The slenderness ratio (λ) is a function of the brace end conditions (k), the

brace length or the brace clear span (L), the second moment of area of the component about axis ii (I_{ii}), and the cross-sectional area of the component (A):

$$\lambda_{ii} = kL\sqrt{\frac{A}{I_{ii}}} = \frac{kL}{r_{ii}} \qquad (7.2)$$

The quantity $\sqrt{I_{ii}/A}$ is known as the radius of gyration (r_{ii}) about axis ii. Data for A, I_{ii}, and r_{ii} for common structural sections can be found in industry handbooks (AISC 1989, AISC 1995, and CISC 1995). The largest value of slenderness ratio for the given cross section is selected for design.

Concentric braces are often described as either slender (large λ), intermediate, or stocky (small λ). The hysteresis loops for braces with different slenderness ratios vary significantly. The area enclosed by a hysteresis loop is a measure of that component's energy dissipation capacity. Loop areas are greater for stocky braces than for slender braces. For small values of λ, the loop shapes resemble those of the material itself.

The ratio of a brace's capacity in tension to that in compression is dependent on its slenderness ratio. The more slender the brace, the larger the ratio. This is evident in the normalized axial load versus normalized axial displacement envelope plots presented in Figure 7.4 for braces with slenderness ratios equal to 40 (stocky), 80 (intermediate), and 120 (slender); tension is positive in this figure. These response envelopes were generated from axial tension-compression displacement cycling of brace components at differing levels of axial displacement. Note that the ratio of tension capacity to compression capacity increases with increasing cyclic axial displacements.

7.2.4.2.1 Slender braces Slender braces can be defined as those braces for which the elastic buckling stress is less than or equal to half the yield stress (F_y). The use of $0.5F_y$ in lieu of F_y is intended to recognize that self-equilibrating compressive residual stresses in rolled shapes may be as large as $0.5F_y$. If the member under axial load is to remain in the elastic range over its entire cross section, a maximum axial stress of only $0.5F_y$ can be applied in the presence of such a residual stress condition. Neglecting the influence of initial out-of-straightness imperfections, the elastic buckling stress (σ_{crr}) is calculated by the Euler buckling equation:

$$\sigma_{crr} = \frac{\pi^2 E}{\left(\frac{kL}{r}\right)^2} = \frac{\pi^2 E}{\lambda^2} \qquad (7.3)$$

where E is Young's modulus. Given these assumptions, the limiting slenderness value for a slender brace is given by:

$$\lambda_{slender} \geq 770 / \sqrt{F_y} \qquad (7.4a)$$

where Fy is the yield stress in kips per square inch (ksi), or

$$\lambda_{slender} \geq 2020 / \sqrt{F_y} \qquad (7.4b)$$

where F_y is in units of MPa. The lower limit from Equation 7.4a is equal to 130 for A36 braces and 110 for Grade 50 braces. Conservative slenderness limits of $720/\sqrt{F_y}$ (AISC 1995) and $1900/\sqrt{F_y}$ (CISC 1995) are adopted in North American seismic regulations.

Sample hysteresis for a slender tubular brace (TS1x1x0.10 or HSS25x25x2.5) with a width-to-thickness ratio of 7.5, an average yield stress (F_y) of 57 ksi (395 MPa), and a slenderness ratio (λ) of 140 is presented in Figure 7.5. The axial load on the slender brace is P, and the corresponding axial displacement is δ_j; in this figure, tension is positive.

Very slender braces have little stiffness in a buckled configuration. Accordingly, the lateral stiffness of a concentrically braced frame with slender compression braces will drop substantially following brace buckling. As these slender braces straighten out from the buckled configuration under tensile loading, they pick up axial stiffness rapidly. The rapid increase in stiffness may produce an impact type of loading that could lead to either brace damage or connection failure.

An axially loaded component loses strength rapidly with repeated inelastic load cycles and does not return to its original straight-line

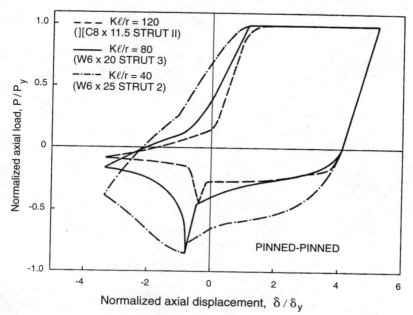

Figure 7.4 Hysteretic envelopes for braces. *(Black et al. 1980)*

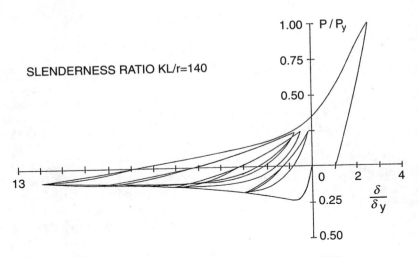

Figure 7.5 Sample hysteresis for a slender brace. *(Jain et al. 1978)*

position upon removal of load. This can be seen in Figure 7.6 from Black et al. (1980) for an I-shaped section with a slenderness ratio of 120. If the brace returned to its straight-line position, the residual lateral displacement (Δ) at zero applied load (P) would be zero. Note the difference between the P–δ relationships for the slender braces of Figures 7.5 and 7.6. These differences can be attributed to the use of different testing protocols and brace cross sections.

Black identified that one of two major causes for the observed decrease in compression brace capacity during inelastic cyclic loading was the residual curvature in the brace that remained following tensile yielding of a previously buckled brace. (The other major cause described by Black et al. was related to the Bauschinger effect.) The residual lateral displacement was treated by Black as a strut with an initial curvature or camber. Black demonstrated by analysis that the axial capacity of a strut reduced with increasing initial curvature (residual lateral displacement) and that reductions were greater for slender struts than stocky struts.

Figure 7.6 captures the key characteristics of the cyclic inelastic behavior of slender braces:

- Loss of axial compression stiffness (P/δ)
- Axial shortening (given by the residual axial displacement δ at zero applied load)
- Loss of tangent stiffness ($=\Delta P/\Delta\delta$) at zero applied load

Axial shortening of braces will distort the frame in which the braces are located, potentially causing substantial damage to the gravity

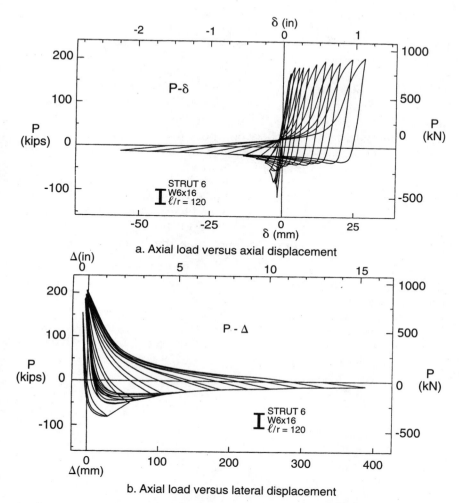

Figure 7.6 Force-displacement relations for a slender brace. *(Black et al. 1980)*

load-resisting system. This problem was identified in the earlier discussion on K-braced frames.

Consider the axial force (P) versus axial displacement relation (δ) for the slender brace of Figure 7.5. The tangent stiffness of the brace at zero axial load is approximately 1700 kips/inch (29.4 kN/mm) in the first loading cycle and approximately 20 kips/inch (0.35 kN/mm) in the loading cycle to δ = 35 mm (1.38 inches). If two such braces formed an inverted-V CBF configuration (see Figure 7.1b), the lateral stiffness of the braced frame near the point of zero lateral displacement may be less than 5 percent of the elastic stiffness of the frame—negating the key benefit of adding the concentric braces to the frame in the first instance.

7.2.4.2.2 Stocky braces Stocky braces can be defined as those braces for which yielding and local buckling dominate the response. Local buckling results in loss of moment capacity at the plastic hinge location; this in turn results in a loss in the axial strength of the member and produces a reduction in the energy dissipation capacity of the brace. The limiting value for λ_{stocky} varies as a function of the material stress-strain relationship, the width-to-thickness ratio of the brace (a measure of the likelihood of local buckling), the residual strains in the brace, and the initial out-of-straightness of the brace. The limiting values for λ_{stocky} are approximately equal to 60 for compact braces composed of A36 steel and 50 for compact braces composed of Grade 50 steel. Sample hysteresis for a stocky tubular brace (TS1x1x0.105 or HSS25x25x2.5) with a width-to-thickness ratio of 7.5, an average yield stress (F_y) of 57 ksi (395 MPa), and a slenderness ratio (λ) of 30 is presented in Figure 7.7; tension is positive in this figure. Note that the hysteresis of the stocky brace is fuller than that of the slender brace.

7.2.4.2.3 Intermediate braces Intermediate braces are those braces for which local buckling phenomena are less critical than inelastic buckling (Khatib et al. 1988). Khatib noted the important distinction between stocky and intermediate braces: in stocky braces, the large lateral displacements initiated by local buckling of a flange (for W sections) or plate (tubular sections) will likely trigger global brace buckling, whereas intermediate braces experience global buckling at an effective

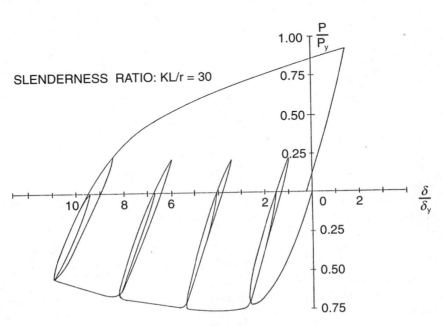

Figure 7.7 Sample hysteresis for a stocky brace. *(Jain et al. 1978)*

buckling stress that is reduced from the nominal values because of residual fabrication stresses. For braces composed of A36 steel, intermediate braces can be characterized as those braces for which:

$$60 < \lambda_{int} < 130 \qquad (7.5)$$

and for Grade 50 steel braces:

$$50 < \lambda_{int} < 110 \qquad (7.6)$$

Sample hysteresis for an intermediate tubular brace with a width-to-thickness ratio of 7.5, an average yield stress (F_y) of 57 ksi (395 MPa), and a slenderness ratio (λ) of 80 is presented in Figure 7.8. The rate of degradation of stiffness and strength in the intermediate brace is less than that of the slender brace but greater than that of the stocky brace.

7.2.4.3 Brace end conditions Earlier studies on the influence of end- or boundary-conditions on the buckled shapes of bracing components focused on brace response in the linearly elastic range. Black and Popov (Black et al. 1980, Popov and Black 1981) extended the scope of these works to determine whether the effective length (kL) approach used for calculating the buckling capacity of an elastic brace could be applied to braces cyclically loaded into the inelastic range.

Consider the normalized data presented in Figure 7.9 for two braces with different slenderness ratios and end-conditions. Figure 7.9a presents inelastic buckled shapes compared with the elastic

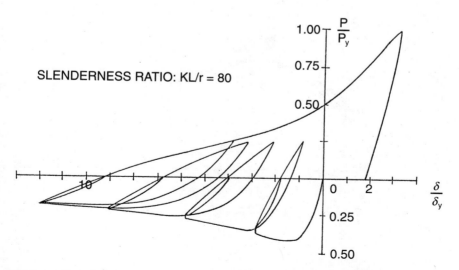

Figure 7.8 Sample hysteresis for an intermediate brace. *(Jain et al. 1978)*

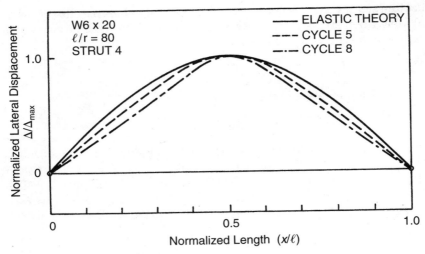

Figure 7.9 Elastic and inelastic buckled shapes for I-shaped beams (a) Pinned-pinned end conditions. *(Black et al. 1980)*

buckled shape for a pin-ended I-shaped brace with a slenderness ratio (kL/r) equal to 80. Figure 7.9b presents inelastic buckled shapes compared with the elastic buckled shape for an I-shaped brace pinned at one end and fixed at the other, with a slenderness ratio of 40. From these data it can be concluded that the inelastic shapes are similar to the elastic shapes. Note that the brace curva-

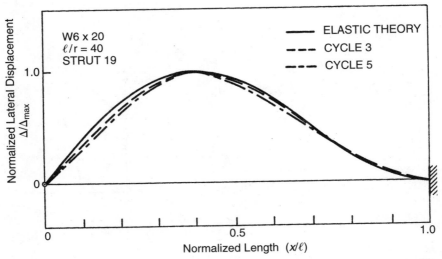

Figure 7.9 Elastic and inelastic buckled shapes for I-shaped beams (b) Pinned-fixed end conditions. *(Black et al. 1980)*

ture tends to concentrate in the plastic hinge regions as the number of inelastic cycles increases (e.g., from cycle 5 to cycle 8 in Figure 7.9a) but that the inflection points for both the elastic and inelastic shapes are essentially identical.

The effect of brace end-restraint on the hysteretic behavior of axially loaded braces was also examined by Black et al. (1980). Normalized force-displacement envelopes were used to compare the hysteretic behavior of braces with identical slenderness ratios but differing lengths. I-shaped, circular tube, and double-angle braces were included in the study. Sample hysteretic envelopes for braces with different end conditions, resulting from the use of an axial displacement history with increasing amplitude, are presented in Figures 7.10a (I-shaped brace) and 7.10b (circular tube brace). Marginally improved performance is realized with the fixed-pinned end conditions for the I-shaped and circular tube braces. There is no discernible difference for the two double-angle braces.

Although the scope of Black's study was limited to two end conditions (pinned-pinned and pinned-fixed), the similarity of the elastic and inelastic deflected shapes (Figure 7.9) and the hysteresis loops (Figure 7.10) for the two cases considered strongly support the extension of the effective length approach to other end conditions for evaluating the cyclic inelastic response of bracing components.

7.2.4.4. Section shape The effect of cross-section shape on the hysteretic response of a bracing member was evaluated by Black et al. (1980). Black studied the behavior of a total of six pin-ended braces: one I-shaped, one T-shaped, two circular tube (CHS), one rectangular tube (TS), and one double-angle; all with a slenderness ratio of 80. Two failures modes, both functions of the section shape and geometry, were shown to influence the results, namely, local buckling and lateral-torsional buckling. Sample hysteretic envelopes from Black et al. (1980) for different braces each with a slenderness ratio equal to 80, resulting from the use of an axial displacement history with increasing amplitude, are presented in Figures 7.11a and 7.11b.

7.2.4.4.1 Section shape efficiency The trends established in Figures 7.10 and 7.11 and other figures in Black et al. (1980) suggest that the most efficient braces are tubular cross-sections with small kL/r, and that improved performance can be achieved by reducing the ratios of b/t (rectangular tube brace), d/t (circular tube brace), and $b_f/2t_f$ (I-shaped brace) ratio. This trend is clearly seen in Figure 7.11a, wherein a reduction in d/t (Strut 14 to Strut 16) results in a larger hysteresis loop area. Black ranked the tested cross-sections in the following descending order of effectiveness for a given slenderness ratio:

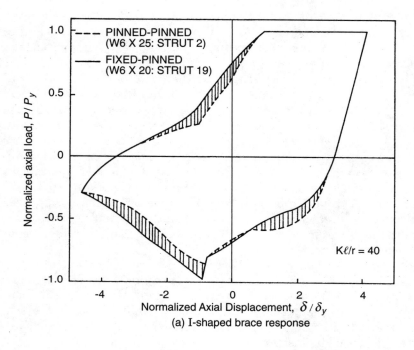
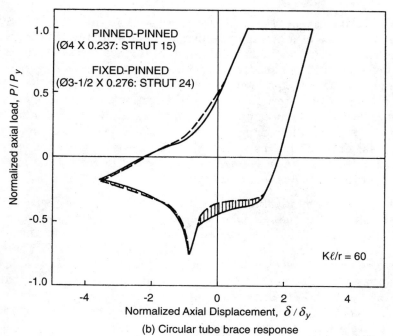

Figure 7.10 Hysteretic curves for braces with different end conditions. *(Black et al. 1980)*

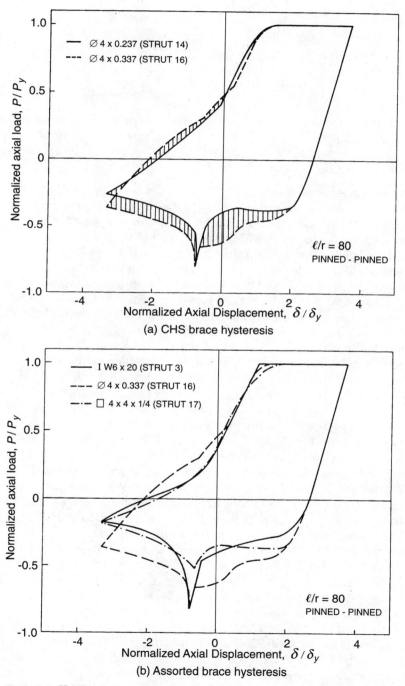

Figure 7.11 Hysteretic curves for braces with different cross-section shapes. *(Black et al. 1980)*

1. Circular tube brace
2. Rectangular tube brace
3. I-shaped brace
4. T-shaped brace
5. Double-angle brace

Black recommended that built-up members not be used as braces for applications in which severe cyclic loading was anticipated unless the members making up the built-up section were adequately stitched together.

7.2.4.4.2 Tubular sections Rectangular tubes are commonly used as seismic braces because of their high efficiency, which is due to their larger radii of gyration and resistance to local buckling for the same cross-sectional area relative to other rolled shapes, such as W and T sections. However, tubular braces are susceptible to failure induced by local buckling and subsequent material fracture (Liu and Goel 1987, Uang and Bertero 1986). Local buckling of compression elements in a tubular brace reduces the brace's plastic moment and consequently its axial compressive strength. Further, the degree and extent of the local buckling at the plastic hinge locations has a major influence on the fracture life of a brace (Tang and Goel 1987). (For a pin-ended brace, plastic hinges will likely form at the midlength of the brace, which is at the point of maximum lateral displacement in the brace. For a fixed-ended brace, plastic hinges will likely form at the brace ends and the brace midlength.) Preventing severe local buckling is the key to precluding premature material fracture.

One strategy for delaying the onset of local buckling in tubular braces is to reduce the width-to-thickness ratio of the brace. Both Tang and Goel (1987) and Uang and Bertero (1986) recommended that the limit on the width-to-thickness ratio (b/t) for rectangular tubes be reduced to $190/\sqrt{F_y}$ ($= 500/\sqrt{F_y}$ in S.I. units) from the value then specified in the AISC Manual of Steel Construction (AISC 1980). Tang and Goel recommended a b/t limit of $95/\sqrt{F_y}$ ($= 250/\sqrt{F_y}$ in S.I. units) for rectangular tube sections. Uang and Bertero (1986) studied the behavior of rectangular Grade 50 concentric braces using data acquired from the earthquake simulator testing of a six-story concentrically braced steel frame. The slenderness (kL/r) and width-to-thickness ratios for the braces studied by Uang ranged between 48 and 61, and 12.7 and 20.5, respectively. All braces in the six-story frame would be classified as intermediate, and the b/t ratios were substantially less than the AISC limit of 26. In evaluating the response of the concentric braces, Uang concluded that those braces with slenderness ratios exceeding 20 and buckling loads less than 80 percent of the nominal squash load ($= AF_y$) fractured due to local buckling that followed global brace buckling. On the basis of limited data, Uang and Bertero (1986) recommended that for Grade 50 tube braces, the b/t ratio be limited to 18 or $125/\sqrt{F_y}$ ($= 330/\sqrt{F_y}$ in S.I. units) and that the slenderness ratio (kL/r) be limited to 68.

7.2.4.4.3 Concrete-filled tubular braces Another promising method for delaying the onset of local buckling in tubular braces is to fill the brace with expansive concrete. Liu and Goel (1987) studied the effects of filling rectangular tubular braces with concrete by evaluating the hysteretic response of similar braces with (1) a high b/t ratio (approximately equal to 30), (2) a low b/t ratio (approximately equal to 14), and (3) a low b/t ratio and concrete infill. Liu and Goel concluded that there was no significant difference in the overall buckling modes of the three specimens prior to plastic hinge formation and local buckling, but that following plastic hinge formation, the braces with the smaller b/t ratio and concrete infill performed substantially better because local buckling was delayed and the strength of the brace remained relatively constant with repeated cycling.

All specimens tested by Liu developed plastic hinges near the midspan and at both ends of the brace after global buckling. Local buckling was observed to be localized within the plastic hinge zones. For the two hollow tubular sections tested by Liu, the compression flange in the brace at the plastic hinge began to buckle inward and the brace webs bulged outward after global brace buckling and plastic hinge formation (see Figure 7.12a). Local buckling in the form of narrow groovelike patterns was concentrated in the middle of each hollow brace and at both ends of each brace. Local buckling was significantly delayed in the brace with the smaller b/t ratio. The bulges in the corners of the braces grew with additional cyclic axial loading leading to the formation of small cracks, which spread quickly into the compression flange and both webs of each brace.

In contrast, the local buckling in the concrete-filled braces followed a different course. Specifically, the flange of each concrete-filled tube buckled outward rather than inward because the concrete prevented the inward local buckling of the compression flange; the zone of local buckling was lengthened to be approximately equal to the width of the tube and its severity reduced (see Figure 7.12b). The addition of the concrete infill to the brace reduced the severity of the local buckling, avoided the concentration of excessive strains in the plastic hinge zones due to local buckling, delayed the onset of material cracking, increased the fracture life of the brace, and minimized the reduction in the moment capacity of the brace because the brace flanges could not substantially deform in the presence of the concrete. The reader is referred to Liu and Goel (1987, 1988) for additional information.

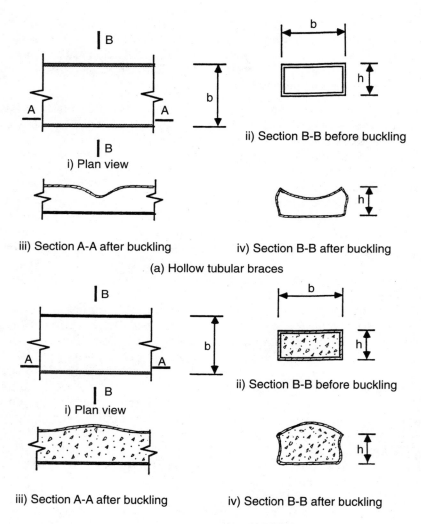

Figure 7.12 Buckled sections in tubular steel braces at plastic hinge locations. (*Liu and Goel 1988*)

7.2.5 CBF design philosophy

A common configuration of the concentrically braced frame (see Figure 7.13) in the United States is the inverted V-braced (chevron-braced) frame of Figure 7.1b. This configuration is both self-contained and amenable to the placement of doorways, windows, and mechanical systems. The following discussion therefore focuses on the design of chevron-braced frames and is parsed into sections addressing bracing

Figure 7.13 Concentrically braced frame building in California. (*Courtesy of Degenkolb Engineers, Inc, San Francisco, California.*)

members, connection design, and design of beams and columns in the braced bays. Proposed improvements to this frame configuration are presented in Section 7.2.5.6.

The concepts presented below are easily extended to other CBF configurations such as the X-braced and split-X-braced frames. These configurations are often used to avoid (or delay) imposing large unbalanced vertical forces at the midlength of the beams in the braced bays.

The inelastic cyclic response of a chevron-braced frame is dependent upon many factors including the following:

- The slenderness and compactness of the bracing members
- The relative axial strengths of the brace in compression and tension
- The strength of the brace connections to the beams and columns
- The degree of lateral restraint provided to the brace-to-beam connection
- The stiffness, strength, and compactness of the beam into which the brace frames

For the purpose of this presentation, it is assumed that the brace connections are sufficiently strong and that the brace-to-beam connection is adequately restrained against lateral-torsional buckling.

7.2.5.1 Lateral stiffness of CBFS Seismic design of structures typically assumes bilinear or trilinear response (Figure 7.14) of the seismic building frame. To achieve such response in a braced frame, negative stiffness must be avoided. Given that bracing members typically exhibit negative postbuckling stiffness, the response of the beam in the braced bay is key to achieving the desired frame hysteresis of Figure 7.14. Accordingly, the design of the bracing members and the beams in a braced bay cannot be uncoupled.

Khatib et al. (1988) conducted an exhaustive study on the postbuckling characteristics of the inverted-V chevron-braced frame. Khatib's study assumed that the beams were pin-connected to the columns, that vertical loads could be ignored, and that the beams and columns were axially inextensible. In Khatib's frame, shown in Figure 7.15a, the braces were denoted as B1 and B2, and the angle of inclination of the braces to the horizontal was θ. Key results from Khatib's study are presented below. For this derivation, tension forces in brace B1, axial elongation of brace B1, compression forces in brace B2, and axial shortening of brace B2 are all considered positive. Any downward movement of the brace-to-beam intersection point is resisted by the vertical stiffness of the beam. The kinematics of the brace-to-beam intersection point are shown in Figure 7.15b.

In the elastic range, the tension force (T) in brace B1 and the compression force (C) in brace B2 are equal in magnitude. The lateral strength of the braced bay at the onset of buckling in brace B2, provided by the braces only, is given by:

$$F_c = 2P_c \cos \theta \tag{7.7}$$

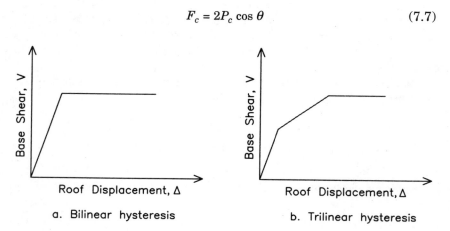

a. Bilinear hysteresis   b. Trilinear hysteresis

Figure 7.14 Idealized frame hysteresis.

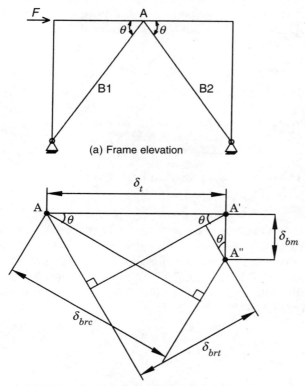

(a) Frame elevation

(b) Kinematics of brace-to-beam intersection point

Figure 7.15 Frame geometry and kinematics of an inverted V-braced CBF. *(Adapted from Khatib et al. 1988)*

where P_c is the elastic buckling load of brace B2. The corresponding unbalanced vertical load applied to the beam (P_{un}) at the brace-to-beam intersection point is:

$$P_{un} = (T - C) \sin \theta \tag{7.8}$$

and equal to zero. The lateral stiffness of the story is equal to:

$$K_s = 2\left(\frac{AE}{L}\right)_{br} \cos^2 \theta = 2k_e \cos^2 \theta \tag{7.9}$$

where k_e is the elastic axial stiffness of one brace, and the lateral stiffness of the beam-column framing is ignored. In loading the frame from $F = 0$ to $F = F_c$, the brace-to-beam intersection point translates from point A to point A' in Figure 7.15b.

Following buckling of brace B2, the forces in the tension and compression braces will generally be unequal, an unbalanced vertical load will be applied to the beam (see Equation 7.8), and the beam will deflect by an amount δ_{bm} (from point A' to point A" in Figure 7.15b). From Figure 7.15b, the displacements f_{brc} and f_{brt} can be calculated as:

$$\delta_{brc} = \delta_t \cos\theta + \delta_{bm} \sin\theta \qquad (7.10)$$

$$\delta_{brt} = \delta_t \cos\theta - \delta_{bm} \sin\theta \qquad (7.11)$$

The increase in tension force in brace B1 is given by:

$$\Delta T = \delta_{brt} k_{br} \qquad (7.12)$$

and the increase in the compression force in brace B2 is:

$$\Delta C = \delta_{brc} k_{bb} \qquad (7.13)$$

where k_{br} and k_{bb} are the tangent tensile stiffness of brace B1 and the tangent post buckling stiffness of brace B2, respectively. In the elastic range, k_{br} equals k_e. For negative values of k_{bb} (see post buckling stiffness of braces in Figures 7.4 through 7.8), the compression force in brace B2 will decrease with increasing values of δ_{brc}. The unbalanced vertical load (P_{un}) applied to the beam is equal to:

$$P_{un} = (\Delta T - \Delta C) \sin\theta \qquad (7.14)$$

and the increase in lateral resistance (ΔF) is equal to:

$$\Delta F = (\Delta T + \Delta C) \cos\theta \qquad (7.15)$$

For positive δ_t and positive δ_{bm}, both δ_{brc} and δ_{brt} will generally be positive. Given that k_{bb} is generally negative following buckling, and k_{br} is positive up to yielding in the tension brace, the compression force in brace B2 will decrease by ΔC, the tension force in brace B1 will increase by ΔT, and the unbalanced vertical load will be equal to:

$$P_{un} = (\Delta T + |\Delta C|) \sin\theta \qquad (7.16)$$

Assuming k_{bb} to be negative and δ_{bm} to be small, the lateral resistance of the frame will increase beyond F_c only if $|k_{bb}|$ is less than $|k_{br}|$.

The tangent stiffness of the story (k_{st}) following buckling of brace B2 is equal to:

$$K_{st} = \frac{\Delta F}{\delta_t} = \cos^2\theta [(k_{br} + k_{bb}) - \frac{(k_{br} - k_{bb})^2 \sin^2\theta}{k_{bm} + (k_{br} + k_{bb}) \sin^2\theta}] \qquad (7.17)$$

where k_{bm} (>0) is the stiffness of the beam associated with a vertical degree of freedom at the brace-to-beam intersection point and the lateral stiffness of the beam-column framing is ignored. For an infinitely stiff beam, Equation 7.17 reduces to:

$$K_{st} = (k_{br} + k_{bb}) \cos^2\theta \qquad (7.18)$$

and for elastic brace buckling ($k_{bb} = 0$):

$$K_s t = \frac{k_{br} \cos^2\theta}{1 + \frac{k_{br}}{k_{bm}} \sin^2\theta} \qquad (7.19)$$

For the design objective of positive story tangent stiffness (to maintain the bilinear or trilinear hysteresis of Figure 7.14), that is, $K_{st} > 0$, the minimum required stiffness of the beam is a function of the brace tangent stiffness values. If k_{br} and k_{bb} are both positive, K_{st} will be positive. If k_{br} is positive and k_{bb} is negative, the beam stiffness required to ensure that $K_{st} > 0$ is:

$$k_{bm} = \frac{-4k_{br}k_{bb} \sin^2\theta}{(k_{br} + k_{bb})} \qquad (7.20)$$

The flexural stiffness of the beam plays an important role in determining the post buckling stiffness of the braced bay. The flexural stiffness of a beam is a function of its second moment of area (I_{bm}) and its boundary conditions. Khatib classified beams as either flexible, intermediate, or stiff. Flexible beams were those that did not satisfy Equation 7.20, assuming complete fixity at their supports. Stiff beams were those that satisfied Equation 7.20, assuming pin-ended supports. Intermediate beams were all those beams not classified as either flexible or stiff.

7.2.5.2 CBF collapse mechanisms To demonstrate the sensitivity of the nonlinear response of a CBF to different design assumptions, consider the single-story, single-bay frame of Figure 7.15a. Assume that the frame has been used in designing a strong-column/weak-beam philosophy and that the framing elements have sufficient shear strength to avoid shear failure. Only flexible and stiff beams are considered.

Idealized, normalized axial force-axial displacement relationships for the tension brace (B1) and stocky (B2-ST), intermediate (B2-INT), and slender (B2-SL) compression braces are shown in Figure 7.16. The piecewise linear relations were estimated from the data of Black et al. (1980).

7.2.5.2.1 Flexible beam collapse mechanism For stocky braces, the tension yield force ($= P_y$) is equal to or slightly larger than the elastic buckling load; that is, P_2 ($= P_c$) is approximately equal to P_4 ($= P_y$) in Figure 7.17a. The maximum unbalance force applied to the beam is equal to $(P_y - P_c) \sin\theta$ and approximately equal to zero. If the beam is flexible, the tensile force in brace B1 remains practically constant and axial

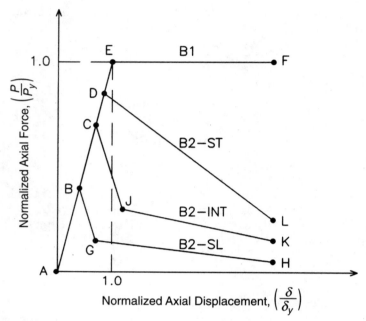

Figure 7.16 Idealized axial force-axial displacement relations.

deformations in the compression brace (δ_{brc}) increase more quickly than the axial deformations in the tension brace (δ_{brt}). This relationship is identified in Figure 7.17a by the dashed lines joining the force-displacement relations for braces B1 and B2. The beam displacements δ_{bm} increase δ_{brc} and decrease δ_{brt} (see Equations 7.10 and 7.11). The resulting story shear force versus lateral displacement relation is shown in Figure 7.17b. (The numbered points in Figure 7.17b relate to the corresponding points in Figure 7.17a).

For intermediate braces, the tension yield force is significantly greater than the elastic buckling load (see Figure 7.17c), and a much larger unbalanced force could be applied to the beam. However, if the beam is flexible, the tension force in B1 remains essentially constant and cannot increase to P_y. Intermediate braces lose strength quickly with increasing deformation (see brace 2 in Figure 7.17c), so the story shear resistance reduces with increasing lateral displacement as shown in Figure 7.17d.

For slender braces, the tension yield force is much larger than the elastic buckling load. If the beam is flexible, the tension force in brace B1 remains essentially constant with increasing lateral displacement. However, given that the resistance of buckled slender braces decreases relatively slowly with increasing lateral displacement (see brace B2 in Figure 7.17e), the story shear resistance (see Figure 7.17f) remains relatively constant with increasing lateral displacement.

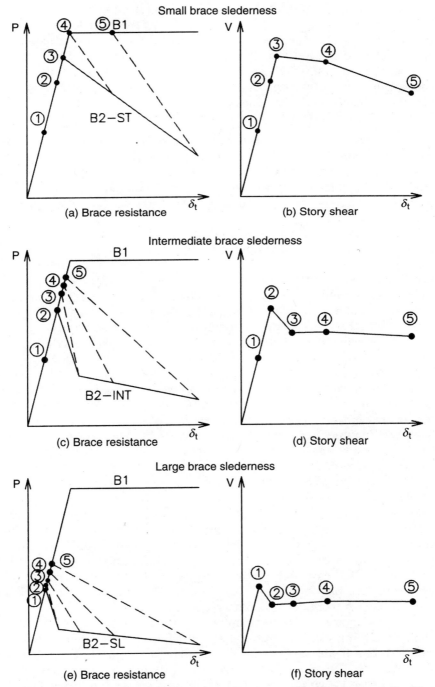

Figure 7.17 Force-displacement relations for an inverted V-braced CBF with flexible beams.

In the discussion above, the beam has been assumed to be strong. Clearly, plastic hinges could form in the beam adjacent to the brace-to-beam intersection point prior to yielding of the tension brace. The effect of replacing the strong beam with a weak beam is that the frame will collapse at a smaller lateral displacement. However, in all three cases, the story shear resistance at incipient collapse is smaller than that at the point of elastic buckling in brace B2 (points 2 in Figures 7.17a, 7.17c, and 7.17e).

7.2.5.2.2 Stiff beam collapse mechanism For the three brace slenderness ratios described below, the assumption that the beam stiffness is large results in small beam displacements (δ_{bm}). Per Equations 7.10 and 7.11, if δ_{bm} is small, the elongation of the tension brace B1 (δ_{brt}) will be approximately equal to the shortening of the compression brace B2 (δ_{brc}).

For stocky braces, the tension yield force ($= P_y$) is equal to or slightly larger than the elastic buckling load (P_c). Because the beam is stiff, the tension force in brace B1 continues to increase, producing an unbalanced force at the midspan of the beam equal to $(P_y - P_c)\sin\theta$. If the beam is strong, the tension force in brace B1 will reach the tension yield force before the beam forms a plastic hinge (point 4 in Figure 7.18b). For larger lateral displacements (segment 4-5 in Figure 7.18b), the lateral resistance provided by the braces decreases slowly because the compression load in brace B2 drops with increasing lateral displacements.

For intermediate braces, the tension yield force is significantly greater than the elastic buckling load (see Figure 7.18c), and a much larger unbalanced force can be applied to the beam. Upon buckling of brace B2 (points 2 in Figures 7.18c and 7.18d), the tension force in brace B1 will increase at k_e kips per inch of brace elongation, and the compression force in brace B2 will decrease at $|k_{bb}|$ kips per inch of brace shortening. The resultant story shear force versus lateral displacement relation is shown in Figure 7.18d. If the absolute value of the tangent post buckling stiffness of the intermediate brace (see segment CJ in Figure 7.16) is larger than the elastic stiffness of the tension brace, the story shear force will decrease with increasing displacement. In the example, $|k_{bb}|$ equals $0.8k_e$ in the segment CJ, and the story shear force increases slowly following brace buckling. Following tension yielding of brace B1 at point 4, the story shear force decreases with increasing displacements at a rate equal to $|k_{bb}|$.

For slender braces, the tension yield force is much larger than the elastic buckling load. A large unbalanced force can be applied to the beam if the beam is strong. The tension force in brace B1 will increase at k_e kips per inch of brace elongation and the compression force in brace B2 will decrease at $|k_{bb}|$ kips per inch of brace shortening. As drawn, the tangent post-buckling stiffness of the slender

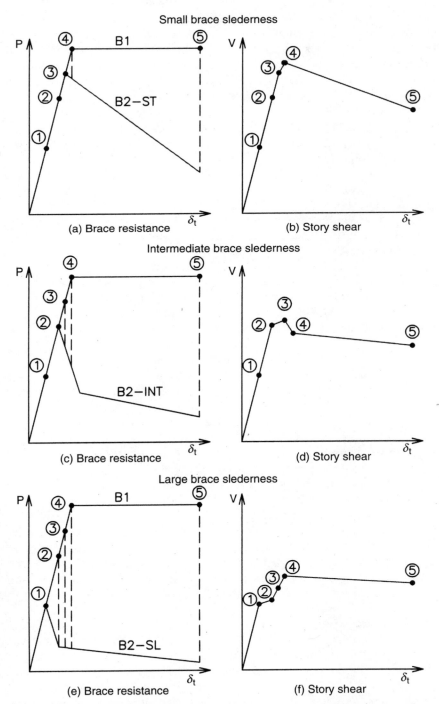

Figure 7.18 Force-displacement relations for an inverted V-braced CBF with stiff beams.

brace is negative, with an absolute value equal to 0.83 times the elastic stiffness of the tension brace. The resultant story shear force versus lateral displacement relation is shown in Figure 7.18f. Upon buckling of brace B2 (point 2), the story shear force increases slowly because k_e is greater than k_{bb}. Following yielding of the tension brace at point 4, the story shear force decreases slowly with increasing displacements.

Although yielding of the stiff beam could precede yielding of the tension brace, substantial changes in the story shear versus displacement relations would not be observed.

The preceding discussion on lateral stiffness of chevron-braced frames highlights the difficulties confronting the designer of braced frames. The choice of beam stiffness, generally not considered by designers, was shown to substantially influence the story shear force versus lateral displacement relation; for example, compare the information presented in Figures 7.17d and 7.18d.

7.2.5.3 Braces The postbuckling seismic response of a chevron-braced frame is extremely sensitive to the relative flexural stiffness of the beam to the axial stiffness of the brace and cannot be predicted by elastic analysis. Braces of intermediate stiffness should likely be avoided unless large force redistributions can be accommodated.

If a design objective is trilinear story shear force-displacement response similar to that shown in Figure 7.14, stiff beams can be used with either slender or intermediate braces (see Figure 7.18). However, it may be impractical to implement stiff beams for braces with typical proportions. Further, the lateral stiffness afforded to a frame by slender braces may be insufficient to adequately control interstory drifts, and the vertical unbalanced force at the brace-beam intersection point may be too large to prevent yielding in the beam for typical brace and beam proportions. The results of Khatib's analysis support the use of stocky braces wherever possible.

Guidance is provided in the AISC LRFD Specification (AISC 1995) for the design of bracing members in CBFs. The AISC manual permits the use of braces with slenderness ratios less than or equal to $720/\sqrt{F_y}$ ($= 1900/\sqrt{F_y}$ in S.I. units); that is, it permits only stocky and intermediate braces per the definitions presented earlier. Information on the design force levels for different types of CBFs can be found in (AISC 1995).

The design strength of a brace, from the AISC LRFD Specification (AISC 1995) is calculated as $0.8\phi_c P_n$ where ϕ_c is the resistance factor for compression components, and P_n is the nominal axial strength of the brace. The design strength is reduced to 80 percent of the nominal design strength ($= \phi_c P_n$) to account for the reduction in strength due to inelastic buckling. The reduction factor 0.8 is a single value of the ratio of C'_r to C_r given by Equation 7.1a, assuming a slenderness ratio of approximately 65. If the strength of a brace is to be used to compute the maximum load it can deliver to other components or systems, the reduction factor of 0.8 should not be used because its use will likely lead to underestimation of brace strength for the first few cycles of seismic loading.

7.2.5.4 Connections Bracing members are typically connected to the beams and columns in the braced bays by gusset plates and either welding or bolting. Given that the CBF design philosophy typically focuses on dissipating energy in the braces, the connections should be designed to remain essentially elastic at all times. Capacity design procedures can be used to achieve this design objective; see Chapter 6 for details.

The AISC LRFD Specification (AISC 1995) adopts capacity design procedures for connection design by requiring that joints in braced frames have a minimum strength greater that the least of

- The design axial tension strength of the bracing member
- The maximum expected axial compression force in the brace
- The maximum force that can be delivered to the brace by the framing system

The first limit should be calculated as the product of the cross-sectional area of the brace and the expected yield stress of the brace material; the factor ϕ should be set equal to 1.0. The second limiting force is calculated in the LRFD Specification by amplifying the design actions due to earthquake shaking by $0.4R$ (greater than 1.0), where R is the response modification factor used to calculate the design earthquake forces. This limit recognizes that most bracing components have reserve elastic compressive strength beyond their nominal design strength. The cap of $0.4R$ ensures that the design force on the connection is less than or equal to the design force that would be used for the identical connection in a nonductile frame that would be designed for a smaller value of R. One can calculate the third limiting force using either plastic or nonlinear static analysis; see (FEMA 1997) for details.

Rules for designing gusset plates are given by AISC (1995). Design rules are provided for two specific cases: the braces buckle in-the-plane of the gusset, and the braces buckle out-of-plane of the gusset. For braces that buckle in the plane of the gusset, the gusset and other parts of the connection must have a flexural strength equal to or greater than the nominal in-plane bending strength of the brace. This rule is intended to prevent hinges forming outside the brace ends in the connection.

For braces that buckle out-of-plane of the gusset, each gusset should be detailed to permit formation of a hinge line in the gusset because it is generally not possible to design the gusset to be stronger than the brace. AISC (1995) requires that the brace terminate a minimum of two times the gusset thickness from the nominal line of unrestrained bending (see Figure 7.19). This rule is based on the work of Astaneh, Goel, and Hanson (1982), which showed an increase in the deformation and energy dissipation capacity of out-of-plane buckling braces when a clear distance of twice the gusset plate thickness was maintained between the brace end and the line of flexural restraint.

7.2.5.5 Columns and beams

7.2.5.5.1 Columns The research of Khatib et al. (1988) demonstrated the sensitivity of the nonlinear response of a CBF to brace slenderness and beam stiffness; relatively minor changes in component stiffnesses can have a substantial impact on hysteretic response. Given this sensitivity and the obvious need to protect columns that resist gravity loads, one reliable but conservative approach for calculating maximum and minimum column loads is capacity design.

The design of a column in a braced bay should account for gravity load actions, seismic actions introduced by overturning moments, and

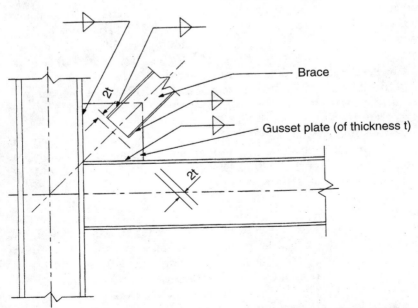

Figure 7.19 Brace-to-gusset plate requirements for out-of-plane buckling of braces. (*AISC 1995.*)

seismic actions that result from the vertical unbalanced force (if any). Calculation of column loads by direct addition of the forces that could be delivered by the braces above the story under consideration is likely conservative for medium- to high-rise CBFs because maximum forces are generally not realized in all bracing members simultaneously. For the lower stories of such buildings, it is appropriate to assume that the times at which individual brace forces are maximized are uncorrelated and to use the square-root-sum-of-the-squares (SRSS) rule for accumulating earthquake-induced brace forces from different stories. An example of this approach to calculate column forces in a 10-story concentrically braced frame is presented by Redwood and Channagiri (1991). For low-rise CBFs and the upper stories of medium- and high-rise buildings, the assumption that the maximum brace forces are uncorrelated may be unconservative, and direct addition of the forces that could be delivered by the braces above the story in question is recommended. The reader is referred to Khatib et al. (1988) for additional information.

7.2.5.5.2 Beams Although beam stiffness plays an important role in the hysteretic response of V and inverted-V CBFs (see Section 7.2.5.2), no clear trends can be established if reasonable beam proportions are to be maintained. The LRFD Specification (AISC 1995) requires the following:

- A beam intersected by braces must be continuous between columns (to allow plastic hinging in the beam and to facilitate force redistribution).
- A beam intersected by V braces must be capable of supporting gravity loads independently of the braces.
- The top and bottom flanges of the beam at the point of intersection of V braces must be laterally braced (in recognition of likely plastic hinge formation in the beam at the intersection point).

7.2.5.6 Improved chevron bracing configurations

The seismic response of concentrically braced frames with traditional geometry is dependent on factors that include brace slenderness and geometry and beam stiffness and strength. Behaviors are complicated by complex interactions between braces, beams, and columns, and optimal relations between the key design parameters have been found to be dependent on the ground motions used for analysis and design (Khatib et al. 1988).

In an attempt to reduce the sensitivity of the response of V- and inverted-V-braced frames to some of the factors noted above and the characteristics of the earthquake ground motions used for analysis and design, Khatib studied the seismic performance of simple variants on the basic inverted-V configuration with the following objectives:

- Reducing the likelihood of forming weak stories.
- Mitigating the effects of postbuckling force redistributions.
- Achieving a trilinear story shear force versus story displacement relation without having to use overly stiff beams and slender braces.

The configuration changes studied by Khatib were limited to different brace arrangements and the addition of vertical struts joining the brace intersection points. Five of alternative configurations to the conventional inverted-V braced frame (termed INVV) studied by Khatib were VREG, XREG, SLITX, STG, and ZIP. See Figure 7.20 for details of these frames.

The V-braced frame (VREG) is a simple variant on the inverted V-braced frame. The potential advantage of this configuration is that gravity loads will pretension the braces and delay the onset of buckling. However, the gravity load forces will typically be substantially smaller than the earthquake-induced forces, and the potential benefit is likely minimal. Khatib et al. (1988) drew the same conclusion and also noted that connecting the braces to rigid framing at the ground-floor level resulted in a concentration of energy dissipation demands in the second story, VREG column axial forces were significantly higher than those in the INVV frame because brace forces are imposed at the top of the columns rather than at the bottom of the columns, and the magnitude of the brace forces due to vertical inertial forces at

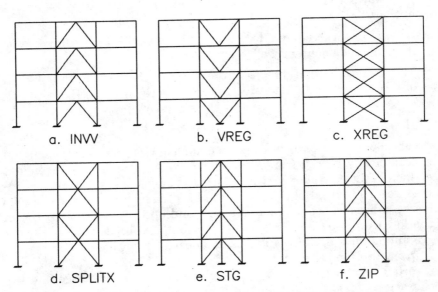

Figure 7.20 Alternative CBF configurations. (*Adapted from Khatib et al. 1988*)

the brace-to-beam intersection point, resulting from the sudden application of the vertical unbalanced force, were greater than the gravity load forces—negating one key benefit of the V-braced frame.

The X-braced frame (XREG) is often used as an alternative to the inverted-V-braced frame; the X-braces in the frame studied by Khatib were not connected at the brace intersection point. Two possible advantages of this configuration are that vertical unbalanced forces are not imposed on the beams in the braced bay and trilinear hysteresis can be achieved in the initial loading cycle if the braces are slender. Design issues identified by Khatib and others include the hysteresis loops for an X-braced frame tend to pinch with inelastic cyclic loading, thereby reducing the tangent stiffness and energy dissipation capacity of the frame. Also, large axial forces are developed in the beam in the braced bay—a beam with an in-plane effective length equal to twice that of the INVV beam.

The cross bracing in the split-X-braced frame (SPLITX) spans two stories with the intent of providing trilinear hysteresis and avoiding the problems identified earlier in this chapter with vertical unbalanced forces at brace-to-beam intersection points. Khatib concluded that using the same braces in each story of a two-story X-bracing module resulted in a concentration of damage in the lower story of the module.

The strut-to-ground (STG) configuration employs vertical bracing components that link the brace-to-beam intersection points at each floor level directly to the foundation. These struts resist the vertical unbalanced forces at the brace-to-beam intersection that develop following brace buckling. The advantages of this configuration identified by Khatib include the following:

- The tension braces can develop their yield strength.
- Additional axial loads in columns due to the vertical unbalanced forces in the beams in the braced bays are avoided.
- Trilinear hysteresis is achieved.

The vertical struts should be designed to remain linearly elastic to effectively serve their intended purpose. These struts may have to be designed for large axial forces if intermediate or slender braces are adopted because of the large difference between the tension yield strength and buckling load of such braces. The size of the resulting struts may resemble column sections.

Analyses by Khatib and others suggest that the braced frame configurations identified above exhibit reduced sensitivity to ground motion characteristics with respect to the conventional chevron (inverted-V)-braced frame. Recognizing the need to develop a new frame configuration capable of better seismic performance, Khatib proposed a new configuration termed the *zipper*. The zipper (ZIP) configuration is similar to the STG configuration, except that the strut in the first story is eliminated.

The intent of the ZIP configuration is to tie all brace-to-beam intersection points together, to force all compression braces in a braced bay to buckle simultaneously, and thereby to distribute the energy dissipation (damage) over the height of the building. Simultaneous brace buckling over the height of a building will produce a single-degree-of-freedom mechanism and result in a more uniform distribution of damage over the height of the building (Whittaker et al. 1990).

Khatib et al. (1988) studied the behavior of a ZIP frame using nonlinear response-history analysis. Comparing the response of the ZIP frame with that of the other frame configurations considered above, Khatib concluded the following:

- The response of the ZIP frame was less sensitive to ground motion characteristics.
- The ZIP frame achieved a more uniform distribution of damage over its height.
- The ZIP frame developed a trilinear story shear force-displacement relation.
- The ZIP frame concept could be successfully implemented with flexible beams and braces of intermediate slenderness.

7.2.6 CBF design example

7.2.6.1 Introduction The sample frame is a three-story concentric braced frame with X-bracing in the third story and split-X bracing in the lower two stories. An elevation of the frame is shown in Figure 7.21a; the beam-to-column connections are assumed to be pinned. Gravity unfactored concentrated dead loads of 250 kN (56.2 kips) and live loads of 100 kN (22.5 kips) are applied to each column at each level, and uniformly distributed dead loads of 15 kN/m (1.11 kips/ft.) and live loads 10 kN/m (0.74 kips/ft.) are applied along the beams. Unfactored lateral seismic loads of 690 kN (155 kips), 460 kN (104 kips), and 230 kN (51.7 kips) are applied at the third, second, and first floor levels, respectively, of the frame. The seismic loads were calculated based on a value of R equal to 3 per the National Building Code of Canada (NRC 1990) for the design of CBFs. For simplicity, only two load cases were considered in this example: 1.0D + 1.0E and 1.2D + 1.6L, where D, L, and E are the dead, live and earthquake loads, respectively.

In this example, the frame is sized first to resist the actions resulting from the application of the above forces using *strength* design and then evaluated and redesigned as necessary to comply with requirements for *ductile* response.

7.2.6.2 Strength design requirements Using strength design in accordance with the AISC LRFD Specification (AISC 1995) and minimum weight as a design objective, the structural members summarized in Table 7.1 and Figure 7-21b were selected. ASTM A572 Grade 50 steel with a specified yield strength of 345 MPa (50 ksi) was used for the beams and columns. ASTM A500 Grade B square structural tubes manufactured to a specified yield strength of 320 MPa (46 ksi) were chosen for the braces. Effective length factors (k) of 0.5 and 1.0 were used to calculate the in-plane and out-of-plane buckling strengths of the third floor braces per standard practice. Some studies on the elastic behavior of X-braced frames have suggested that the tension brace can provide some resistance against *out-of-plane* buckling of the compression brace and that a value of k less than one could be used for out-of-plane buckling. The error of this assumption for the seismic design of braced frames was identified in Section 7.2.2 wherein it was noted that the tension brace can lose stiffness in the vicinity of zero frame displacement due to repeated buckling, therefore, the value of k for out-of-plane buckling should be taken as 1.0.

Note that the engineer would likely select identical column and beam sizes for this frame to simplify the detailing of the frame and its connections. Such a rationalization of structural sizes has not been undertaken herein.

7.2.6.3 Design for ductile response In this section, the brace, column, and beam sizes obtained from strength design for the imposed gravity and earthquake actions are evaluated and modified as necessary to guarantee ductile response of the frame. The member sizes required for ductile frame response are presented in Figure 7-21c.

7.2.6.3.1 Braces It has been established that the degree of ductile behavior in a CBF depends largely on the slenderness ratios (kL/r) and width-to-thickness ratios (b/t) of the braces. Therefore, a logical first check is whether these ratios remain within the permissible limits for the members selected using strength design. The values presented in Table 7.1 show that the width-to-thickness ratios of all braces, and the slenderness ratio of some braces, exceed the code-mandated limits. For example, the HSS178x178x6 (TS7x7x0.25 in U.S. units) third-story braces have a length of 8732 mm (343 in.), cross-sectional area and width of 4250 mm^2 (6.58 sq. inches) and 177.8 mm (7.00 in.) respectively, wall thickness of 6.35 mm (0.25 in.), radius

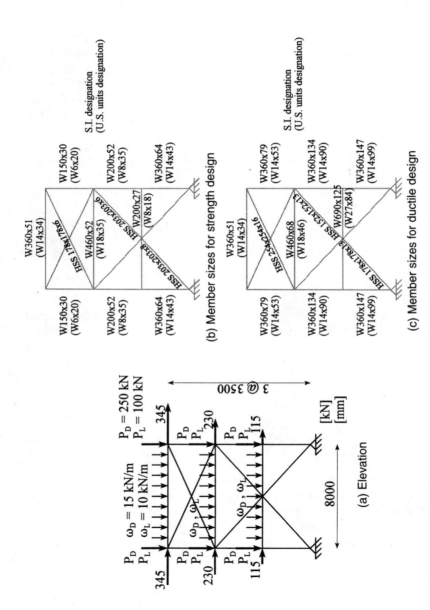

Figure 7.21 Example of three-story concentrically braced frame.

Table 7.1 Strength Design Data for Three-Story Braced Frame

Element	Story	Strength design		$\dfrac{KL}{r}$	$\dfrac{b}{t}$
		S.I. Shapes	U.S. Shapes		
Columns	3	W150x30	W6x20		
	2	W200x52	W8x35		
	1	W360x64	W14x43		
Beams	3	W360x51	W14x34		
	2	W460x52	W18x35		
	1	W200x27	W8x18		
Braces	3	HSS178x178x6	TS7x7x.25	67.1	25.0
	2	HSS203x203x6	TS8x8x.25	66.4	29.0
	1	HSS203x203x8	TS8x8x.313	125.5	22.6

of gyration of 69.6 mm (2.74 in.), and yield strength of 320 MPa (46 ksi). For these braces:

$$kL/r = 8732/69.5 = 125.6 > \frac{1890}{\sqrt{320}}$$

$$b/t = (177.8 - 2(6.35))/6.35 = 27 > \frac{288}{\sqrt{320}}$$

and new braces must be selected. Note that the limiting values of $1890/\sqrt{F_y}$ (slenderness ratio) and $288/\sqrt{F_y}$ (width-to-thickness ratio) are the metric equivalents of $720/\sqrt{F_y}$ and $110/\sqrt{F_y}$, respectively, adopted in the AISC LRFD Specification (AISC 1995). The corresponding limits in the Canadian Limit States Design of Steel Structures (CSA 1994) are $1900/\sqrt{F_y}$ and $330/\sqrt{F_y}$, respectively, where F_y is in units of MPa

The problem of excessive slenderness in the third-story braces could be resolved if the brace configuration in that story were changed from X-bracing to chevron (inverted-V) bracing. However, some seismic codes (CSA 1994, AISC 1995) penalize inverted-V-braced frames by requiring them to be designed for 50 percent greater seismic forces.

Further, the increase in the seismic loads must typically be applied to the entire frame (in the direction under consideration) because most seismic codes require the design of mixed structural systems to be based on the design loads associated with the least ductile of the framing systems that make up the seismic framing system.

New brace sections that comply with the width-to-thickness and slenderness ratio limits are presented in Table 7.2. Substantially heavier braces are needed for ductile response design than for strength design.

The loss in compression strength of the braces due to repeated buckling, identified in Section 7.2.3 as the ratio C'_r/C_r, should be checked next. In the Canadian seismic code (CSA 1994), this reduction need not be considered if the tension member has enough reserve capacity to compensate for this loss of compression strength. Checking the second story HSS152x152x12 brace:

Factored design forces: $C_f = 913$ kN $\qquad T_f = 611$ kN

Calculated resistance: $C_r = 985$ kN $\qquad T_r = 1932$ kN $(= \phi A F_y)$

Nominal resistance: $C_n = 1159$ kN $\qquad T_n = 2147$ kN $(= A F_y)$

Slenderness factor: $\lambda^* = \dfrac{kL}{r}\sqrt{\dfrac{F_y}{\pi^2 E}} = \sqrt{\dfrac{5320}{56.1}} = \sqrt{\dfrac{320}{\pi^2(200000)}}$

Reduced compression resistance: $C'_r = \dfrac{C_r}{1 + 0.35\lambda^*} = \dfrac{985}{1 + 0.35(1.21)} = 692$ kN

That is, after a few cycles of buckling, the compression strength of the brace will drop from 985 kN to 692 kN, and this loss of brace strength

Table 7.2 Brace Member Sizes for Ductile Response in the Three-Story Braced Frame Example

Story	Ductile Response		$\dfrac{kL}{r}$	$\dfrac{b}{t}$	C_f (kN)	C_r (kN)	C'_r (kN)	T_f (kN)	T'_f (kN)	T_r (kN)
	S.I. Sizes	U.S. Sizes								
3	HSS254x254x16	TS10x10x.625	90.9	16.0	393	2240	1590	360	1010	4160
2	HSS152x152x13	TS6x6x.50	94.7	12.0	913	985	692	611	904	1930
1	HSS178x178x13	TS7x7x.50	79.9	14.0	1130	1400	1030	710	1080	2300

must be absorbed by the tension member if the design is to remain unchanged:

$$T'_f = T_f + (C_r - C'_r) = 611 + (985 - 692) = 904 \ kN < T_r = 1932 \ kN$$

Another check is whether the compression braces along any line of framing resist at least 30 percent, but no more than 70 percent of the design shear force assigned to the line of bracing. This requirement is an attempt to balance the resistance of the compression and tension braces across the breadth and width of the story. For frames with the same number of compression and tension braces, and designed for slenderness limits described earlier, this is generally not a concern. The LRFD Specification provides an exception to this requirement if the braces are sufficiently strong to provide essentially elastic response in the design earthquake.

Results for the other braces are presented in Table 7.2.

7.2.6.3.2 Brace connections To maximize the energy dissipation capacity in a braced frame, the brace connections should be designed to be stronger than the members they connect so that the bracing members can yield and buckle. Connections should therefore be designed to resist the nominal tensile strength ($= AF_y$) of the braces. In addition, standard gusset plate design practice is to provide a clear distance of approximately twice the gusset plate thickness between the end of the brace and the adjacent members to allow a plastic hinge to form in the gusset plate without excessive local straining of that plate (see Section 7.2.5.4).

It would be prohibitive to design the third-story brace connection to resist AF_y of the brace given that the brace was increased in size to meet slenderness limits—the third-story braces would remain elastic for forces 5.5 times the minimum design forces. In this instance, the options set forth in the LRFD Specification and reproduced in Section 7.2.5.4, should be used to calculate the design forces for the connection.

7.2.6.3.3 Columns Column forces for design can be estimated using capacity design principles. Such an approach is likely appropriate for low-rise frames wherein all braces may attain their capacity simultaneously (see Section 7.2.5.5). Using a capacity design approach, the columns in the braced bay should be designed to remain elastic for gravity load actions and the forces delivered by the braces assuming that the braces achieve their maximum nominal tensile strength (= AF_y) or compressive strength (i.e., with $\phi = 1.0$, and ignoring the compression capacity reduction factor described above). Both compressive and tensile forces introduced by the braces should be considered to calculate maximum axial compressive and tensile forces. The column sizes needed to satisfy this requirement are presented in Table 7.3.

Table 7.3 Column Sizes for Ductile Response in the Three-Story Braced Frame Example

Story	Resolved brace forces		Column forces		C_f (kN)	Ductile design		C_r (kN)
	$C_u \sin\theta$ (kN)	$T_u \sin\theta$ (kN)	E (kN)	D (kN)		S.I. Shapes	U.S. Shapes	
3	896	1164	1664	310	1974	W360x79	W14x53	2040
2	649	1272	3832	620	4452	W360x134	W14x90	4590
1	923	1516	3817	930	4747	W360x147	W14x99	5040

7.2.6.3.4 Beams Floor beams in braced frame buildings serve multiple purposes: to support gravity load effects, to act as collector elements to transfer seismic inertial forces into the braced bays, and to facilitate load redistribution in the braced bays. Inadequate design of collector beams and their connections transferring seismic forces into the vertical lateral force resisting system will likely compromise the response of the entire building. Much attention must be paid to designing a complete load path to the foundation for the seismic inertial forces.

The effect of load redistribution due to brace buckling and yielding should be considered for the design of beams in the braced bays. To ensure ductile frame response, the resultant forces on the beams in the braced bays can be calculated though the use of capacity design principles. Elastic design using these forces will produce the strong beams described by Khatib et al. (1988).

Generally, braces will have different strengths in tension and compression. Internal force redistributions will occur following brace buckling (see Section 7.2.5.2). For the X-braced (third story) and split-X-braced (first and second stories) configurations presented in this example, two free-body diagrams of the beams can be prepared (Figure 7.22). Using the first free-body diagram (Figure 7.22c), for a beam supported at midlength by the intersecting braces, the axial load in the beam can be calculated as:

$$\Sigma F_x = 0 = 2P + (T_{i+1} + C_{i+1}) \cos\theta_{i+1} - (T_i + C_i) \cos\theta_i$$

where the nominal member forces in stories i and $i+1$ are considered. Recognizing that braces in adjacent stories may not reach their maximum strengths simultaneously, Redwood and Channagiri (1991) suggested that only 75 percent of the brace strengths in story $i+1$ be considered for beam design:

$$P = 0.5(T_i + C_i) \cos\theta_i - 0.5(0.75)(T_{i+1} + C_{i+1}) \cos\theta_{i+1}$$

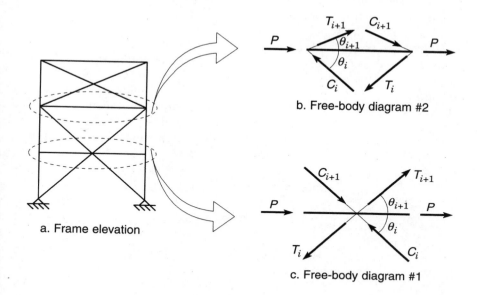

Figure 7.22 Free-body diagram for calculating beam actions.

In the second free-body diagram (Figure 7.22b), the beam spans the full width of the braced bay, and brace buckling and yielding will produce an internal redistribution of forces. Although the earthquake loads are symmetrically applied to the frame, this beam acts as a load-transfer member. The axial force in the beam is given by:

$$P = -0.5(T_i - C_i)\cos\theta_i - 0.5(T_{i+1} - C_{i+1})\cos\theta_{i+1}$$

where the value of P should be taken as the maximum value calculated using either C_r or C_r' for the compression braces above and below the beam.

When these axial forces are combined with the moments acting on the beams due to gravity and other seismic actions, the adequacy of the beams can be checked with the standard beam-column design equations.

For the first free-body diagram, and considering the first floor beam ($i=1$), the nominal tension (T_i) and compression (C_i) forces are 2300 and 1400 kN, respectively, in story i, and 1930 and 985 kN, respectively, in story $i+1$. The brace angle θ is 41.2° in both stories. The design axial force in the beam is:

$$P = 0.5(2300 + 1400)\cos 41.2° - 0.5(0.75)(1930 + 985)\cos 41.2° = 569 \text{ kN}$$

This beam must be sized to simultaneously resist the flexural moments resulting from the same load condition including gravity load actions. The unbalanced vertical force (= 125 kN) due to brace buckling can be calculated by summing the forces in the vertical direction. For this beam, the gravity-load moment is equal to 120 kN-m, and the earthquake-induced moment is equal to 250 kN-m for a design moment of 375 kN-m. Assuming that the beam is adequately braced to prevent lateral-torsional buckling and global member buckling, the W200x27 (W8x18) beam initially sized by strength design is checked using the beam-column code interaction equation required by the AISC LRFD Specification (AISC 1995):

$$\frac{P_u}{\phi P_n} + \frac{8}{9}\left(\frac{M_{ux}}{\phi_b M_{nx}}\right) = \frac{569}{1000} + \frac{8}{9}\left(\frac{375}{87}\right) = 0.56 + 3.84 = 4.4 > 1.0$$

and a significantly larger beam (W690x125 or W27x84) is required for ductile response.

For the second free-body diagram, and considering the second floor beam ($i=2$), the nominal tension strength (T) and compression strengths (C_r, C_r') are 1930 kN and (985 kN, 692 kN), respectively, in story i, and 4160 kN and (2240 kN, 1590 kN), respectively, in story $i+1$. The brace angle θ is 41.2° in the second story and 23.6° in the third story. The maximum design axial force in the beam is:

$$P = -0.5(1930 - 692)\cos 41.2° - 0.5(4160 - 1590)\cos 23.6° = -1643 \text{ kN}$$

Assuming a gravity load moment of 120 kN-m, the W460x52 (W18x35) sized above using strength design is checked with the LRFD Specification and found to be inadequate, and a larger beam (W460x68 or W18x46) is required.

The size of the third-floor beam is checked through a procedure identical to that outlined above for the beams at the first and second floors. The maximum axial design force in the beam is calculated as:

$$P = -0.5(4160 - 1590)\cos 23.6° = -1177 \text{ kN}$$

The W360x51 (W14x34) beam sized using strength design is checked for a moment of 120 kN-m and an axial force of 1177 kN, and it is just adequate to ensure ductile response.

7.2.6.3.5 Other considerations The design example has served to illustrate some of the key steps in the design of a ductile CBF. Other considerations for the structural engineer would include analysis of alternate CBF configurations, column splice details, and column axial forces for design.

The weight of the design of the example CBF for ductile response is substantially greater than that for the design for strength alone.

Given the significant increase in weight (and cost), alternative ductile configurations would likely be considered by most design professionals. One alternative configuration is the inverted-V or chevron configuration. For such a configuration, lateral loads 50 percent greater than those required for the design of the X-braced or the split-X braced frame were considered per standard practice (AISC 1995). The column, beam, and brace sizes resulting from a strength design of this chevron-braced frame are presented in Table 7.4. To ensure ductile response, the slenderness and width-to-thickness ratios of the braces must be reduced to allowable values, and the columns and beams should be sized using capacity procedures. Larger brace and column elements will be required. Substantially larger beam sizes will be required to resist the unbalanced vertical forces due to brace buckling—forces that are generally much greater for chevron-braced frames than for split-X-braced frames: 592 kN for HSS178x178x13 chevron braces in the first story versus 125 kN for the split-X braces identified earlier. The 50 percent increase in design seismic forces for chevron braced frames (AISC, 1995) indirectly addresses the substantial unbalanced vertical

Table 7.4 Strength Design Data for Three-Story Chevron-Braced Frame Example

Element	Story	Strength design	
		S.I. Shapes	U.S. Shapes
Columns	3	W150x30	W6x20
	2	W200x46	W8x31
	1	W310x97	W12x65
Beams	3	W310x39	W12x26
	2	W200x46	W8x31
	1	W200x52	W8x35
Braces	3	HSS178x178x6	TS7x7x.25
	2	HSS203x203x8	TS8x8x.31
	1	HSS254x254x8	TS10x10x.31

forces that can develop in chevron braced frames and the consequent need for stronger beams in such frames.

Particular attention must be paid to column splice details. Complete penetration groove welds are preferred for column splices because partial penetration groove welds perform poorly under cyclic loading (Bruneau and Mahin 1990). However, at a minimum, partial penetration groove-welded column splices should be designed for the lesser of 150 percent of the calculated maximum tension force or 50 percent of the tensile strength of the column cross section.

The use of capacity procedures to calculate column forces is appropriate for low-rise buildings and the upper stories of medium- and high-rise buildings but may be too conservative in other instances because the braces in each story will likely not yield or buckle simultaneously. An example of an alternate square-root-sum-of-the-squares (SRSS) is illustrated in Figure 7.23 using the eight-story CBF analyzed by Redwood and Channagiri (1991). At the fourth floor, the use of the SRSS procedure results in a design axial force of 5859 kN rather than the design axial force of 7412 kN that is obtained through capacity procedures. For high-rise framed structures, the use of the SRSS procedure will generally result in substantially smaller column loads and foundation design forces.

7.3 Eccentrically braced frames

7.3.1 General

The eccentrically braced frame (EBF) is a hybrid lateral force-resisting system composed of two conventional framing systems: the moment-resisting frame (Chapter 8) and the concentrically braced frame (Section 7.2). The EBF serves to combine many of the individual advantages of each conventional framing system and minimizes their respective disadvantages. Specifically, eccentrically braced frames possess high elastic stiffness, stable inelastic response under cyclic lateral loading, and excellent ductility and energy dissipation capacity.

The eccentrically braced frame has recently emerged as the braced framing system of choice in regions of high seismicity. Research on the behavior of EBFs commenced in earnest in the mid-1970s (Roeder and Popov 1977, Roeder and Popov 1978) and continued apace through the 1980s (Engelhardt and Popov 1989a, 1989b, 1992; Kasai and Popov 1986a, 1986b, 1986c; Ricles and Popov 1987; Whittaker, Uang, and Bertero 1987). Early applications of the EBF date back to the early 1980s (Merovich et al. 1982).

Provisions for the analysis, design, and detailing of EBFs have been introduced into seismic regulations and guidelines in the United States. Sample regulations and guidelines include the Structural

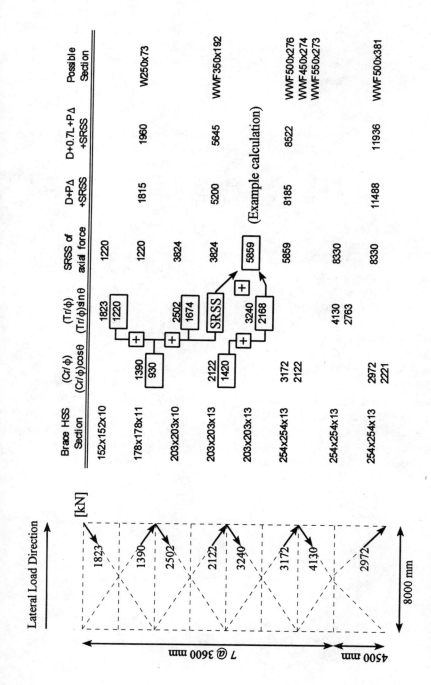

Figure 7.23 SRSS estimates of column forces in an eight story CBF analyzed by Redwood and Channagiri (*1991*)

Engineers Association of California (SEAOC) Recommended Lateral Force Requirements (SEAOC 1996), the Uniform Building Code (ICBO 1994), the NEHRP Recommended Provisions for the Development of Seismic Regulations for New Buildings (BSSC 1997), the AISC LRFD Specification (AISC 1995), and Limit States Design of Steel Structures (CSA 1994).

7.3.2 Development of EBFs

The high elastic stiffness of the concentrically braced steel frame and the ductility and stable energy dissipation capacity of the moment-resisting frame are characteristics of the EBF. The key distinguishing feature of an EBF is that at least one end of each brace is connected so as to isolate a segment of beam called a *link*. Common EBF arrangements are illustrated in Figure 7.24; in each figure the links are identified by the link length e. The three EBF arrangements are termed the D-braced frame (Figure 7.24a), the split-K-braced frame (Figure 7.24b), and the V-braced frame (Figure 7.24c). Elevations of D-braced and split-K-braced EBFs in the San Francisco Museum of Modern Art are presented in Figures 7.25a and 7.25b, respectively.

Sample hysteresis from the first story of a six-story split-K EBF dual system tested by Whittaker et al. (1987) on an earthquake simulator is presented in Figure 7.26. The hysteresis curves plot base shear force versus interstory drift in the first story for the major cycle of excitation for each of five earthquake simulations; the tests listed in the legend identify the earthquake record (e.g., Taft) and the peak ground acceleration (e.g., 66 for 0.66g). The hysteresis is stable and well rounded, indicative of much energy dissipation. The energy history from the Taft-66 test of the six-story EBF is presented in Figure 7.27. From this figure it is evident that the shear links dissipated in excess of 90 percent of the earthquake input energy.

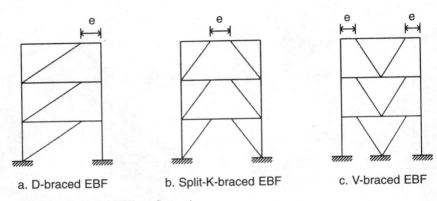

a. D-braced EBF   b. Split-K-braced EBF   c. V-braced EBF

Figure 7.24 Common EBF configurations.

Design of Ductile Braced Frames

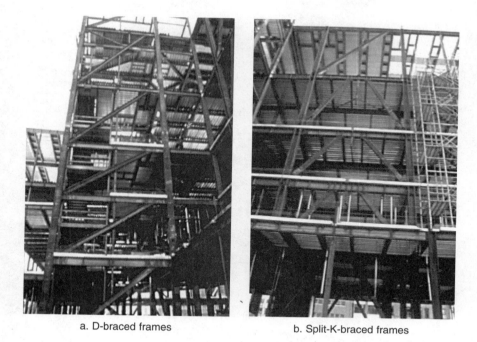

a. D-braced frames

b. Split-K-braced frames

Figure 7.25 Eccentrically braced frames in the San Francisco Museum of Modern Art. *(Courtesy of Forell/Elsesser Engineers, Inc.)*

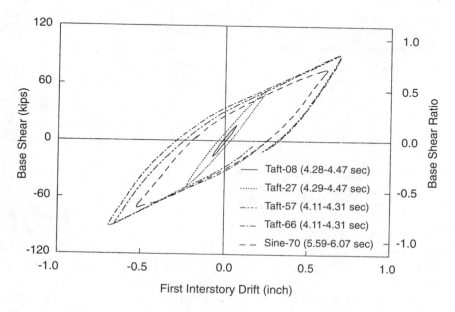

Figure 7.26 EBF earthquake simulation hysteresis. *(Whittaker et al. 1987)*

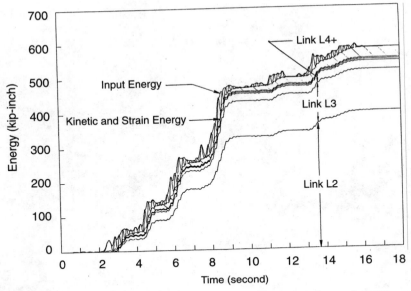

Figure 7.27 EBF earthquake simulation energy histories. *(Whittaker et al. 1987)*

7.3.3 EBF design philosophy

The EBF design concept is simple: restrict the inelastic action to the links, and design the framing around the links to sustain the maximum forces that can be delivered by the links. Design using this strategy should ensure that the links act as ductile seismic fuses and preserve the integrity of the surrounding seismic framing.

This concept of capacity design provides an elegant means of limiting forces in selected framing components. For design using linear static procedures (ICBO 1994, BSSC 1994), links are typically proportioned for forces substantially smaller than those calculated by analysis for elastic response. Other components in the seismic framing system are then designed for the forces generated by the fully yielded and strain-hardened links. That is, all other components are designed for the capacity of the links.

Explicit capacity design methods are not common in seismic regulations and guidelines for steel construction in the United States. Exceptions include the requirement to design connections of joined components to be stronger than the individual components and the strong-column/weak-girder strategy. Inelastic actions in the connections are precluded in the former case, and inelastic actions are limited to the beam in the latter case. Key to the success of this approach for the design of EBFs is the deformation capacity of the links. Links must be properly designed and detailed to achieve the large requisite plastic deformations.

Simple relationships between frame shear force and link shear force can be developed for common EBF configurations (see Figure 7.28) for the purpose of preliminary design. These relationships depend only on frame geometry and are independent of whether the link response is elastic or inelastic. Links can be proportioned using linear frame analysis, and design actions in components surrounding the links can be calculated using equilibrium concepts. For the split-K-braced EBF of Figure 7.28a, and assuming that the moment at the center of the link is equal to zero, the link shear force (V_L) can be estimated as:

$$V_L = \frac{Ph}{L} \tag{7.21}$$

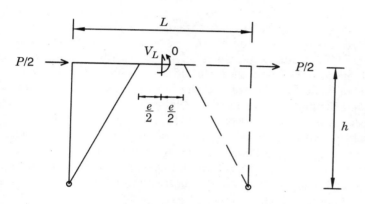

a. Split-K-braced EBF

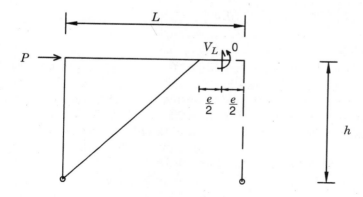

b. D-braced EBF

Figure 7.28 Component actions in EBFs.

where P is the lateral load, h is the story height, and L is the width of the braced bay. Similarly, for the D-braced EBF of Figure 7.28b, the link shear force can be estimated as:

$$V_L = \frac{Ph}{(L - \frac{e}{2})} \qquad (7.22)$$

where e is the link length and the other terms are defined above.

7.3.4 EBF frame geometry

Commonly used EBF configurations are shown in Figure 7.24. Most framing arrangements are composed of links that are located at only one end of each brace, with a nominally concentric connection at the other end. For most framing arrangements, the horizontal component of the brace force must be resisted as axial force in the beam segment outside the link. To avoid excessive axial forces in these beam segments, small brace-to-beam angles should be avoided. A minimum included angle of 35 degrees between the eccentric brace and the beam is recommended (Engelhardt and Popov 1989a).

Simple diagonal bracing (see Figure 7.24a) is often used in narrow bays of bracing that are typically found around elevator cores. Ideally, this framing arrangement should be used in symmetrically opposing bays to maintain overall symmetry in the response of the building frame. For wider bays, the framing configurations shown in Figures 7.24b and 7.24c are preferred. The split-K configuration of Figure 7.24b is particularly advantageous because of its symmetrical configuration and because the links are not directly connected to the columns—thus avoiding the problems associated with developing full moment connections to the column (see Chapter 8).

7.3.5 Kinematics of the EBF

The relationship between story plastic drift (Δ_p) and link plastic rotation (γ_p) can be simply derived. Consider the EBFs of Figure 7.24. The relations between story plastic drift and link plastic rotation for these three frames are presented in Figure 7.29 (AISC 1995). Recognizing that the elastic deformations in the framing outside of the links are small compared with the plastic deformations, the framing outside the links can be assumed to be rigid. For the D-braced frame of Figures 7.24a and 7.29a, the story plastic drift is related to the link plastic rotation as follows:

$$\Delta_p = \frac{\gamma_p e h}{L} \qquad (7.23)$$

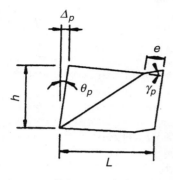

a. D-braced frame

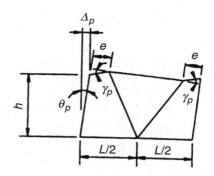

b. V-braced frame

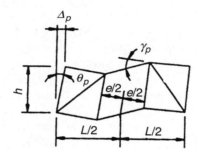

c. Split-K-braced frame

Δ_v = Plastic story drift
e = Link length
h = Story height
L = Column to column distance
θ_p = Plastic story drift angle, radians = Δ_p/h
γ_p = Link rotation angle, radians.

Figure 7.29 Frame drift and link rotation relations. *(AISC 1995)*

where the variables are defined in the figure. For the split-K-braced frame of Figure 7.24b and Figure 7.29c, the story plastic drift is related to the link plastic rotation by:

$$\Delta_p = \frac{\gamma_p \, eh}{L} \qquad (7.24)$$

and for the V-braced frame of Figures 7.24c and 7.29b, the story plastic drift is related to the link plastic rotation by:

$$\Delta_p = \frac{2\gamma_p\, eh}{L} \tag{7.25}$$

The reduced link rotation per unit frame drift of the V-braced frame may be advantageous in cases where link rotation capacity controls the design of the EBF.

Link rotation demand grows quickly as the link length decreases with respect to the frame dimension L. Figure 7.30 plots link rotation demand versus e/L for the split-K-braced EBF. The limiting case of e/L equal to 1.0 corresponds to a moment-resisting frame; that is, γ/θ approaches 1.0 as e/L approaches 1.0. Although large link rotation demands can be realized by short (shear) links, Figure 7.30 demonstrates that links should not be too short, or else the link rotation demands will become excessive—even for a shear-yielding link.

Procedures for estimating total seismic drift in frames are well established (FEMA 1997). For global displacement ductilities of 3 to 5, frame plastic drifts will be substantially greater than frame elastic drifts. For the purpose of preliminary design, the frame elastic drifts can be ignored, and the frame plastic drift can be set equal to the estimated drift. Given this information and limiting link rotations (see Section 7.3.7.3) and recognizing that the frame bay dimensions h and L often cannot be substantially altered by the engineer, the designer can use Equations 7.23 through 7.25 to select the link length e.

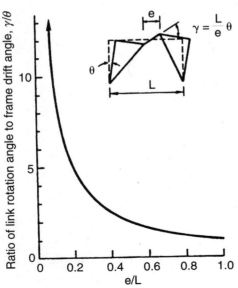

Figure 7.30 Link rotation demand. *(Engelhardt and Popov 1989a)*

7.3.6 Link behavior and length

The length and geometry of the links in an EBF will dictate the behavior of the frame. Short links are shear-critical and long links are flexure-critical. Architectural concerns with the use of short links are similar to those associated with the use of concentric braces. Long links have architectural and planning advantages by providing significant openings in the frame for doors, windows, and mechanical equipment.

The link in an EBF is designed to act as a ductile seismic fuse, dissipating much of the energy input to the building by the earthquake. Because the link is designed to protect the remainder of the frame from overload by limiting the maximum forces that can be developed in the EBF, key to the design of an EBF is an understanding of the inelastic behavior of its links. The critical factor affecting the inelastic behavior of a link is its length. Link length controls the yielding mechanism and the ultimate failure mode. For short links, shear dominates the link response; for longer links, flexure controls link response.

Figure 7.31 illustrates typical distributions of actions (axial force P, bending moment M, and shear force V) in the braces, beams, and links of a D-braced EBF and a split-K-braced EBF under lateral load. The links shown in Figure 7.31 are subjected to high shear forces, high end moments, and low axial force. For very short links, the link shear force reaches the plastic shear capacity V_p ($= 0.55 dt_w F_y$) before the end moments reach the plastic moment capacity M_p ($= ZF_y$), and the link yields in shear, forming a shear hinge or shear link. For longer links,

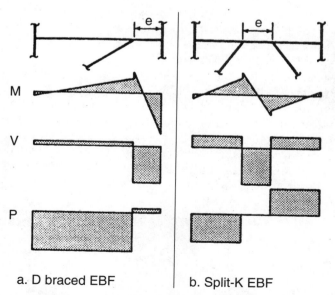

a. D braced EBF b. Split-K EBF

Figure 7.31 Distribution of forces in EBFs. *(Engelhardt and Popov 1989a)*

the end moments reach M_p, forming flexural hinges before shear yielding can occur; these links are termed flexural links.

Material strain hardening can substantially influence link behavior, and both shear and flexural yielding can occur over a wide range of link lengths. Rules have been developed to guide the designer's selection of W sections to ensure shear yielding. The following equations for the link length (e) can be used to classify links:

Shear (short) links:
$$e \leq \frac{1.6 M_p}{V_p} \tag{7.26}$$

Intermediate links:
$$\frac{1.6 M_p}{V_p} < e < \frac{2.5 M_p}{V_p} \tag{7.27}$$

Moment (long) links:
$$e \geq \frac{2.5 M_p}{V_p} \tag{7.28}$$

For the purpose of this textbook, a link length exceeding the threshold set in Equation 7.26 will be considered to be a long link.

The behaviors of short and long links have been exhaustively studied. The shear yielding of short links is preferred to the flexural yielding of long links (Engelhardt and Popov 1989a). In a shear link not subjected to gravity loads between its ends, the shear force is constant along the length of the link (see Figure 7.31), and inelastic shear strains are fairly uniformly distributed over the length of the link. This permits the development of large inelastic link deformations without the development of excessively high local strains. In long links, flexural yielding dominates the response, and very high bending strains are required at the link ends to produce large inelastic link deformations. For reference, a well-detailed short link can likely sustain link chord rotations of 0.1 radian (Kasai and Popov 1986a, Whittaker et al. 1987), whereas the maximum chord rotation of the corresponding long link is likely less than 0.02 radian (Engelhardt and Popov 1989a).

The ultimate failure modes of short and long links are quite different. Inelastic web shear buckling is the ultimate failure mode for short links. This mode of buckling can be delayed by the addition of web stiffeners. Simple rules were developed by Kasai and Popov (1986a) relating stiffener spacing and maximum link inelastic rotation (γ_p) up to the onset of web buckling:

$$a = 29 t_w - \frac{d}{5} \quad \text{for } \gamma_p = \pm 0.09 \text{ rad} \tag{7.29a}$$

$$a = 38 t_w - \frac{d}{5} \quad \text{for } \gamma_p = \pm 0.06 \text{ rad} \tag{7.29b}$$

$$a = 56 t_w - \frac{d}{5} \quad \text{for } \gamma_p \leq \pm 0.03 \text{ rad} \tag{7.29c}$$

where a is the distance between equally spaced stiffeners, d is the beam depth, and t_w is the web thickness. For intermediate values of γ_p, the required stiffener spacing can be interpolated. These rules have been adopted with only minor modifications in current seismic regulations in the United States (see Section 7.3.8.1).

The experiments of Engelhardt and Popov demonstrated that the rotation capacity of long links is appreciably smaller than that for short links. The primary failure mode of long links was fracture of the link flange at or near the link flange-to-column groove weld; other failure modes included severe flange buckling and lateral-torsional buckling of the link. Although the ancillary failure modes can be delayed by the addition of stiffeners within and around the link, the concentration of bending strains at the link ends generally produces flange fracture at relatively low inelastic rotations. Further, the calculation of the maximum inelastic rotation capacity of a long link is unreliable.

The behavior of short links is clearly superior to that of long links. Accordingly, the prudent course of action for a designer is to limit the maximum link length to e given by Equation 7.26.

7.3.7 Link strength and deformation calculations

7.3.7.1 Yield strength Typically, one uses capacity design to size the framing components that deliver force to and from links in an EBF. The starting point in this process is the calculation of the shear yielding strength of the link. This strength may be taken as:

For $e < \dfrac{2M_p}{V_p}$:
$$V_y = V_p \tag{7.30}$$

For $e \geq \dfrac{2M_p}{V_p}$:
$$V_y = \dfrac{2M_p}{e} \tag{7.31}$$

where Equations 7.30 and 7.31 correspond to shear-yielding and flexural-yielding links, respectively. The shear strengths presented in Equations 7.30 and 7.31 are based on the assumption of perfect plasticity (that is, no strain hardening), negligible moment-shear interaction, and small axial force in the link. If the axial force in a link is not small (say, $P > 0.15P_y$), the shear yielding strength must be reduced using shear-axial (for Equation 7.30) and moment-axial (for Equation 7.31) interaction equations such as those presented in Chapter 3 and in the LRFD Specification (AISC 1995). A loss in link rotation capacity will also result if the axial force in a link is high. The reader is referred to Kasai and Popov (1986a) for additional information.

For EBFs including links attached to columns (see Figure 7.24a and 7.24c), evaluation by linear analysis will generally show large bending moments at the column end of the link because of the relatively high flexural stiffness of the column. For this case, the link should not be sized for these large moments but rather for their shear yielding strength. Studies by Kasai and Popov (1986a) have shown that early flexural yielding at one end of a link has little effect on either the shear strength or rotation capacity of the link. In this case, the large elastic moment at the column end of the link redistributes to the other end of the link, and general yielding of the link does not occur until the link shear force reaches the values given by Equations 7.30 and 7.31.

7.3.7.2 Ultimate strength The braces, columns, and beam segments outside the links are intended to remain essentially elastic during earthquake shaking. This objective can be realized by the careful application of capacity design procedures using the maximum link forces as the basis for design. The maximum strength of a link will exceed the strengths given by Equations 7.30 and 7.31 because of the following:

- Material strain hardening
- Actual material strength exceeding nominal values
- The use of dual-certified steels
- The presence of composite floor systems

The combination of the first three may produce maximum link strengths in excess of 200 percent of the nominal values given above. Composite floor systems may further increase the maximum link resistance by upward of 50 percent (Whittaker et al. 1987). Yield-line procedures can be used to estimate the contribution of a composite slab to the strength of a link.

7.3.7.3 Link deformation capacities Early research work on EBFs focused on the response of short links. These studies determined that a well-stiffened shear link can achieve cyclic plastic rotations (γ) of ± 0.10 radian and a monotonic plastic rotation of $+0.20$ radian. By comparison, testing of long links by Engelhardt and Popov (1992) suggest that a well-stiffened long link can achieve cyclic plastic rotations (γ) of between ± 0.015 to ± 0.09 radian and a monotonic plastic rotation of between $+0.03$ and $+0.12$ radian.

The most current standard for the seismic design of eccentrically braced frames at the time of this writing is the AISC LRFD Specification (AISC 1995). The Specification limits the rotation of the link relative to the remainder of the beam to 0.09 radian for links having clear lengths of $1.6M_p/V_p$ or less and 0.030 radian for links having clear lengths of $2.6M_p/V_p$ or greater, where one calculates the link rotation angle using displacements equal to $0.4R$ times the displacements determined using the specified design base shear. (To calculate the specified design base shear, divide the elastic spectral demand is divided by the response modification factor R.) Linear interpolation is used between these rotation limits. Other seismic standards (e.g., the Uniform Building Code [ICBO 1994]) include similar limits on the link rotation angle.

7.3.8 Link details

7.3.8.1 Spacing of link stiffeners It is well established that the addition of web stiffeners to bare steel links will delay the onset of web shear buckling and increase the link rotation capacity. The occurrence of such buckling in a short link is considered as failure of the link because tearing of the web occurs soon after buckling (Kasai and Popov 1986a). Kasai and Popov developed criteria (Equations 7.29a, b, and c) for stiffener spacing as a function of link rotation demand for links yielding in shear (short links). These relationships were developed using both plastic plate buckling theory and experimental data, and they apply up to the onset of inelastic web buckling.

As links transition from short to long, the types of local instability change from those related to shear (shear buckling of the web) to those related to flexure (flange buckling and lateral-torsional buckling). Testing by Engelhardt and Popov suggest that flange buckling cannot be as successfully delayed by the use of transverse web stiffeners. However, flange buckling within a stiffened long link is likely not as serious an event as shear web buckling, although link strength may decrease quickly with increasing plastic hinge rotations. For longer links, the dominant form of instability will probably be lateral-torsional buckling. Transverse stiffeners placed near the link ends appear to provide restraint against lateral-torsional buckling and improve link behavior. Similar observations were made in tests by Hjelmstad and Popov (1983). Tests on long links also identified the need for stiffening outside the link in the brace connection panel. The need for such stiffening is dependent on a number of factors including link length, flange b/t ratio, and the included angle between the brace and the beam. Engelhardt and Popov (1992) proposed a conservative solution, which was to provide a partial depth stiffener at a distance of $1.5b_f$ beyond the end of the link.

The AISC LRFD Specification (AISC 1995) writes the following rules for stiffening links in eccentrically braced frames, most of which are based on the research findings of Engelhardt, Kasai, and Popov:

- Full depth web stiffeners shall be provided on both sides of a link's web at the diagonal brace ends of the link. Limits are set on the combined width and thickness of the stiffeners.
- Links of lengths less than or equal to $1.6M_p/V_p$ shall be provided with intermediate web stiffeners spaced at intervals not exceeding $30t_w - 0.2d$ for a link rotation angle of 0.09 radian and $52t_w - 0.2d$ for link rotation angles of 0.03 radian or less. Linear interpolation shall be used for link rotation angles between 0.03 and 0.09 radian.
- Links of length greater than $1.6M_p/V_p$ and less than $5M_p/V_p$ shall be provided with intermediate web stiffeners placed at a distance of $1.5b_f$ from each end of the link.
- Links of length between $1.6M_p/V_p$ and $2.6M_p/V_p$ shall be provided with intermediate web stiffeners meeting the requirements of the second and third rules above.
- Intermediate web stiffeners are not required in links of lengths greater than $5M_p/V_p$.

7.3.8.2 Intermediate link stiffener details Intermediate stiffeners are required to be full depth and should be equally spaced along the length of the link. For links 25 inches (635 mm) in depth or greater, the stiffeners are required on both sides of the web; for links less than 25 inches in depth, the stiffeners are required on one side only. Link stiffeners must be fillet welded to the link web; these fillet welds shall have a strength equal or exceeding the nominal vertical tensile strength of the stiffener. The fillet welded connection of the link stiffener to the link flange must be designed to transfer a force equal to 25 percent of the tensile strength of the stiffener.

7.3.8.3 Link end transverse bracing Lateral support must be provided at each end of a link to stabilize both the link and the eccentric bracing at its connection to the link. The function of the transverse bracing is to restrain the link and the beam segments outside the link against lateral-torsional buckling and to hold the end of the eccentric brace in the vertical plane of the frame. Any out-of-plane movement of the brace end will produce twisting of the beam and the link when the brace is in compression. The presence of a composite deck cannot be counted upon to provide adequate lateral support at the link ends. Transverse beams of a similar depth to the links are recommended for the perpendicular bracing elements.

The AISC LRFD Specification writes that lateral supports must be provided at both the top and bottom flanges of the link at the ends of the link. The end lateral supports must be designed for a force equal to 6 percent of link flange nominal strength calculated as $b_f t_f F_y$. This recommendation is based on the work of Engelhardt and Popov, who measured end lateral support forces of approximately $0.06b_f t_f F_y$.

7.3.8.4 Link-to-column connections

The AISC LRFD Specification (AISC 1995) provides special requirements for the connection of links to columns (D-braced EBFs and V-braced EBFs). The Specification recommends that only short links (i.e., $e \leq 1.6M_p/V_p$) be connected directly to columns unless it can be demonstrated that the link-to-column connection can develop the required inelastic rotation of the link. This requirement acknowledges that longer links may develop flexural plastic hinges at their ends and that the standard moment connection details may be incapable of sustaining the corresponding plastic hinge rotations.

The link-to-column flange connection is required to be an all-welded connection with the link flanges having complete penetration welded joints to the column and the link webs being welded to the column such that the axial, shear, and flexural strength of the web can be developed. A sample detail for this connection is presented in Figure 7.32. Bolted web connections cannot be used in lieu of the welded web connection because large cyclic shear force demands on a

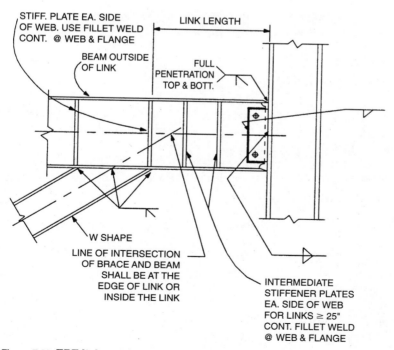

Figure 7.32 EBF link-to-column connection details. *(AISC 1995)*

bolt group may produce bolt slippage, resulting in failure of the link flange complete penetration welds due to high localized deformations.

Link-to-column web connections are not recommended. The maximum link rotation angle for such a connection should be limited to 0.015 radians for any length of link (AISC 1995). In this all-welded connection, the link flanges must be complete penetration welded to continuity plates in the column, and the link web-to-column web joint weld must be capable of developing the axial, shear, and flexural strength of the link web.

7.3.8.5 Other design rules for links The AISC LRFD Specification (AISC 1995) writes other rules for links in EBFs. The key rules are summarized below:

- Links are required to be compact sections using the same criteria as adopted for beams in special moment-resisting frames.
- The specified yield stress of steel used for links shall not exceed 50 ksi in order to obtain sufficiently ductile behavior. (There is little test data on links composed of high-strength steels, and high-strength steels generally have lower ductility than Grade 36 and Grade 50 steel.)
- The web of a link shall be single thickness without doubler plate reinforcement and without openings.

7.3.9 Frame design outside links

7.3.9.1 Beam segments and eccentric braces The EBF design objective is to dissipate much or all of the earthquake-induced energy in the links. Link shear force will generate axial forces and moments in the eccentric brace and the beam segment outside the link. To achieve the design objective, the beam segments and eccentric braces outside the links should be designed to remain elastic for the maximum forces that can be developed by the links. The estimate of maximum link force should include all likely sources of link overstrength, including composite slab effects, increased material yield strength, and strain hardening.

The AISC LRFD Specification (AISC 1995) requires that the minimum moment, axial, and shear strength of the beam segment outside the link be sufficient to resist the forces generated by 1.25 times the nominal shear strength of the link. The same rule is adopted for the design of the eccentric braces. The eccentric brace and the beam segment outside the link should be checked as beam-columns for the specified minimum forces. Further, the beam must not be spliced at or adjacent to the connection between the eccentric brace and the beam.

Earlier design procedures for EBFs (SEAOC 1990) required that nominal strength of eccentric braces be equal to or greater than 1.5 times the axial force corresponding to the nominal strength of the link beam. Brace design for a maximum link strength of 150 percent of the nominal link strength was considered an extreme loading condition, and ϕ was set equal to 1.0 to avoid an overly conservative design (Engelhardt and Popov 1989b).

The LRFD Specification requires that the design strength of the eccentric brace and the beam segment outside the link exceed the forces generated by 1.25 times the nominal strength of the link; for such design, ϕ factors must be used to reduce the nominal strength to the design strength. For axial force in the brace and the beam segment, ϕ generally equals 0.85, and the reserve strength factor is approximately equal to 1.5 ($\approx 1.25/0.85 = 1.47$); for moments, the reserve strength factor is approximately equal to 1.4 using a value of ϕ equal to 0.9.

Adopting the link reserve strength factor of 1.25, the required strength of the eccentric brace and beam segment outside the link can be taken as the forces generated by the following link shear and end moments (AISC 1995), assuming that the link end moments are equal when the link achieves its maximum strength:

For $e \leq 2M_p/V_p$: link shear = $1.25V_p$ (7.32a)

link end moment = $e(1.25M_p)/2$ (7.32b)

For $e \geq 2M_p/V_p$: link shear = $2(1.25M_p)/e$ (7.32c)

link end moment = $1.25M_p$ (7.32d)

For links attached to columns with lengths less than $1.3M_p/V_p$, Kasai and Popov (1986b) note that link end moments may not be equal when the link reaches its maximum strength. The following relations for link shear and link end moments can be used (AISC 1995):

For $e \leq 1.3M_p/V_p$: link shear = $1.25V_p$ (7.32e)

link end moment at column = $0.8M_p$ (7.32f)

link end moment at brace end = $e(1.25V_p) - 0.8M_p$ (7.32g)

At the connection of the eccentric brace, link, and beam segment outside the link, the intersection of the brace and beam centerlines must be either in the link or at the end of the link (see Figures 7.32 and 7.33) because an intersection outside the link will increase the moment generated in the brace and beam segment.

7.3.9.2 Link-to-eccentric brace connections The connection between the eccentric brace and the link beam should be designed for the nominal strength of the brace. The connection must not extend into the link. If the brace-to-link beam connection is designed as a pin, the beam segment outside the link must be adequate to resist the entire link end moment (see Equations 7.32a–f), and the connection detailed to transfer the maximum brace force to the link beam and beam segment.

A sample pinned connection detail for a tubular eccentric brace is presented in Figure 7.33. Note the use of bent or two welded plates to stiffen the brace gusset plate and mitigate the likelihood of plate buckling such as that observed during testing of a full-scale six-story EBF in Japan (Engelhardt and Popov 1989b).

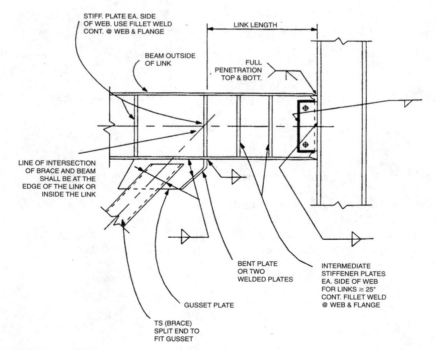

Figure 7.33 EBF brace-to-link connection. *(AISC 1995)*

If the eccentric brace is designed to resist a portion of the link end moment, the connection shall be designed as fully restrained per the provisions of the LRFD Specification. An example of a fully restrained wide flange eccentric brace connection is presented in Figure 7.32.

7.3.9.3 Columns Inelastic behavior in columns should generally be avoided because such response may compromise the stability of the gravity-load resisting system. Column hinging at the base may be unavoidable if a fixed-base detail is adopted, and this connection should be detailed to prevent failure of both the column and the brace-to-column connection.

The LRFD Specification (AISC 1995) requires columns in braced bays to have the strength to resist the sum of gravity-load actions and the moments and axial loads generated by 1.25 times the nominal strength of the links. For medium- and high-rise EBFs, the accumulation of earthquake-induced axial loads corresponding to 125 percent of the nominal strength of each link may be too conservative because it is unlikely that all links will reach their maximum strengths simultaneously. In such cases, the SRSS procedure presented in Section 7.2.6.3 for design of columns in CBFs should be considered.

7.3.9.4 Beam-to-column connections If the EBF configuration is such that a link is not adjacent to a column (e.g., a split-K-braced EBF), the beam-to-column connections away from the link can be designed as a pin in the plane of the beam web, provided that the connection provides some restraint against torsion in the beam. The LRFD Specification (AISC 1995) requires the design torsional moment to be calculated by considering forces equal to 1.5 percent of the nominal axial tensile strength of the flange applied in opposite directions on each flange.

7.3.10 EBF design example

7.3.10.1 Introduction The sample frame is a three-story eccentrically braced frame with split-K bracing in all stories. An elevation of the frame is shown in Figure 7.34a. The beam-to-column connections were assumed to be pinned.

Gravity unfactored concentrated dead loads of 250 kN (56.2 kips) and live loads of 100 kN (22.5 kips) are applied to each column at each level, and uniformly distributed dead loads of 15 kN/m (1.11 kips/ft.) and live loads 10 kN/m (0.74 kips/ft.) are applied along the beams. See Figure 7.21 for details. Unfactored lateral seismic loads of 518 kN (116 kips), 345 kN (78 kips), and 173 kN (38.8 kips) are applied at the third, second, and first floor levels of the frame. The seismic loads were calculated using a value of R equal to 4 per the National Building Code of Canada (NRC 1990) for the design of EBFs. For simplicity, only two load cases were considered in this example: 1.0D + 1.0E and 1.2D + 1.6L, where D, L, and E are the dead, live, and earthquake loads, respectively.

Buildings identical in all aspects except for their lateral framing systems will generally have different fundamental periods. Consequently, their lateral load-resisting frames would be designed for different valves of base shear. The CBF design example (see Section 7.2.6) and the EBF design example use the same elastic spectral demand. This approach was chosen because the intent is to illustrate how frame design can proceed given an elastic spectral demand. The results of the CBF and EBF design examples should not be used to gauge the relative cost or merits of the two framing systems.

7.3.10.2 Strength design requirements The AISC LRFD Specification (AISC 1995) was used to size the frame members for strength. Link plastic shear strengths were calculated per the appropriate EBF provisions in the Specification. Link flange and web compactness ratios satisfied the requirements of the Specification. The sizes of the structural members required by strength design are summarized in Table 7.5 and Figure 7.34b. ASTM A572 Grade 50 steel with a specified yield strength of 345 MPa (50 ksi) was used for the beams, columns, and links. ASTM A500 Grade B square structural tube manufactured to a specified yield strength of 320 MPa (46 ksi) was used for the eccentric braces. The beams, each composed of one link and two beam segments, were assumed to be continuous between the columns. The effective length factor (k) was set equal to 1.0 for the eccentric braces.

7.3.10.3 Design for ductile response Seismic standards that contain design provisions for EBFs generally require that one can evaluate the beam segments outside the links, eccentric braces, and columns in the braced bay using capacity design procedures based on the forces associated with the maximum expected strengths of the links. Such a procedure is required by the AISC LRFD Specification and was described in Section 7.3.9. The procedure presented in the LRFD Specification was used to evaluate the strength design of Section 7.3.10.2.

7.3.10.3.1 Braces and beam segments outside the link The required design combined axial and moment strength of the eccentric brace must

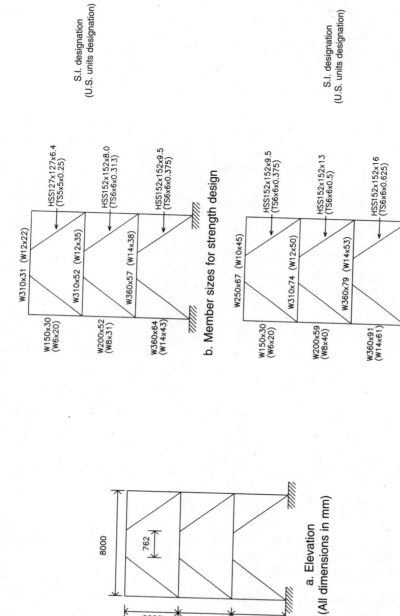

Figure 7.34 Example of three-story eccentrically braced frame.

Table 7.5 Strength Design Data for Three-Story Split-K Eccentrically Braced Frame Example

Element	Story	Strength design		Link Length
		S.I. Shapes	U.S. Shapes	
Columns	3	W150x30	W6x20	–
	2	W200x52	W8x31	–
	1	W360x64	W14x43	–
Beams/Links	3	W310x31	W12x22	30"
	2	W310x52	W12x35	30"
	1	W360x57	W14x38	30"
Braces	3	HSS127x127x6.4	TS5x5x0.25	–
	2	HSS152x152x8.0	TS6x6x0.313	–
	1	HSS152x152x9.5	TS6x6x0.375	–

exceed the axial forces and moments generated by 1.25 times the nominal shear strength of the link, where the nominal shear strength is calculated as $0.6F_y(d - 2t_f)t_w$ (AISC 1995). Similarly, the required design strength of the beam segment outside the link must exceed the forces generated by 1.25 times the nominal strength of the link.

None of the three beams sized for strength (see Table 7.5) had sufficient strength to resist the moments and axial forces generated by 125 percent of the corresponding link shear strength. Four options, and combinations thereof, were considered in order to meet the requirements of the Specification:

- Select a new beam size with a similar value of V_p but a much larger value of M_p.
- Reduce the length of the link, thereby reducing the link end moment for a given V_p.

- Replace the pin-ended brace-to-link connection with a moment connection, attracting part of the link end moment into the brace.
- Detail the brace-to-link connection so that the brace and link beam centerlines intersect inside the link, introducing a moment of the opposite sign to the link end moment.

The first option was selected for this design example. The new beam sizes, listed in Table 7.6, satisfied the requirements for flange and web compactness.

None of the braces sized for strength (see Table 7.5) had sufficient strength to resist the axial forces generated by 125 percent of the corresponding link shear strength. The braces sizes required for ductile response are listed in Table 7.6.

7.3.10.3.2 Columns The required strength of columns is determined from the following equations, except that the earthquake-induced moments and axial loads introduced into the column at the connection

Table 7.6 Member Sizes for Ductile Response in the Example Three-Story Split-K Eccentrically Braced Frame

Element	Story	Ductile Response		Link Length
		S.I. Shapes	U.S. Shapes	
Columns	3	W150x30	W6x20	–
	2	W200x59	W8x40	–
	1	W360x91	W14x61	–
Beams/Links	3	W250x67	W10x45	30"
	2	W310x74	W12x50	30"
	1	W360x79	W14x53	30"
Braces	3	HSS152x152x9.5	TS6x6x0.375	–
	2	HSS152x152x13	TS6x6x0.500	–
	1	HSS152x152x16	TS6x6x0.625	–

of the brace must exceed those generated by 1.25 times the nominal strength of the link:

$$1.2D \pm 1.0E + 0.5L + 0.2S$$

$$0.9D \pm 1.0E$$

where D is the dead-load action, E is the earthquake-induced action, L is the live-load action, and S is the snow-load action. This requirement increased the size of the columns in the lower two stories of the example frame; see Table 7.6 for the new sizes.

7.3.10.3.3 Other considerations The design of the example frame is not complete. One must design connections using capacity procedures similar to those described above. Link rotation demands must be evaluated by analysis, and the links detailed to sustain the calculated deformations.

The use of capacity procedures to calculate column forces was discussed in Section 7.2.6.3. For the three-story example frame, the use of summed maximum link forces to calculate earthquake-induced actions in columns is appropriate. However, this approach may be too conservative in the lower stories of high-rise frames because it is unlikely that each link in a braced bay will yield simultaneously. The reader is referred to Redwood and Channagiri (1991) for an alternate force calculation procedure.

References

1. Astaneh-Asl, A.; Goel, S.C.; and Hanson, R.D. 1982. *Cyclic Behavior of Double Angle Bracing Members with End Gusset Plates.* Report No. UMEE 82R7. August. Department of Civil Engineering. The University of Michigan. Ann Arbor, Michigan,
2. AISC. 1995. *Load and Resistance Factor Design.* Volume 1, 2nd Edition. American Institute of Steel Construction. Chicago, IL.
3. AISC. 1989. *Manual of Steel Construction, Allowable Stress.* 9th Edition. American Institute of Steel Construction. Chicago, IL.
4. AISC. 1980. *Manual of Steel Construction, Allowable Stress.* 8th Edition. American Institute of Steel Construction. Chicago, IL.
5. Black, R.G.; Wenger, W.A.; and Popov, E.P. *Inelastic Buckling of Steel Struts Under Cyclic Load Reversal.* Report No. UCB/EERC-80/40. Berkeley: Earthquake Engineering Research Center. University of California.
6. Bruneau, M., and Mahin, S.A. 1990. "Ultimate Behavior of Heavy Steel Section Welded Splices and Design Implications." *Journal of Structural Engineering.* Vol. 116, No. 18: 2214–2235. August. ASCE.
7. BSSC. 1995. *NEHRP (National Earthquake Hazards Reduction Program) Recommended Provisions for the Development of Seismic Regulations for New Buildings.* Report FEMA 222A. Building Seismic Safety Council. Federal Emergency Management Agency. Washington, DC.
8. CISC. 1995. *Handbook of Steel Construction.* 6th Edition. Canadian Institute of Steel Construction. Willowdale, Ontario, Canada.
9. CSA. 1994. *Limit States Design of Steel Structures.* CAN/CSA-S16.1-M94. Canadian Standard Association. Rexdale, Ontario, Canada.

10. Engelhardt, M.D., and Popov, E.P. 1989a. *Behavior of Long Links in Eccentrically Braced Frames*. Report No. UCB/EERC-89/01. Berkeley: Earthquake Engineering Research Center. University of California.
11. Engelhardt, M.D., and Popov, E.P. 1989b. "On Design of Eccentrically Braced Frames." *Earthquake Spectra*. Vol. 5, No. 3. August.
12. Engelhardt, M.D., and Popov, E.P. 1992. "Experimental Performance of Long Links in Eccentrically Braced Frames." *Journal of Structural Engineering*. Vol. 118, No. 11: 3067–3088. November. ASCE.
13. FEMA. 1997. *Guidelines for the Seismic Rehabilitation of Buildings*. Report FEMA 273. Federal Emergency Management Agency. Washington, D.C.
14. Hjelmstad, K.D., and Popov, E.P. 1983. *Seismic Behavior of Active Beam Links in Eccentrically Braced Frames*. Report No. UCB/EERC-83/24. Berkeley: Earthquake Engineering Research Center. University of California.
15. Ikeda, K., and Mahin, S.A. 1984. *A Refined Physical Theory Model for Predicting the Seismic Behavior of Braced Frames*. Report No. UCB/EERC-84/12. Berkeley: Earthquake Engineering Research Center, University of California.
16. ICBO. 1970, 1976, 1994. *Uniform Building Code*. International Conference of Building Officials. Whittier, CA.
17. Jain, A.K., Goel, S.C., and Hanson, R.D. 1978. *Hysteresis Behavior of Bracing Members and Seismic Response of Braced Frames with Different Proportions*. Report No. UMEE 78R3. July. Ann Arbor: Department of Civil Engineering. The University of Michigan.
18. Kasai, K., and Popov, E.P. 1986a. *A Study of Seismically Resistant Eccentrically Braced Frames*. Report No. UCB/EERC-86/01. Berkeley: Earthquake Engineering Research Center. University of California.
19. Kasai, K., and Popov, E. P. 1986b, "General Behavior of WF Steel Shear Link Beams." *Journal of the Structural Division*. Vol. 112, No. 2: 362–382. February. ASCE.
20. Kasai, K., and Popov, E.P. 1986c. "Cyclic Web Buckling Control for Shear Link Beams." *Journal of the Structural Division*. Vol. 112, No. 3: 505–523. March. ASCE.
21. Khatib, I.F., Mahin, S.A., and Pister, K.S. 1988. *Seismic Behavior of Concentrically Braced Steel Frames*. Report No. UCB/EERC-88/01. Berkeley: Earthquake Engineering Research Center. University of California.
22. Liu, Z., and Goel, S.C. 1988. "Cyclic Load Behavior of Concrete-Filled Tubular Braces." *Journal of the Structural Division*. Vol. 114, No. 7: 1488–1506. July. ASCE.
23. Liu, Z., and Goel, S.C. 1987. *Investigation of Concrete-Filled Steel Tubes under Cyclic Bending and Buckling*. Report No. UMCE 87-3. Ann Arbor: Department of Civil Engineering. The University of Michigan.
24. Merovich, A.T., Nicoletti, J.P., and Hartle, E. 1982 "Eccentric Bracing in Tall Buildings." *Journal of the Structural Division*. Vol. 108, No. 9. September. ASCE.
25. Nonaka, T. 1989. "Elastic-Plastic Bar under Changes in Temperature and Axial Load." *Journal of the Structural Engineering*. Vol. 115, No. 12: 3059–3075. December ASCE.
26. Nonaka, T. 1987. "Formulation of Inelastic Bar under Repeated Axial and Thermal Loadings." *Journal of the Engineering Mechanics*. Vol. 113, No. 11: 1647–1664. November. ASCE.
27. NRC. 1990. *National Building Code of Canada*. National Research Council. Ottawa, Ontario, Canada.
28. Popov, E.P., and Black, R.G. 1981. "Steel Struts under Severe Cyclic Loadings." *Journal of the Structural Division*. Vol. 107, No. ST9: 1857–1881. September. ASCE.
29. Redwood, R.G., and Channagiri, V.S. 1991. "Earthquake-Resistant Design of Concentrically Braced Steel Frames." *Canadian Journal of Civil Engineering*. Vol. 18, No. 5: 839–850.
30. Ricles, J.M., and Popov, E.P. 1987. *Dynamic Analysis of Seismically Resistant Eccentrically Braced Frames*. Report No. UCB/EERC-87/07. Berkeley: Earthquake Engineering Research Center. University of California.
31. Roeder, C.W., and Popov, E.P. 1978. "Eccentrically Braced Frames for Earthquakes." *Journal of the Structural Division*. Vol. 104, No. 3: 391–412, March. ASCE.

32. Roeder, C.W., and Popov, E.P. 1977. *Inelastic Behavior of Eccentrically Braced Steel Frames under Cyclic Loadings.* Report No. UCB/EERC-77/18. Berkeley: Earthquake Engineering Research Center. University of California.
33. SEAOC. 1974, 1990, 1996. *Tentative Lateral Force Requirements.* Seismology Committee, Structural Engineers Association of California. Sacramento/San Francisco/Los Angeles, CA.
34. Tang, X., and Goel, S.C. 1987. *Seismic Analysis and Design Considerations of Braced Steel Structures.* Report No. UMCE 87-4. June. Ann Arbor: Department of Civil Engineering. The University of Michigan.
35. Uang, C.M., and Bertero, V.V. 1986. *Earthquake Simulation Tests and Associated Studies of a 0.3-Scale Model of a Six-Story Concentrically Braced Steel Structure.* Report No. UCB/EERC-86/10. Berkeley: Earthquake Engineering Research Center. University of California.
36. Wakabayashi, M. 1986. *Design of Earthquake Resistant Buildings.* New York: McGraw-Hill.
37. Whittaker, A.S.; Uang, C.M.; and Bertero, V.V. 1987. *Earthquake Simulation Tests and Associated Studies of a 0.3-Scale Model of a Six-Story Eccentrically Braced Steel Structure.* Report No. UCB/EERC-87/02. Berkeley: Earthquake Engineering Research Center. University of California.
38. Whittaker, A.S.; Uang, C.M.; and Bertero, V.V. 1990. *An Experimental Study of the Behavior of Dual Steel Systems.* Report No. UCB/EERC-88/14. Berkeley: Earthquake Engineering Research Center. University of California.

Chapter 8

Design of Ductile Moment-Resisting Frames

8.1 Introduction

Moment-resisting frames (also called moment frames) are, in their simplest form, rectilinear assemblages of beams and columns, with the beams rigidly connected to the columns. Resistance to lateral forces is provided primarily by rigid frame action—that is, by the development of bending moments and shear forces in the frame members and joints. By virtue of the rigid beam-to-column connections, a moment frame cannot displace laterally without bending the beams and columns. The bending rigidity and strength of the frame members is therefore the primary source of lateral stiffness and strength for the entire frame.

Steel moment-resisting frames have been popular in many regions of high seismicity for several reasons. First, moment frames have been viewed as highly ductile systems. Building code formulae for design earthquake forces typically assign the largest force reduction factors (and therefore the lowest lateral design forces) to moment-resisting frames, reflecting the opinion of code writers that moment-resisting frames are among the most ductile of all structural systems. Second, moment frames are popular because of their architectural versatility. There are no bracing elements present to block wall openings, providing maximum flexibility for space utilization. A penalty for this architectural freedom results from the inherent lateral flexibility of moment-resisting frames. Compared with braced frames, moment frames subjected to lateral loads generally require larger member sizes than those required for strength alone to keep the lateral deflections within the code-mandated drift limits. The inherent flexibility of moment frames may also result in greater drift-induced nonstructural damage under earthquake loading than with other stiffer systems.

These perceptions regarding the expected performance of steel moment frames in earthquakes were significantly challenged by the 1994 Northridge (Los Angeles) earthquake in the United States, and by the 1995 Hyogo-ken Nanbu (Kobe) earthquake in Japan. In both earthquakes, steel moment frames did not perform as well as expected. Brittle failures were observed at beam-to-column connections in modern steel moment frame structures, challenging the assumption of high ductility and demonstrating that our understanding of steel moment-resisting frames is incomplete.

This chapter provides an overview of ductile steel moment frames. Section 8.2 discusses some basic concepts of overall frame behavior. Sections 8.3 and 8.4 cover basic concepts of column and panel zone behavior and design. Beam-to-column panel zones influence both the elastic and inelastic behavior of steel moment frames. Section 8.5 discusses beam-to-column connections. Typical North American practice for beam-to-column moment connections prior to the 1994 Northridge earthquake is presented. Observations of connection damage from this earthquake are also reviewed together with a discussion on some of the major reasons for this damage. Finally, some of the moment-connection strategies developed since the Northridge earthquake are presented. Current moment frame design practice is addressed in Section 8.6, followed by a short example.

Throughout this chapter, the AISC Seismic Provisions (1992) are used when necessary to illustrate how principles of ductile design have been implemented in codes. Numerous code documents have introduced specific detailing requirements for earthquake-resistant steel structures since the first comprehensive provisions formulated in code language appeared in 1988 (SEAOC 1988). Emphasis on the AISC requirements here is intended to help focus the discussion (note that although steel frames damaged by the Northridge earthquake were generally designed per the Uniform Building Code, ductility detailing requirements of the AISC Seismic Provisions are compatible with those of the 1988 edition of the Uniform Building Code). At the time of this writing, intensive research is under way worldwide in response to the unexpectedly poor performance of steel moment frames in the 1994 Northridge and 1995 Hyogo-ken Nanbu earthquakes. Most building codes are being revised in response to these earthquakes and will likely evolve further. The reader is encouraged to keep abreast of new developments in this field.

8.2 Basic response of ductile moment-resisting frames to lateral loads

8.2.1 Internal forces during seismic response

A steel moment-resisting frame is composed of three basic components: beams, columns, and beam-column panel zones. These are illustrated in Figure 8.1 for a simple two-story, single-bay moment frame. Beams span the clear distance from face-of-column to face-of-column, L_b, and columns are divided into a clear span portion, h_{ci}, and a panel zone region of height, h_{pzi}. The panel zone is the portion of the column contained within the joint region at the intersection of a beam and a column. This definition is useful when one is considering sources of elastic and inelastic deformations, as well as possible plastic hinge locations.

In traditional structural analysis, moment frames are often modeled as line representations of horizontal and vertical members, with the lines intersecting at dimensionless nodes. Such models do not explicitly consider the panel zone region, and they provide an incomplete picture of moment frame behavior. Design of ductile moment frames requires explicit consideration of the panel zone region. (Inelastic behavior and design of the panel zone are addressed in Section 8.5.)

Figure 8.1 also shows qualitatively the distribution of the bending moment, shear force, and axial force in a moment frame under lateral load. These internal forces are shown for the beam, clear span portion of the column, and the column panel zone, and they do not include gravity load effects. The beams exhibit high bending moments, typically under reverse curvature bending, with maximum moments occurring at the member ends. The shear and axial force in the beam are generally much smaller and less significant to the response of the beam as compared with bending moment, although they must be considered in design. Similarly, the clear span portions of the columns are typically subjected to high moments, with relatively low shear forces. Axial forces in columns, both tension and compression, can be significant because of overturning moments on the frame. Finally, the column panel zone is subject to high moments, high shear forces due to a severe moment gradient, and possibly high axial forces.

The qualitative distribution of internal forces illustrated in Figure 8.1 is fundamentally the same for both elastic and inelastic ranges of behavior. The specific values of the internal forces will change as elements of the frame yield and internal forces are redistributed. The basic patterns illustrated in Figure 8.1, however, remain the same.

276 Chapter Eight

Inelastic step-by-step response-history analysis is needed to obtain exact values for the internal forces in moment frames, but this analytical complexity can be avoided if capacity design principles are integrated into the design process along with the conventional elastic analyses.

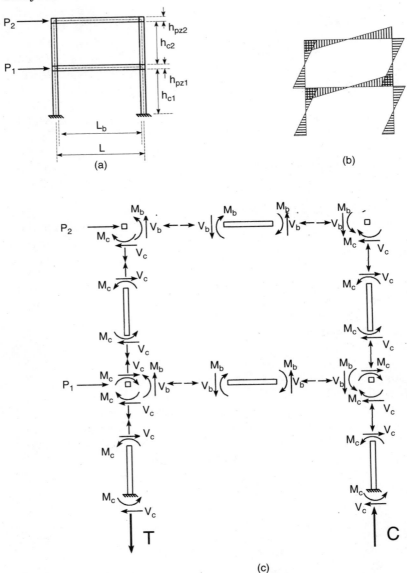

Figure 8.1 Ductile moment resisting frame: (a) Geometry considering finite dimensions of members, (b) Typical moment diagram under lateral loading, and (c) Corresponding member forces on beams, columns, and panel zones.

Plastic analysis of moment frames was described in Chapters 4 through 6. It was shown that, depending on the relative strength of beams and columns framing into a joint, different plastic collapse mechanisms can develop and that, as described in Chapter 6, the development of plastic hinges in the beams is the superior mechanism (see Figure 6.10). In actual frames, however, as opposed to frames studied using simple plastic analysis, strain hardening makes possible yielding of more than one component at any given joint. In an example sequence of events, the panel zone may yield first, but still exhibit significant postyielding stiffness because of strain hardening and other effects described in Section 8.4. As a result, greater forces can be applied at the joint, and other framing members, such as a beam, may reach their plastic capacities. Thus, the beam, column, and even panel zone could contribute to the total plastic deformation at the joint, depending on their relative yield strengths and yield thresholds. A structural component considerably weaker than the others framing into the joint will have to provide alone the needed plastic energy dissipation, whereas components of comparable strength would share this burden.

Once identified, those structural components expected to dissipate hysteretic energy during an earthquake must be detailed to allow development of large plastic rotations, without significant loss of strength. Only those components and connection details capable of providing cyclic plastic rotation capacities in excess of the demands should be used to ensure satisfactory seismic performance.

8.2.2 Plastic rotation demands

Estimates of the plastic rotation demands for a given moment frame are typically obtained by inelastic response-history analyses. Results from such analyses are sensitive to modeling assumptions and vary when different ground motion records are considered. The amount of the plastic energy dissipated by beams, panel zones, and columns will also be a function of the design philosophy adopted.

For those reasons, general expectations of plastic rotation demand for generic moment frames are based on the synthesis of observations from past analytical studies. Prior to the Northridge earthquake, the largest plastic rotations expected in beams alone (in the absence of panel zone plastic deformations) were expected to be 0.02 radian (Tsai 1988, Popov and Tsai 1989), although some studies reported values as high as 0.025 radian (Roeder et al. 1989). Smaller plastic rotation demands are obviously expected in flexible frames whose design is governed by compliance to code-specified drift limits.

An approximate way to estimate the plastic rotation demands in a frame is to examine its plastic collapse mechanism at the point of maximum drift. For example, if the beam sway mechanism shown in

Figure 6.10 develops in a frame designed in compliance with the code-specified interstory drift limit, the maximum plastic hinge rotations in a beam can be estimated as Δ_e/h, where h is the story height, and Δ_e is the inelastic interstory drift. Inelastic interstory drifts are approximately related to those one calculates using design-level forces, Δ_c, by simple relationships such as $\Delta_e = R\,\Delta_c$ (CSA 1994) or $\Delta_e = C_d\,\Delta_c$ (BSSC 1994) where R is a seismic force reduction factor per the CSA code respectively and C_d is a deflection amplification factor (see Chapter 9 for more details). Typically, for code-specified drift limits, the use of these relationships produces plastic rotation demands of approximately 0.02 radian. This procedure is conservative because a large percentage of the total frame drift occurs elastically before plastic hinges form, provided that the method to calculate Δ_e and the seismic hazard characterization are accurate.

After the Northridge earthquake, the required connection plastic rotation capacity was increased to 0.03 radian for new construction and 0.025 radian for postearthquake modification of existing buildings (SAC 1995b). This target rotation was a consensus value developed following the earthquake based on analysis of code-compliant moment frames using ground motion histories recorded during the earthquake (e.g., Bertero et al. 1994). Although this rotation capacity may exceed real earthquake demands on most structural connections, it will likely remain as the target value until substantial research demonstrates that lower values are acceptable.

8.2.3 Lateral bracing and local buckling

Selected structural members must be able to reach and maintain their plastic moment through large plastic rotations that permit hysteretic dissipation of earthquake-induced energy. The engineer must therefore delay local flange and web buckling, and lateral-torsional buckling, to prevent premature failures due to member instability.

For that reason, only compact structural shapes should be used for structural members expected to develop plastic hinges. For example, AISC (1992) limits the flange width-to-thickness ratios, $b_f/2t_f$, of W shapes to $52/\sqrt{F_y}$, for F_y in ksi (which roughly corresponds to the $145/\sqrt{F_y}$ limit in CSA 1994, for F_y in MPa). Moreover, lateral bracing to both flanges of these members should be provided at each plastic hinge location and spaced at no more than $2500 r_y/\sqrt{F_y}$, with F_y in ksi and where r_y is the member's radius of gyration about its weak axis, in inch (or $980 r_y/\sqrt{F_y}$ for MPa and mm units). This requirement recognizes that top and bottom flanges will alternatingly be in compression during an earthquake and accounts for some uncertainty in the location of plastic hinges under various load conditions. Local buckling of flanges and webs and lateral-torsional buckling will unavoidably develop at very large plastic rotations (at least in commonly used structural shapes), but compliance with the above requirements will slow the progressive loss in strength and help ensure good inelastic energy dissipation. This topic is further discussed in Chapter 10.

8.3 Ductile moment-frame column design

8.3.1 Axial forces in columns

Column buckling is not a ductile phenomenon and must be prevented. Columns should therefore be designed to remain stable under the maximum forces they can be subjected to during an earthquake. These forces will generally exceed those predicted by elastic analysis using code-specified earthquake loads, but may be difficult to estimate. As an upper bound, with some allowance for strain hardening effects, one can obtain maximum axial forces using capacity design principles (as described in Chapter 6, Figure 6.8). However, during an earthquake, plastic hinges do not form simultaneously at all stories, but rather develop in only a few stories at a time, often in a succession of waves traveling along the height of the building. As a result, the capacity design approach may be conservative, particularly in multistory buildings.

There is no agreement on what constitutes a proper alternative method to capture the maximum axial force acting on a column during earthquake shaking. Some codes typically resort to an additional load case, with higher specified earthquake loads to be considered only for the design of columns. For example, in the 1992 AISC Seismic Provisions, a seismic force multiplier of $2R/5$ is used. Application of the SRSS technique in conjunction with capacity design principles, as presented in Chapter 7 for ductile concentric braced frames, is another method to estimate more realistic maximum column axial forces.

8.3.2 Considerations for column splices

Typically, the bending moment diagram for the beams and columns will show a point of inflection somewhere along the length of the member. Frequently, for preliminary design, the points of inflection are assumed to be at midlength of the members. Although this is a convenient assumption, it is important to recognize that the location of the inflection points will vary significantly. This is particularly true as yielding occurs in the frame during an earthquake and bending moments are redistributed within the frame. Even though the basic pattern of bending moments remains the same, the location of inflection points can shift substantially from the locations indicated by an elastic frame analysis.

Assumptions regarding the location of inflection points can substantially impact the design of column splices. A designer may elect to locate a column splice near an inflection point based on elastic frame analysis (or slightly lower than midheight to provide convenient site-welding conditions) and design the splice for a relatively small bending moment, based on those same elastic frame analysis results. This would be an error because the possibility of significant bending moments at the splice location must be considered, regardless of the results of elastic analysis.

Tests have showed partial penetration welds in thick members to be brittle under tensile loads (Popov and Stephen 1977, Bruneau et al. 1987, Bruneau and Mahin 1991). For example, a standard partial penetration splice detail frequently used in seismic regions, shown in Figure 8.2a, was tested for the largest column sizes that could be accommodated in a 17,800 kN (4,000 kips) capacity universal testing machine (Bruneau and Mahin 1991). This specimen, fabricated from A572 Grade 50 steel, was tested in flexure instead of tension to permit consideration of the largest specimen for which cross section could be kept whole; cutting away part of the section would have released some of the lock-in residual stresses. The test setup is illustrated in Figure 8.2b. As shown in Figure 8.2c, the moment-curvature relationship remained practically linear up to a value corresponding to approximately 60 percent of the nominal plastic moment of the smaller column section at the splice, at which the weld fractured in a brittle manner (Figure 8.2d).

For the above reasons, partial penetration welded joints in column splices are viewed apprehensively. Therefore, seismic codes typically require splices subjected to net tension forces to be designed for no less than half of the column axial cross-sectional plastic strength, or 150 percent of the required splice strength calculated by analysis.

8.3.3 Strong-column/weak-beam philosophy

Structural frames can dissipate a greater amount of hysteretic energy when plastic hinges develop in the beams rather than in the columns (see Figure 6.10). This beam-sway mechanism enhances overall seismic resistance and prevents development of a soft-story (column-sway) mechanism in a multistory frame. Frames in which measures

Design of Ductile Moment-Resisting Frames 281

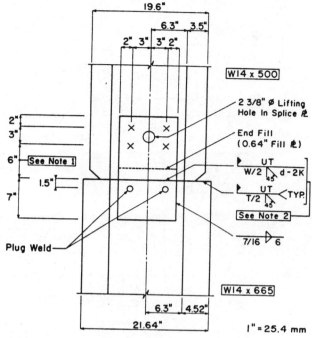

NOTES:
1. Provide 6" <u>as specified</u> instead of 3" as on standard detail.
2. A certified inspector is to be present during construction and perform ultrasonic testing of the weld.

Figure 8.2 Test column splice with partial penetration welds in thick members: (a) Splice detail.

are taken to promote plastic hinges in the beams rather than in the columns are said to be strong-column/weak-beam (SCWB) frames. The alternative is weak-column/strong-beam (WCSB) frames.

Many codes and design guidelines (e.g., AISC 1992, SEAOC 1990) have moved toward the SCWB philosophy by requiring that, at a joint, the sum of the columns' plastic moment capacities exceed the sum of the beams' plastic moment capacities, in which case:

$$\sum M_{pcr} = \sum Z_{cr} F_{yc} = \sum Z_c \left(F_{yc} - \frac{P_{uc}}{A_g} \right) \geq \sum Z_b F_{yb} \quad (8.1)$$

or exceed the sum of the beam moments that correspond to the strength of the panel zone, in which case:

$$\sum M_{pcr} = \sum Z_{cr} F_{yc} = \sum Z_c \left(F_{yc} - \frac{P_{uc}}{A_g} \right) \geq V_n d_b \frac{H}{(H - d_b)} \quad (8.2)$$

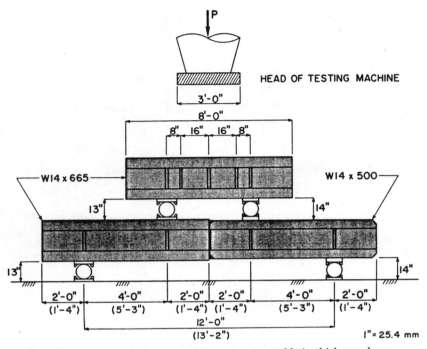

Figure 8.2 Test column splice with partial penetration welds in thick members: (b) Test setup.

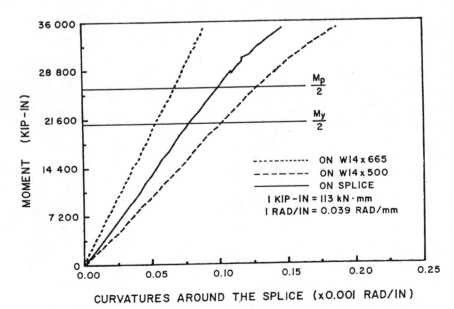

Figure 8.2 Test column splice with partial penetration welds in thick members: (c) Moment-curvature results.

Figure 8.2 Test column splice with partial penetration welds in thick members: (d) Splice after brittle fracture.

where A_g is the gross area of the column, F_{yb} is the beam nominal yield strength, F_{yc} is the column nominal yield strength, H is the average of the story heights above and below the joint, P_{uc} is the required axial strength in the column, V_n is the code-prescribed nominal shear strength of the panel zone (per the Krawinkler equation presented in Section 8.4), Z_b is the beam plastic section modulus, Z_c is the column plastic section modulus, and Z_{cr} is the plastic modulus reduced to account to the presence of axial force (see Chapter 3). These equations are based on simple moment equilibrium at the joint (as in Figure 8.1). The concept embedded in Equation 8.2 will become clear after a reading of the material presented in Section 8.4.

Conceptually, however, the above requirements are insufficient to prevent column plastic hinging at beam-to-column joints. First, to achieve that objective, *probable* capacities instead of *nominal* capacities must be considered in the above equations, taking into account the impact of strain hardening, larger-than-specified beam yield strength, and other factors contributing to reserve strength. Second, a correction factor must be introduced in the above equations to reflect that the ratio of the column moments acting at the top and bottom faces of a joint varies greatly during an earthquake because of the movement of each column's inflection point.

8.3.4 Effect of axial forces on column ductility

Some codes (e.g., CSA 1994) refrain from promoting the SCWB philosophy out of concern that the capacity design approach is difficult to implement in low-rise buildings and hard to justify in zones of moderate and low seismicity. Other seismic codes that promote SCWB design also recognize that prevention of column flexural yielding may be difficult in some applications and permit the design of WCSB frames in some instances. For example, according to AISC (1992), the SCWB requirement can be waived provided that the maximum axial load acting on all columns is less than $0.30P_y$ where P_y equals $F_{yc}A_g$, F_{yc} is the column nominal yield strength, and A_g is the gross area of the column. Engineers taking that route must recognize that plastic hinges may form in columns of WCSB frames and recognize the deleterious impact of axial forces on the rotation capacity of columns.

Surprisingly, there is a paucity of research results on the effect of axial loads on the ductility of steel columns. Adherence to the above strong-column/weak-beam philosophy may partially explain this situation. Popov et al. (1975) showed that the cyclic behavior of W-shaped columns is a function of the applied load to yield load ratio, P/P_y, and the magnitude of interstory drifts. In those tests, for specimens braced to prevent lateral buckling about their weak axis, sudden failure due to excessive local bucking and strength degradation were observed when P/P_y exceeded 0.5. The aforementioned limit of $0.30P_y$ ($0.40P_y$ in some other codes) is historically tied to this series of tests.

The adequacy of the existing code limits has been challenged by Schneider and Roeder (1992); test results showed that moment-resisting steel building frames designed according to the WCSB philosophy suffered rapid strength and stiffness deterioration when the columns were subjected to axial loads equal to approximately $0.25\ P/P_y$.

Columns subject to plastic deformations should be compact sections and be laterally braced in accordance with the requirements for plastic design. This requires lateral bracing at each plastic hinge location and a maximum brace spacing of $2500 r_y/\sqrt{F_y}$ in U.S. units ($980 r_y/\sqrt{F_y}$ in S.I. units), as discussed earlier for beams.

8.4 Panel zone

The satisfactory seismic response of a ductile moment-resisting frame depends on the adequate performance of its beam-column joints. For multistory building frames, in which beams connected to columns are expected to develop their plastic moment, the designer must prevent undesirable beam-column joint failures. In steel structures, doing so requires measures to avoid column flange distortion, column web yielding and crippling, and panel zone failure. This section mostly focuses on the behavior and design of ductile panel zones, but matters relevant to the first and second failure modes are first addressed.

8.4.1 Flange distortion and column web yielding/crippling prevention

The addition of continuity plates (i.e., stiffeners joining the beam flanges across the column web) can effectively prevent flange distortion and column web yielding/crippling. Examples of continuity plates are shown in Figure 8.3. When beams reach their plastic moment at the column face (Figure 8.4a), the beam flanges apply large localized forces to the columns (Figure 8.4b). The beam flange in tension pulls on the column flange. In absence of continuity plates, and if otherwise unrestrained, the column flange would bend under that pulling action, with greater deflections in column flanges of low stiffness and small thickness (Figure 8.4c). However, the column flange is not free to deflect because the beam flange framing into it is rigid in its plane (Figure 8.4d). Because deformations of the connected elements must be compatible, stresses concentrate in the beam flange where column flange is stiffest, that is, near the column web (Figures 8.4e and 8.4f).

In tests of connections without continuity plates, localized cracking originated in the beam flange weld at the column centerline and rapidly propagated across the entire flange width and thickness. To prevent this type of failure, most seismic codes require the addition of continuity plates if the maximum expected beam flange force exceeds the factored flange strength of ϕR_n where:

$$R_n = 6.25 t_{cf}^2 F_{yf} \tag{8.3}$$

where t_{cf} is the column flange thickness, F_{yf} is the column flange nominal yield strength and ϕ is equal to 0.9. This equation is based on

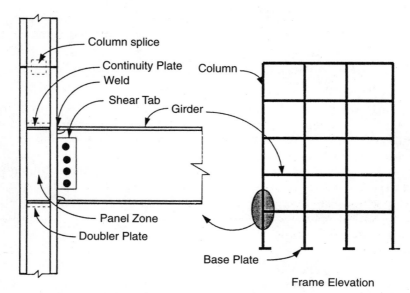

Figure 8.3 Fundamental elements of a ductile moment resisting frame. (*From* Interim Guidelines: Evaluation, Repair, Modifications, and Design of Steel Moment Frames, *SAC Joint Venture, 1995, with permission*)

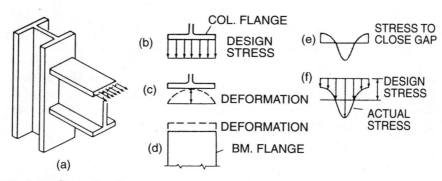

Figure 8.4 Stress distribution in welded beam flange at column face in absence of column continuity plates (stiffeners). (*From* Journal of Constructional Steel Research, *Vol. 8, E.G. Popov*, Panel Zone Flexibility in Seismic Moment Joints, *1987, with permission from Elsevier Science Ltd., Kidlington, U.K.*)

yield line analyses by Graham (1959). Note that AISC (1992) specifies that maximum expected beam flange force to be taken as 1.8 $A_f F_y$, where A_f is the flange area of the connected beam, and F_y is the nominal strength of the beam. This value assumes a strain-hardened beam moment 30 percent greater than the nominal plastic moment, and it assumes that the bolted beam web is ineffective in transferring moment. Thus, if only the beam flanges can effectively transfer the maximum beam moment at the connection, and assuming that the flanges-only plastic modulus, $Z_f (\approx A_f d$, where d is the beam depth), is approximately 70 percent of a beam's plastic modulus, Z, the maximum expected beam flange force becomes:

$$T_{max} = \frac{M_{max}}{d} = \frac{1.3 M_p}{d} = \frac{1.3(ZF_y)}{d} = \frac{1.3\left(\frac{Z_f}{0.7}\right)F_y}{d} \approx \frac{1.8 A_f d F_y}{d} = 1.8 A_f F_y \quad (8.4)$$

Other codes (e.g., CSA 1994) arrive at a similar result by specifying a reduced column flange resistance instead of magnified beam flange forces for seismic applications.

Nonetheless, AISC (1992) suggests that designers use continuity plates even when the above requirement is satisfied because continuity plates were used in nearly all cyclic tests (prior to 1994) that exhibited satisfactory ductile behavior. To avoid the stress concentration depicted in Figure 8.4 in the highly stressed welded region, interim design guidelines released following the Northridge earthquake (SAC 1995b) also recommend the use of continuity plates in all ductile moment frame connections.

When the beam flange applies compression to the column flange, column web yielding must be prevented, as would normally be done in nonseismic applications, using the traditional equations for bearing resistance:

$$B_r = (5k + N)\, t_{cw} F_{yw} = (5k + t_{bf})\, t_{cw} F_{yw} \quad (8.5)$$

where k is the distance from the outer face of the column to the web toe of the fillet, F_{yw} is the yield strength of the column web, and t_{bf} and t_{cw} are the beam flange and column web thicknesses, respectively. Resistance to web crippling must also be checked, using:

$$B_r = C t_w^2 \left[1 + 3\left(\frac{N}{d}\right)\left(\frac{t_w}{t_f}\right)^{1.5}\right] \sqrt{\frac{F_{yw} t_f}{t_w}} \quad (8.6)$$

where C is equal to 135 (ASCE 1993) in U.S. units, and 300 (CSA 1994) in S.I. units (not a direct conversion of units, because the two codes use different ϕ factors) and B_r is the bearing resistance.

Note that seismic design codes generally do not require consideration of strain hardening in the beam flange in compression because web-crippling is not a brittle failure mode. Also note that doubler plates are frequently used instead of continuity plates when increases in web crippling or web yielding resistance are necessary.

8.4.2 Forces on panel zones

The panel zone of a beam-column joint is the rectangular segment of the column web surrounded by the column flanges (left and right vertical boundaries) and the continuity plates (top and bottom horizontal boundaries). Typically, the panel zone is simultaneously subjected to axial forces, shears, and moments from the columns and beams, as shown in Figure 8.5.

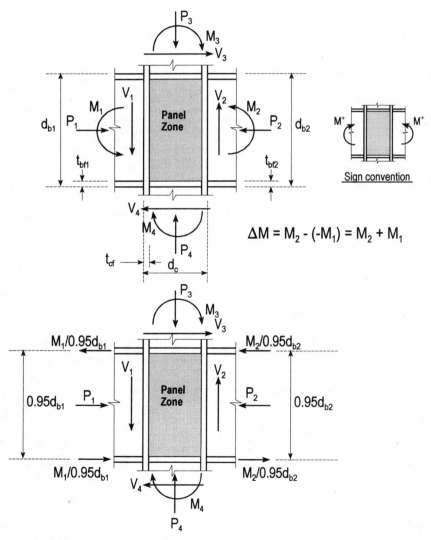

$$\Delta M = M_2 - (-M_1) = M_2 + M_1$$

Figure 8.5 Moment, shear force, and axial forces acting on the panel zone of a ductile moment resisting frame subjected to lateral loading.

Resolving equilibrium on the free-body diagram of Figure 8.5 and taking the forces shown acting on the face of the panel as positive, the horizontal shear acting in the panel zone can be calculated as:

$$V_w = \frac{M_1}{0.95 d_{b1}} + \frac{M_2}{0.95 d_{b2}} - V_c \tag{8.7}$$

where d_{b1} and d_{b2} are the depths of beam 1 and 2, respectively, and $0.95\, d_{b1}$ and $0.95\, d_{b2}$ are approximations for the lever arm of the beam flange forces resulting from the applied moments, as shown in Figure 8.5. V_c is the subassembly equilibrating shear given by:

$$V_c = \frac{M_1 \left(\frac{L_1}{L_{b1}}\right) + M_2 \left(\frac{L_2}{L_{b2}}\right)}{h} \tag{8.8}$$

where h is the average of story heights above and below the joint, L_i is the total span length of beam i measured center-to-center of the columns to which it connects, and L_{bi} is the clear span length of beam i equal to the distance from face-to-face of the columns (i.e., deducting half of the column width at each end of the beam) as shown in Figure 8.1. When member forces are available from computer analysis, one can obtain an estimate of V_c by averaging the column shears at the edges of the panel zone:

$$V_c = \frac{V_3 + V_4}{2} \tag{8.9}$$

This approximation is usually conservative because it gives smaller values of V_c and thus higher values of V_w.

The above equations show that the critical loading condition for the panel zone occurs when it is subjected to large unbalanced moments from the beams framing into the columns. Large shear forces will develop in the panel zones of interior columns participating in a sway frame collapse mechanism (of the type shown in Figure 6.10a) when the beams on all sides of such a panel zone reach their plastic moment. In fact, the panel zone shear in that case is substantially greater than the shear in the adjacent columns and beams, and the possibility of panel zone yielding must be considered.

If Equation 8.8 is substituted into Equation 8.7, the panel zone shear, V_w, can be shown to depend only on beam moments M_1 and M_2. In other words, the magnitude of the unbalanced moment, $\Delta M = M_1 + M_2$, controls the force demand on the panel zone. Different philosophies regarding the magnitude of ΔM to be considered for design have been developed in the past. Tsai and Popov (1990b) reported three such philosophies: strong panel zones, intermediate-strength panel zones, and minimum-strength panel zones. For strong panel zone design, $\Delta M = M_{p1} + M_{p2} = \Sigma M_p$, following capacity design principles (SEAOC 1980). For intermediate strength panel zone design, $\Delta M = \Sigma M_p - 2M_g$, where M_g is the moment due to gravity loads. Assuming this moment to be to 20 percent of M_p, the design requirement becomes $\Delta M = \Sigma 0.8 M_p$ (Popov 1987, Popov et al. 1989). For minimum-strength panel zone design, in an allowable stress design perspective, $\Delta M = \Sigma(M_g + 1.85\ M_e\ ) < \Sigma 0.8 M_p$, where M_e is the beam moment obtained when the specified earthquake loads are acting alone, and 1.85 is a factor chosen to further reduce the design force on the panel zone and promote a greater energy dissipation by panel zone yielding.

The strong panel zone philosophy was used prior to 1988 in the United States, together with a panel zone shear strength of $0.55 F_y A_w$, where A_w is the column web area (SEAOC 1980). The intermediate strength and minimum strength approaches are indirect means to obtain weaker panel zones that will yield sooner and respectively dissipate a greater percentage of the total hysteretic energy. Despite the lack of a sound theoretical basis, the latter two approaches were adopted by many codes and guidelines in the United States (e.g., SEAOC 1988, AISC 1992) after 1988 to be used in conjunction with the panel zone shear strength equation described in section 8.4.5 below. Only a few studies have investigated the consequences of these various design approaches in terms of the relative levels of plastic deformation in beams and panel zones (Popov et al. 1989, Tsai and Popov 1990b, Tsai et al. 1995). These indicated larger panel zone inelastic demands and interstory drifts in frames designed per the minimum-strength panel zone approach.

8.4.3 Behavior of panel zones

Studies of panel zone inelastic behavior started in the 1970s and included the work of Krawinkler et al. (1971, 1973, 1975, 1978), Fielding and Huang (1971), Fielding and Chen (1973), and Becker (1975). Tests of large-scale specimens clearly revealed the dominance of shear distortions on panel zone behavior. Krawinkler et al. (1971) visually captured this phenomenon using photogrammetric techniques, as shown in Figures 8.6c and 8.6e, at large shear strains for the specimens shown in Figure 8.6a and 8.6b. These tests also demonstrated that panel zones, when carefully detailed to avoid column web yielding and crippling, as well as column flange distortion, can exhibit excellent hysteretic energy dissipation characteristics in shear, up to large inelastic deformations. Typical results from cyclic inelastic testing are presented in Figures 8.6e and 8.6f, expressed in terms of the unbalanced beam moment ($\Delta M = M_1 + M_2$) versus average panel zone shear distortions (γ_p, also termed shear strains or shear deformations in the literature).

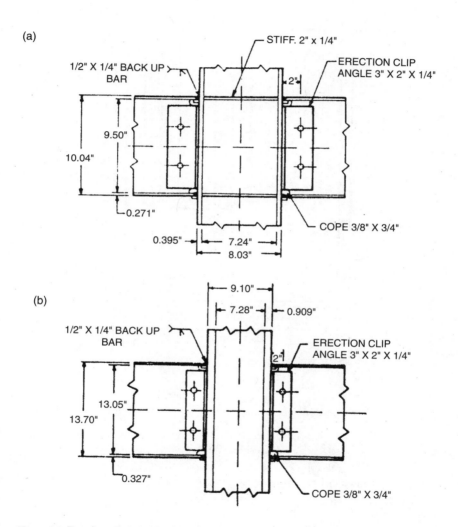

Figure 8.6 Panel zone deformation experimental results: (a) Connection details of specimen A, (b) specimen B. Column in specimen A is W200x36 section (W8x24 in U.S. units) with flanges milled to simulate W360x101 (W14x68 in U.S. units), and column in specimen B is W200x100 (W8x67 U.S. units) to simulate W360x339 (W14x228 U.S. units). (*Parts a-g from Earthquake Engineering Research Center Report UCB/EERC 71-7, "Inelastic Behavior of Steel Beam-to-Column Subassemblages" by H. Krawinkler et al., 1971, with permission from the author*)

292 Chapter Eight

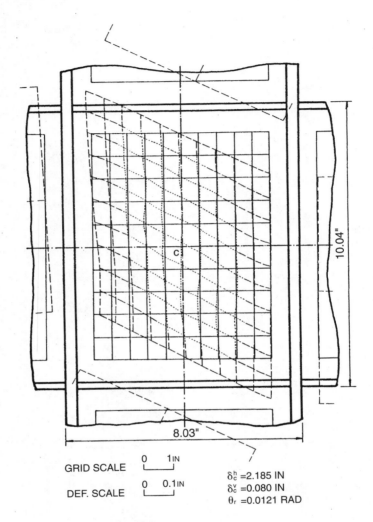

SPECIMEN A-2 LP 9

Figure 8.6 Panel zone deformation experimental results: (c) Deformation pattern in panel zone of specimen A.

Examination of these hysteretic loops shows that panel zones exhibit considerable reserve strength beyond first yield, with a steep strain-hardening slope. This results from the complex state of stress that develops inside the panel zone as shear stresses are progressively increased. Typically, yielding starts in the middle of the panel, consis-

Design of Ductile Moment-Resisting Frames 293

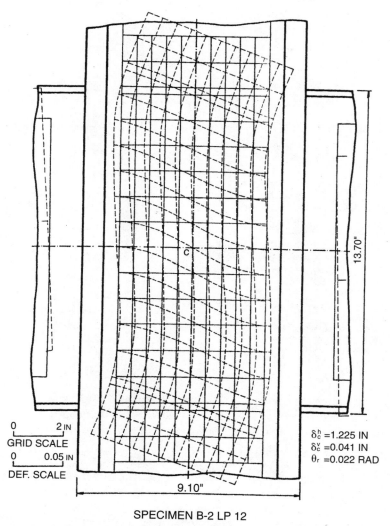

Figure 8.6 Panel zone deformation experimental results: (d) Deformation pattern in panel zone of specimen B.

tently with elastic theory, and progresses approximately in a radial manner over the entire panel zone as the unbalanced moment further increases. As a result, shear distortion is largest at the center of the panel and smallest at the corners. Once the web is fully yielded, the panel zone stiffness depends in a complex manner on the panel aspect ratio, d_c/d_b per Figure 8.5, and the stiffness of its surrounding

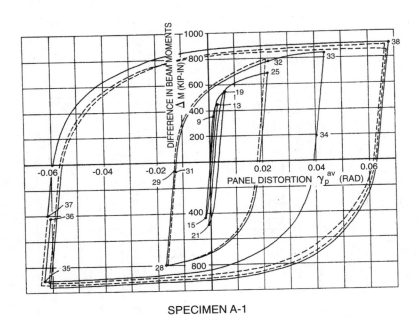

Figure 8.6 Panel zone deformation experimental results: (e) ΔM versus γ_p diagram for specimen A.

elements, such as the column flanges and the webs of the connecting beams. These factors, together with strain hardening of the web in shear, produce the considerable post-yield stiffness observed during tests (see Figures 8.6e and 8.6f).

The column axial load also has an impact on the behavior of the panel zone. In the presence of axial stress, the onset of shear yielding in the panel zone is hastened, in accordance with the Von-Mises yield criterion. Nonetheless, experiments have shown that the ultimate shear strength of the panel is not substantially affected by column axial loads; column flanges were observed to provide axial load resistance when the panel yielded in shear. This redistribution is possible when the column flanges remain elastic during panel zone yielding. Ultimately, at large shear strains, the column flanges will in turn develop their full plastic flexural capacity, in a state of combined flexure and axial force. When that occurs, large kinks in column flanges may develop, producing large strains in or near the welds connecting the beam flanges to the column, and possibly joint fracture. For this reason, researchers have recommended that the maximum shear dis-

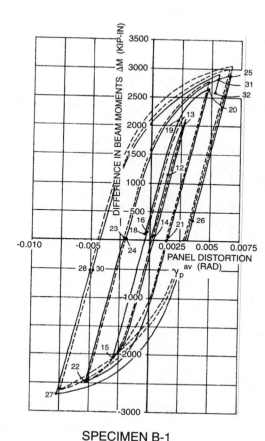

SPECIMEN B-1

Figure 8.6 Panel zone deformation experimental results: (f) ΔM versus γ_p diagram for specimen B.

tortion in a panel zone, γ_{max}, be limited to four times the shear yield distortion, γ_y (Krawinkler et al. 1971).

8.4.4 Modeling of panel zone behavior

Formulation of a simple model that captures the complex behaviors described above remains elusive. Elastic stiffness and yield threshold are relatively simple matters, but modeling postyield stiffness, which was observed to vary considerably from specimen to specimen, is particularly difficult. Krawinkler et al. (1971) proposed a model "...simple enough to permit its inclusion into practical computer programs..." at the "...sacrifice [of] accuracy in modeling actual boundary conditions."

296　Chapter Eight

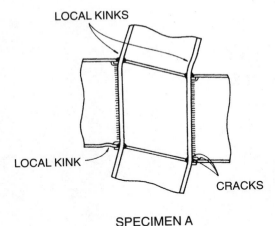

Figure 8.6 Panel zone deformation experimental results: (g) Effects of excessive panel zone distortions.

The model proposed, presented in Figure 8.7a, consists of an elastic-perfectly plastic column web surrounded by four rigid sides connected by springs at the corners.

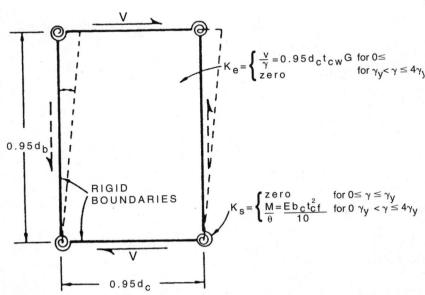

$$K_e = \begin{cases} \dfrac{V}{\gamma} = 0.95 d_c t_{cw} G & \text{for } 0 \leq \gamma \leq \gamma_y \\ \text{zero} & \text{for } \gamma_y < \gamma \leq 4\gamma_y \end{cases}$$

$$K_s = \begin{cases} \text{zero} & \text{for } 0 \leq \gamma \leq \gamma_y \\ \dfrac{M}{\theta} = \dfrac{E b_c t_{cf}^2}{10} & \text{for } 0\, \gamma_y < \gamma \leq 4\gamma_y \end{cases}$$

Figure 8.7 Panel zone behavior: (a) Mathematical model. (*Parts a-d from* Engineering Journal, *3rd Quarter 1978*, Shear in Beam-Column Joints in Seismic Design of Steel Frames *by H. Krawinkler, with permission from the American Institute of Steel Construction*)

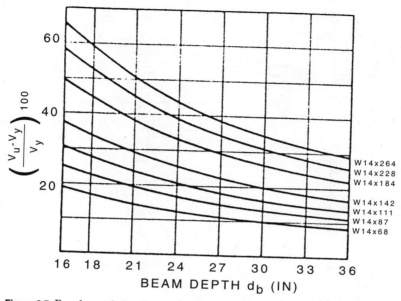

Figure 8.7 Panel zone behavior: (b) Example of ultimate strength per Krawinkler model, V_u, compared with Von Mises yield strength, V_y.

These springs mostly capture the effect of the column flanges on panel zone behavior and neglect other behaviors. In the elastic range, the stiffness of the panel zone is approximately:

$$K_e = \frac{V}{\gamma} = \frac{1}{\dfrac{1}{0.95 d_c t_{cw} G} + \dfrac{d_b^2}{24 E I_{cf}}} \qquad (8.10)$$

where G is the shear modulus, E is the modulus of elasticity, I_{cf} is the moment of inertia of a single column flange, t_{cw} is the column web thickness, and all other terms have been defined previously. Recognizing that the flange typically contributes approximately only 10 percent of the total elastic stiffness, one can ignore the second term in the denominator, which results in the following simpler expression for the elastic stiffness:

$$K_e = \frac{V}{\gamma} = 0.95 d_c t_{cw} G \qquad (8.11)$$

In the postyield range, the panel zone shear stiffness is taken as zero while the spring stiffness is taken as:

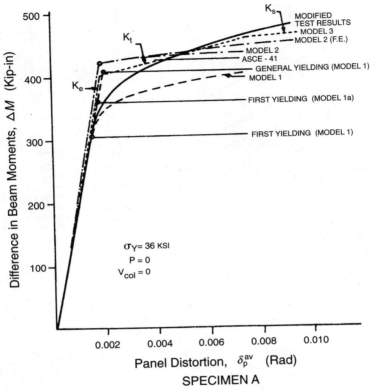

Figure 8.7 Panel zone behavior: (c) Experimental versus theoretical panel zone shear (expressed in terms of ΔM) for specimen A.

$$K_s = \frac{M}{\theta} = \frac{E b_c t_{cf}^2}{10} \tag{8.12}$$

where θ is the concentrated spring rotation, and b_c and t_{cf} are the width and thickness of the column flange, respectively. This definition of K_s cannot be proven through the use of simple models. Krawinkler et al. (1971) report that finite element analyses have been used to determine the concentrated column flange rotation at each corner corresponding to this model. The post yielding stiffness of the panel is thus given by:

$$K_t = \frac{V}{\gamma} = \left[\frac{4M}{0.95 d_b}\right]\frac{1}{\gamma} = \left[\frac{4}{0.95 d_b}\right]\frac{M}{\theta} = \frac{1.095 b_c t_{cf}^2 G}{d_b} \tag{8.13}$$

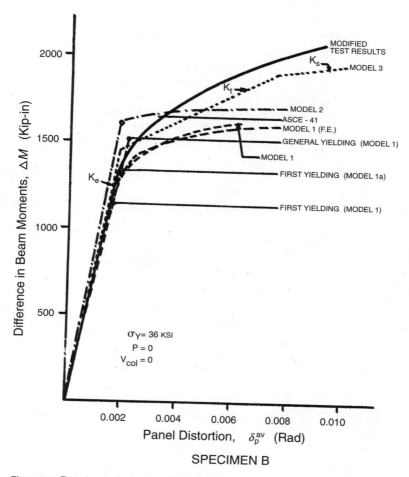

Figure 8.7 Panel zone behavior: (d) Experimental versus theoretical panel zone shear (expressed in terms of ΔM). Specimen B (Krawinkler model is identified as Model 3 on that figure).

using static equilibrium on the panel and knowledge that γ is equal to θ for this model. This equation is reasonable over the range $\gamma_y < \gamma < 4\gamma_y$, where γ_y is the shear yield distortion. Hence, the panel zone shear strength, reached at an angle of distortion of $4\gamma_y$, is:

$$V_u = K_e \gamma_y + 3K_t \gamma_y = V_y \left(1 + 3 \frac{K_t}{K_e}\right) = 0.55 F_y d_c t_{cw} \left(1 + \frac{3.45 b_c t_{cf}^2}{d_b d_c t_{cw}}\right) \quad (8.14)$$

The ratio of the second term over the first term inside the parenthesis represents the increase in panel zone shear resistance beyond that predicted by the Von-Mises criterion. Heavy columns with large flanges will benefit more from the higher resistance provided by this second term, as illustrated in Figure 8.7b. However, tests to date have been conducted on specimens scaled to represent moderate size columns, such as those indicated in Figure 8.6.

Note that the above model fails to check whether the flange flexural plastic capacity is reached before the shear deformation reaches $4\gamma_y$. It also does not consider many other effects that influence panel zone inelastic behavior, such as shear strain hardening and true boundary conditions (in particular, plastic hinges in column flanges can be closer than $0.95\,d_b$ for different boundary conditions). However, given that this model was found to capture the few available experimental results reasonably well (as shown in Figures 8.7c and 8.7d where this model is called Model 3), it has been adopted in many seismic codes.

8.4.5 Design of panel zone

Until the Northridge earthquake, inelastic panel zone action was generally considered to be desirable for energy dissipation. By comparing the behavior of frame subassemblies tested to identical interstory drift levels, Krawinkler et al. (1971) observed that specimens exhibited greater energy dissipation when panel zone shear yielding occurred in combination with beam flexural yielding. When the panel zones tested by Krawinkler did not yield, greater beam flexural plastic rotations were necessary to reach the same interstory drifts, and the beams suffered more of the inelastic local buckling and lateral-torsional buckling that typically develop at large hysteretic flexural deformations, and thus exhibited more strength degradation. It was therefore suggested that "controlled" inelastic panel zone deformations would improve the overall seismic behavior of steel frames, particularly because the cyclic shear hysteretic behavior of well-designed panel zones does not exhibit strength degradation. Designers were also advised to consider panel shear deformations when calculating drifts.

The panel zone design equation typically implemented in North American codes is:

$$V_u = 0.60 F_y d_c t_p \left(1 + \frac{3 b_c t_{cf}^2}{d_b d_c t_p}\right) \quad \text{or} \quad 0.55 F_y d_c t_p \left(1 + \frac{3 b_c t_{cf}^2}{d_b d_c t_p}\right) \quad (8.15)$$

depending on the code (AISC 1992 or CISC 1994, with ϕ factors of 0.75 or 0.9, respectively), where t_p is the thickness of the panel zone including doubler plates if any, and all other terms have been defined previously. When beams of different sizes frame into the column, it is conservative to use the largest of the beam depths for d_b. In nonseismic applications, the AISC code (1993) decreases the strength given by Equation 8.15 to as low as 70 percent of the calculated value when the axial load exceeds 75 percent of the column plastic axial strength (i.e., 0.75 P_y); some researchers have argued that further reductions are necessary to properly account for the effect of axial forces (Chen and Liew 1992). However, in seismic applications, such high axial loads are rarely found in the columns of ductile moment frames.

When the panel zone of a column has insufficient strength, doubler plates can be added locally to increase the column web thickness; this has proven to be an economical solution in North America. To be considered effective in seismic applications, doubler plates must be placed next to the column web, fillet welded along the plate width, and welded to the column flanges to develop the design shear strength of the doubler plate.

In addition to traditional web slenderness limits, seismic design codes typically require that panel zone thickness be at least:

$$t_z \geq \frac{(d_z + w_z)}{90} \qquad (8.16)$$

to prevent premature local buckling under large cyclic inelastic shear deformations. In this empirical equation, d_z is the panel zone depth between the continuity plates, w_z is the panel zone width between the column flanges, and t_z is the panel zone thickness. If doubler plates are used to increase the thickness of the panel zone, their individual thickness must also satisfy the above equation. Note that t_z can be taken as the sum of the panel zone and doubler plate thicknesses only if the doubler plates are connected to the panel zone with plug welds in a manner to preclude independent buckling of these individual elements.

Finally, note that one should consider panel zone deformations when calculating frame deformations. However, designers have typically neglected panel zone flexibility when conducting analyses with line representations of frames. In such models, finite joint sizes are ignored, structural members are modeled by line elements at their centers of gravity, and the flexible lengths of beams and columns are taken as the center-to-center distances between their intersection points. In more exact models, finite joint sizes are considered, member flexibility is derived from the free lengths between the faces of columns and beams, and the flexibility of panel zones is included. For the types and geometries of frames typically used in buildings, the error obtained through use of the simpler model has been reported to be negligible, particularly in view of all other uncertainties involved in the process (Wakabayashi 1986, Englekirk 1994). The engineer should nonetheless beware of instances when design conditions and/or frame geometry would make this error significant.

8.5 Beam-to-Column Connections

The seismic response of a ductile moment frame will be satisfactory only if the connections between the framing members have sufficient strength to permit attainment of the desired plastic collapse mechanism, sufficient stiffness to justify the assumption of fully rigid behavior typically assumed for analysis, and adequate detailing to permit development of the large cyclic inelastic deformations expected during an earthquake without any significant loss of connection strength. Beams, panel zones, and to some extent, columns can dissipate seismic energy through plastic cyclic rotations, but connection failure is not acceptable. From that perspective, bolts and welds are considered to be nonductile elements that must be designed with sufficient strength to resist the maximum forces that can develop in the connected elements. Even though bolts and, to some extent, welds are capable of plastic deformations, their small size and limited ductility generally make those deformations ineffective at the structural level.

Moment frames acquired their excellent reputation as seismic framing systems following the San Francisco 1906 earthquake. However, even though the few midrise steel buildings constructed at that time weathered the earthquake well, one must recognize that the heavily riveted moment connections of that era bear little resemblance to current seismic moment connections. Examples of connections used in the first half of the 1900s are shown in Figure 8.8 for comparison with the standard modern connections illustrated later in this chapter. The oft-stated "excellent performance of steel moment frames in past earthquakes" was biased, to some degree, by the track record of buildings with details that became obsolete in the 1960s when high-strength bolts and welding became the preferred fastening methods in seismic regions. It is the behavior of these modern moment connections that is addressed here.

8.5.1 Knowledge and practice prior to the 1994 Northridge earthquake

The welded moment connection details widely used in many North American seismic regions (notably California) during the 25 years preceding the Northridge earthquake are shown at the top of Figure 8.9. Although the simple plastic theory formulated in the first chapters of this book would suggest that full-penetration groove welds are required in both flanges and the web of a beam to create a connection capable of resisting the beam's plastic moment, by the 1960s the building industry was already frequently using an alternative

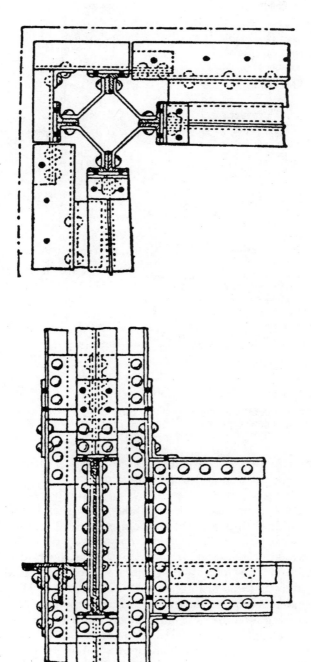

Figure 8.8 Examples of frame connections (a) at turn-of-the-century. (*From* Journal of Constructional Steel Research, *Vol. 10, William McGuire, Introduction to Special Issue, 1988, with permission from Elsevier Science Ltd., U.K.*)

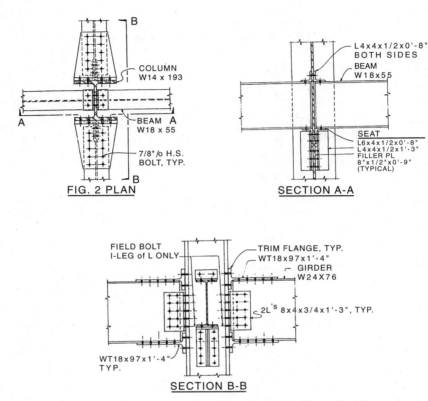

Figure 8.8 Examples of frame connections (b) in the 1930s. (*From* Steel Tips—Structural Steel Construction in the '90s *by F.R. Preece and A.L. Collin, with permission from the Structural Steel Education Council*)

more economic (easier to construct) connection detail with fully welded flanges and a bolted web connection.

The first tests to investigate the cyclic plastic behavior of moment connections were conducted in the 1960s (Popov and Pinkney 1969). Various popular details were considered, as shown in Figure 8.9, and specimens with welded flanges and bolted web connections showed superior inelastic behavior compared with the cover plated moment connection and the fully bolted moment connection alternatives. Typical hysteretic loops are presented in Figure 8.10. The fully bolted detail was considered less desirable because slippage of the bolts during cyclic loading produced a visible pinching of the hysteretic loops and because tensile rupture occurred along a net section between bolt holes.

Further tests in the 1970s (Popov and Stephen 1970) compared the relative performance of the commonly used welded flanges-bolted web detail and fully welded connections. Sample results are shown in Figure 8.11. Both details were significantly stronger than predicted by

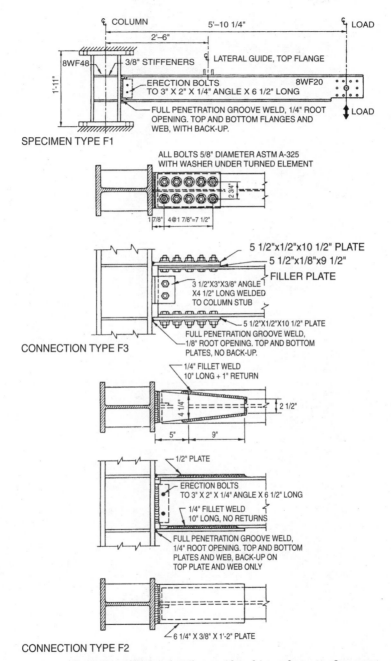

Figure 8.9 Typical connection details considered in early tests of moment connections by Popov and Pinkney. (*From ASCE* Journal of the Structural Division, *Vol. 95*, Cyclic Yield Reversal in Steel Building Connections *by E.P. Popov and R.B. Pinkley, 1969, with permission from the American Society of Civil Engineers*)

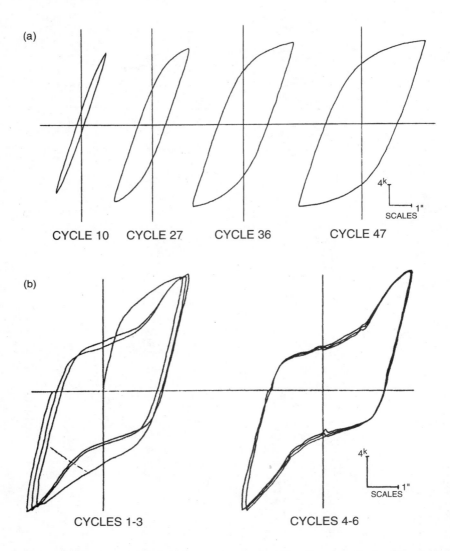

Figure 8.10 Examples of hysteretic behavior obtained in Popov and Pinkney's experiments for (a) specimen type F1, (b) Specimen type F3—See Figure 8.9. (*From* ASCE Journal of the Structural Division, *Vol. 95,* Cyclic Yield Reversal in Steel Building Connections *by E.P. Popov and R.B. Pinkley, 1969, with permission from the American Society of Civil Engineers*)

the simple plastic theory (with F_y=36 ksi), as clearly shown in Figure 8.11, and the fully welded connection exhibited more ductile behavior (Figures 8.11a versus 8.11b). The moment connections with bolted webs were also reported to fail abruptly, and their ductility was more erratic (Popov 1987, in a retrospective of past research). Nonetheless, connections with bolted web were judged to be sufficiently ductile and reported to be less costly to fabricate. It is interesting to note that Popov and Stephen (1972) also concluded that "The quality of workmanship and inspection is exceedingly important for the achievement of best results."

Further studies on frame subassemblies (Krawinkler et al. 1971, Bertero et al. 1973, and Popov et al. 1975) investigated the effect of panel zone and column plastic hinging and helped make the welded flange-bolted web detail a prequalified moment connection provided that it was detailed according to predetermined rules. This standard connection is illustrated in Figure 8.12, although some aspects shown on that detail (such as the supplemental fillet welds along part of the web tab) were actually implemented only in the late 1980s (ICBO 1988). This figure also summarizes some of the doubler plate details described in Section 8.4.5. Note that self-shielded flux-cored arc welding was commonly used, with E70T-4 or E70T-7 electrodes as the filler metal, there was no specified notch toughness requirement for the filler metal.

For a number of years, nearly all beam-to-column connections in structural systems designated as ductile moment-resisting frames were detailed to be able to transfer the nominal plastic moment of the beams to the columns (Roeder and Foutch 1995). As a result, relatively modest column and beam sizes were sufficient in those moment frames to provide the necessary seismic resistance. However, over the years, as a result of the cost premium commanded by full moment connections compared with shear connections, many engineers concluded that it was economically advantageous to limit the number of bays of framing designed as ductile moment-resisting frames. In the extreme, prior to the Northridge earthquake, some engineers routinely designed buildings having only four single-bay ductile moment frames (two in each principal direction, with each in a different plane to provide torsional resistance). This trend developed at the expense of a dramatic loss in structural redundancy, which can be argued to be a nonnegligible reduction in overall structural safety, particularly in the event of construction defects. Moreover, considerably deeper beams, columns with thicker flanges, and bigger foundations were needed in these single-bay ductile moment frames than in the multibay ones previously used to resist the same seismically induced forces.

In that regard, some more recent tests on beam-to-column subassemblies added valuable data to the existing knowledge base and provided an opportunity to investigate potential size effects. In particular, tests by Tsai and Popov (1988, 1989) indicated that some

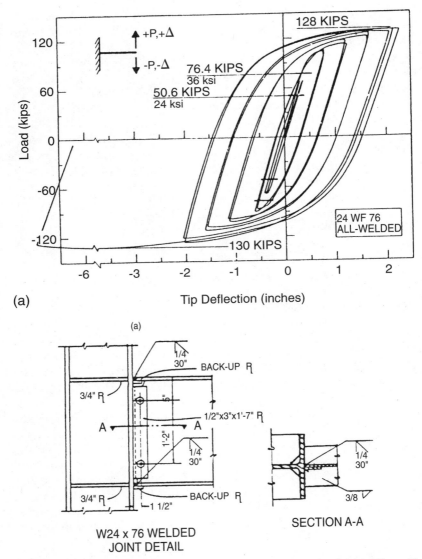

Figure 8.11 Hysteretic behavior of typical connection details having (a) fully welded webs. Results from tests conducted in early 1970s. (*Parts a and b from* Journal of Constructional Steel Research, *Vol. 8, E.G. Popov,* Panel Zone Flexibility in Seismic Moment Joints, *1987, with permission from Elsevier Science Ltd., U.K.*)

prequalified moment connections in ductile moment frames with W460 and W530 beams, equivalent to W18 and W21 in U.S. units and thus similar in depth to those tested by Popov and Stephen (1971), were not as ductile as expected when the web accounted for a substantial portion

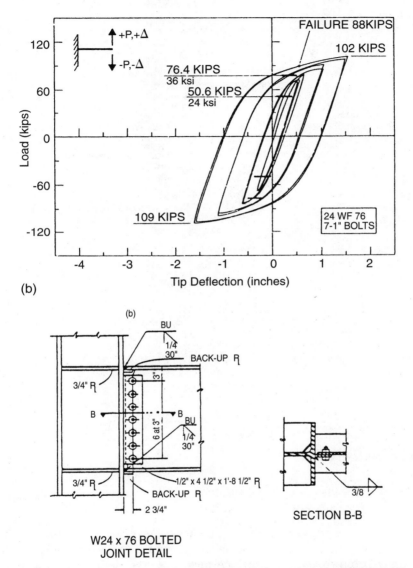

Figure 8.11 Hysteretic behavior of typical connection details having (b) bolted webs. This connection is otherwise identical to the one shown in Figure 8.11a. Results from tests conducted in early 1970s.

of the beam's plastic moment capacity. As shown in Figure 8.13, specimens with the welded flanges-bolted web connections (specimens 3 and 5) failed abruptly before developing adequate plastic rotations. These specimens were constructed by a commercial fabricator, and the welds had been inspected ultrasonically and found to be satisfactory.

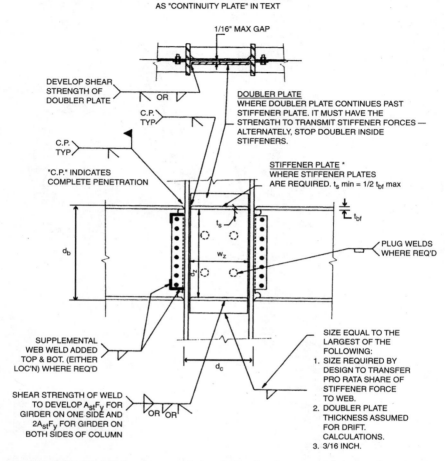

Figure 8.12 Prequalified moment-resisting frame detail in use prior to Northridge earthquake.

The use of bolts with twist-off ends for tension control in the beam web (specimens 17 and 18) or the use of supplemental web welds (specimens 13 and 14) improved hysteretic performance and delayed abrupt failure. It is noteworthy that two specimens with bolted webs failed prior to reaching M_p (even though they were supplied from a commercial fabricator), and two other specimens with fully welded flanges and webs exhibited significant ductility (specimens 9 and 11), as shown in Figure 8.13.

Further to these findings, the prequalified welded flange-bolted web connection detail was modified in the late 1980s for beams having a

Design of Ductile Moment-Resisting Frames 311

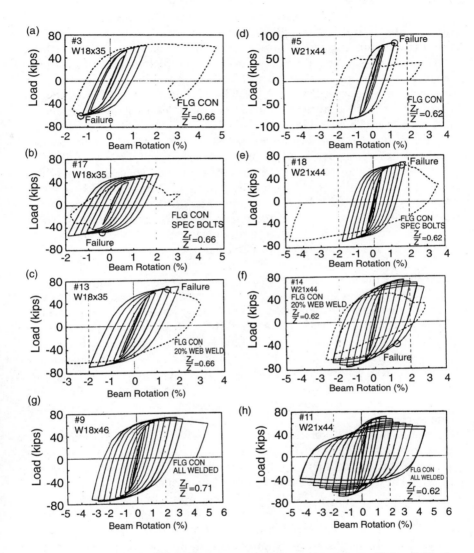

Figure 8.13 Hysteretic loops for moment-resisting frame connections with low Z_f/Z values and different beam web connection methods: (a) W460x52 (W18x35) beam with bolted web, (b) W460x52 beam with tension-control bolts (special bolt whose ends twist off upon reaching specified bolt tension), (c) W460x52 beam with bolted web and 20 percent supplementary weld, (d) W530x66 (W21x44) beam with bolted web, (e) W530x66 beam with tension-control bolts, (f) W530x66 beam with bolted web and 20 percent supplementary weld, (g) W460x68 (W18x46) beam with fully welded web, and (h) W530x66 beam with fully welded web. *(From* Engineering Journal, *2nd Quarter 1989,* Performance of Large Seismic Steel Moment Connections under Cyclic Loads *by E.P. Popov and K.C. Tsai, with permission from the American Institute of Steel Construction)*

ratio Z_f/Z less than 0.7, where Z_f is the plastic modulus of the beam flanges alone, and Z is the plastic modulus of the entire beam section. For those beams, supplemental welds on the bolted web shear tabs were required (i.e., in addition to the usual complete penetration single-bevel groove welds on the beam flanges and the bolted shear tab for the web), as shown in Figure 8.12. The supplemental welds were also required to have a minimum strength of 20 percent of the nominal flexural strength of the beam web.

Given that those new requirements were supported by only limited test data, Engelhardt and Husain (1993) conducted additional tests to investigate the effect of Z_f/Z on rotation capacity using slightly deeper beams than those tested by Tsai and Popov (W460 to W610 shapes, equivalent to W18 to W24 in U.S. units). Interestingly, some of the specimens tested by Engelhardt and Husain showed a disturbing lack of ductility, even though all specimens had been constructed by competent steel fabricators using certified welders, and all welds had been ultrasonically tested by certified inspectors. Some specimens exhibited almost no ductile hysteretic behavior (e.g., Figures 8.14a and 8.14d) while others behaved in a ductile manner until a sudden rupture developed in the connection (e.g., Figures 8.14b and 8.14e). The amount of hysteretic behavior developed prior to failure bore no relationship to Z_f/Z. Three specimens suffered sudden fracture at the weld-to-column interface at the beam bottom flange (such as the specimen shown in Figure 8.14a); the remaining specimens suffered gradual fracture at the same location (three specimens), at the top flange (one specimen), or through the bottom beam flange outside the weld (one specimen).

Engelhardt and Husain also compared their results with past experimental data. Assuming that connections must have a beam plastic rotation capacity of 0.015 radian to survive severe earthquakes, they found that none of their seven specimens could provide this rotation capacity (Figure 8.15), nor could most connections in tests conducted by other researchers. As a result of these observations, Engelhardt and Husain expressed concerns about the welded flange-bolted web detail commonly used in ductile moment frames in severe seismic regions.

And then the Northridge earthquake happened.

8.5.2 Damage during the Northridge earthquake

On January 17, 1994, an earthquake of moment magnitude 6.7 struck the Los Angeles area. The epicenter of the earthquake was at Northridge in the San Fernando valley, 32 km northwest of downtown Los Angeles. This earthquake caused over $20 billion in damage,

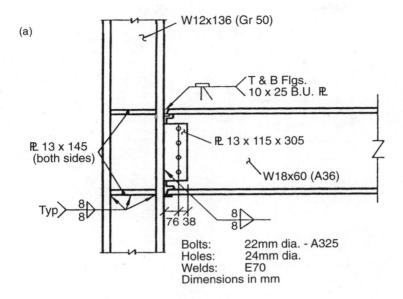

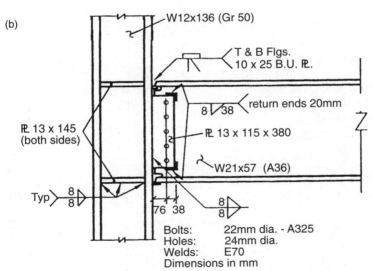

Figure 8.14 Engelhardt and Husain's tests: (a) Specimen 4 details, (b) Specimen 7 details. (*Parts a-e courtesy of M.D. Engelhardt, Dept. of Civil Engineering, University of Texas, Austin.*)

becoming the most costly disaster ever to strike the United States (EERI 1995). Structural and nonstructural damage to buildings and infrastructure was widespread and considerable, but there were no reports of significant damage to steel building structures immediately

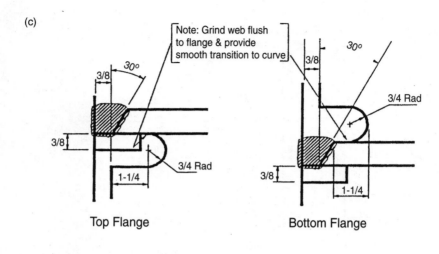

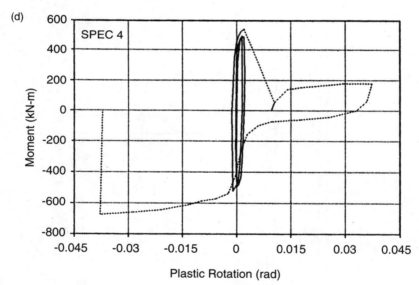

Figure 8.14 Engelhardt and Husain's tests: (c) Typical weld and cope details, (d) Resulting moment versus plastic rotation hysteretic curves for Specimen 4

following the earthquake (Moehle 1994). This should not come as a surprise. Inspectors, as well as reconnaissance teams dispatched by various engineering societies and research centers following a major earthquake can report only readily visible damage not obstructed by nonstructural elements. Careful inspection of a building's steel frame requires the removal of architectural finishes (cladding, ceiling pan-

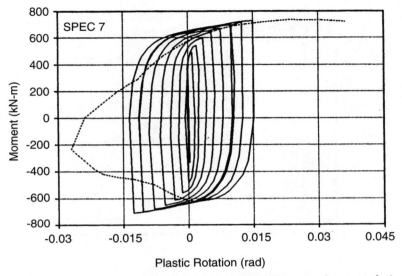

Figure 8.14 Engelhardt and Husain's tests: (e) Resulting moment versus platic rotation hysteric curves for Specimen 7.

els, etc.) and of the fireproofing material covering the steel members—an expensive and time-consuming process. Given that no steel building collapsed or exhibited noticeable signs of structural distress (Tremblay et al. 1995, EERI 1996), the discovery of critical but nonfatal damage was precluded without authority to expose part of the structure.

However, in the months following the earthquake engineers discovered important damage to steel structures, including a large number of beam-to-column connections fractures. Initially, damage was often found accidentally, while engineers were trying to resolve nonstructural problems reported by owners following the earthquake. In one case, for example, beam-to-column connection fractures would have remained hidden if not for complaints by occupants about persisting elevator problems. The structural engineer noticed that the building was leaning in one direction and requested that some connections be exposed. Informal discussion of such problems within the profession led other structural engineers to recognize the potential significance of the problem and to require random inspection of joints in various steel structures. This led to the discovery of more failures. Connection fractures were found in buildings of various vintages and heights (1 to 27 stories), including new buildings under construction at the time of the earthquake (SAC 1995a, Youssef et al. 1995). For example, in a steel building still under construction at the time of the Northridge

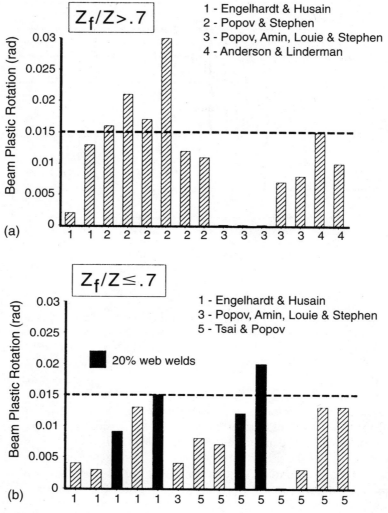

Figure 8.15 Engelhardt and Husain's comparison of beam plastic rotations obtained in past test for (a) specimens with $Z_f/Z > 0.70$ and (b) specimens with $Z_f/Z \leq 0.70$. *(Courtesy of M.D. Engelhardt, Dept. of Civil Engineering, University of Texas, Austin)*

earthquake, one that had apparently survived the earthquake intact, random inspection revealed severe fractures in nearly all beam-to-column connections in one moment-resisting frame. Typically, in the damaged connections of that building, the column flange fractured at the level of the full-penetration weld of the beam's bottom flange to the column, and the crack propagated horizontally a short distance into the column web and then vertically toward the other flange of the same beam (Figure 8.16).

Figure 8.16 Examples of Northridge fractures propagating through column flanges: (a) Column without stiffener, with fracture propagating into column web and vertically toward top flange.

Within two months, more than a dozen buildings with brittle failures of beam-to-column moment connections attributable to the Northridge earthquake had been reported. This became a rather delicate issue given that most buildings in which fractures were discovered were still occupied after the earthquake. A first special AISC task committee meeting allowed researchers and practicing engineers to meet and exchange information (AISC 1994). Tentative provisions for the repair of observed damage were formulated, and although many potential causes for the problem could be identified, failures could not be conclusively explained.

Figure 8.16 Examples of Northridge fractures propagating through column flanges: (b) Close-up view of fracture shown in Figure 8.16a.

Three months following the earthquake, approximately 50 steel buildings were known to have suffered moment frame damage, based on records from the Los Angeles Department of Building and Safety. By the end of 1994, more than 100 had been identified, but the actual number of buildings with damaged moment frames was suspected to be higher given that some owners disallowed inspection of their buildings (SAC 1995a, SAC 1995b). For perspective, approximately 500 buildings with steel moment frames were located where severe ground shaking occurred during that earthquake. Lessons from the Northridge earthquake also prompted engineers to suspect that damage to steel moment frames might have occurred in previous earth-

Figure 8.16 Examples of Northridge fractures propagating through column flanges: (c) Column with partial stiffener, with fracture through column flange.

quakes, and remained hidden. In the San Francisco Bay Area, hit by the Loma Prieta earthquake in October 1989 (EERI 1990), this suspicion has been confirmed and more buildings with damaged connections are being discovered as inspection opportunities arise (Rosenbaum 1996).

Various types of damage were discovered during the surveys conducted following the Northridge earthquake. Cracks that developed at or near beam bottom flanges were most frequently reported. Figure 8.17 summarizes the various types of fractures observed in that case (types 1 to 8). Most frequently, cracking initiated near the steel backup bar, in the root pass of the weld. Those cracks either

Figure 8.16 Examples of Northridge fractures propagating through column flanges: (d) Close-up view of fracture shown in Figure 8.16c.

remained within the weld material, propagating through part or all of the flange weld (type 1 and 2 respectively), or spread into the adjacent base metal (types 3 to 6). Cracks in the adjacent steel propagated into the column flange either vertically (types 3 or 4, depending on whether a piece of the column was completely pulled out in the process) or horizontally by fracturing the entire column flange (type 5) and sometimes a significant portion of the column web (type 6). In some cases, cracks that extended into the column web ruptured the entire column section horizontally or were found to bifurcate and propagate vertically toward the other flange of the beam in which it initiated. In a few instances, cracking initiated at the weld toe and propagated through the flange heat-affected zone (type 7), and at least one case of lamellar tearing of a column flange has been reported (type 8) (Bertero et al. 1994). Examples of such damage are shown in Figures 8.18, 8.19 and 8.20.

These cracks and fractures were frequently reported in the absence of similar damage to the top flange. In a few buildings, there were

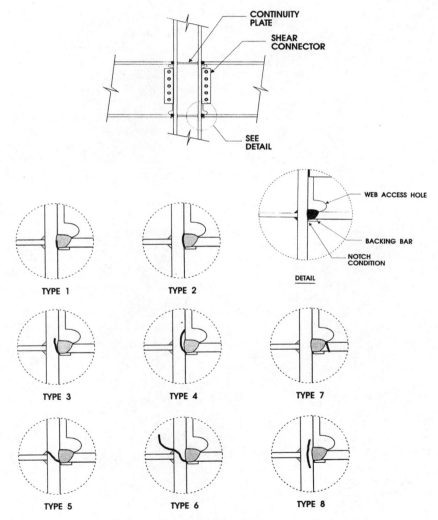

Figure 8.17 Typical welded flange and bolted web beam-to-column connection in moment resisting frames, with close-up view of notch-condition at backing bar, and eight types of reported Northridge fractures. *(Courtesy of R. Tremblay, Dept. of Civil Engineering, Ecole Polytechnique, Montreal, Canada)*

322 Chapter Eight

(a)

(b)

Figure 8.18 Four examples of bottom flange welds fractures. In case (a), for a fracture located near the face of a box column, a business card is dropped in to illustrate that the fracture passes completely through the weld. *(Parts a to c are courtesy of M.D. Engelhardt, Dept. of Civil Engineering, University of Texas, Austin.)*

Design of Ductile Moment-Resisting Frames 323

(c)

(d)

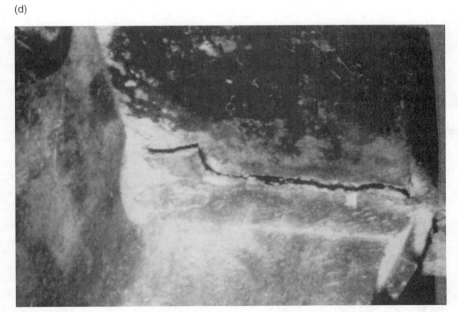

Figure 8.18 Continued. (*Part d is courtesy of David P. O'Sullivan, EQE International, San Francisco.*)

324　Chapter Eight

(a)

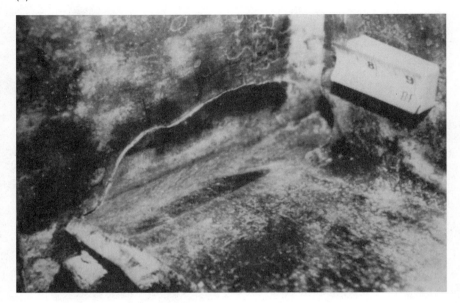

(b)

Figure 8.19 Two examples of divot fracture at beam bottom flange. *(Courtesy of David P. O'Sullivan, EQE International, San Francisco)*

Design of Ductile Moment-Resisting Frames 325

(a)

(b)

Figure 8.20 Examples of fractured columns, with fractures propagating from near the beam bottom flange weld (on the right side in both cases) to the column flange and into the column web. *(Part a courtesy of M.D. Engelhardt, Dept. of Civil Engineering, University of Texas, Austin. Part b courtesy of David P. O'Sullivan, EQE International, San Francisco.)*

instances of weld damage at the beam top flanges without damage to the corresponding bottom flange welds, but generally, both flanges were found to have suffered damage when cracks were found in the top flange welds. Only a few instances of base metal fractures adjacent to beam top flanges were reported, but other such failures may have been left undetected in many cases because floor slabs frequently obstruct inspection at that location (Youssef et al. 1995).

The damage reported above was sometimes accompanied by severe damage to the beam's shear tabs, with vertical net section fractures between the bolt holes over part of the height of the web connector (this occurred after the beam flange fractured completely). Note that gravity load resistance could be seriously jeopardized when complete rupture of such shear tabs follows flange fracture. Finally, in a few instances, panel zone yielding was also observed (Youssef et al. 1995).

Given that the above damage was reported in buildings having widely different characteristics, attempts were made to correlate damage statistics to beam depth, beam span, steel grade, design details, shear connection type, weld process, composite-beam behavior, material, and construction quality. These studies have proven inconclusive (Youssef et al. 1995).

Although no steel buildings collapsed during the Northridge earthquake, the discovery of these unexpected failures forced the structural engineering community to reexamine its design, detailing, and construction practice for steel moment frames. A sense of urgency was fueled by the recognition that the Northridge earthquake is certainly not the largest earthquake expected to occur in North America and that steel frames could be subjected to larger inelastic deformation demands in future earthquakes. To develop short-term and long-term solutions, extensive research activities have been initiated by federal agencies and private industries. More notable is a coordinated research effort initiated through a joint venture of the Structural Engineers Association of California (SEAOC), the Applied Technology Council (ATC), and the California Universities for Research in Earthquake Engineering (CUREe). The SAC Joint Venture combines the efforts of practicing engineers, code writers, industry representatives, and researchers who share either a professional or a financial interest in the resolution of the problems in beam-to-column connections that arose as a result of the Northridge earthquake. This venture has already published important documents reporting findings to date (SAC 1995a, 1995b, 1997).

8.5.3 Causes for failures

Numerous factors have been identified as potentially contributing to the poor seismic performance of the pre-Northridge steel moment connections, and failures may have been caused by different combinations of those factors. Yet, after much debate and deliberation, the professional engineering community had not reached a consensus on that topic at the time of this writing. Thus, in the post-Northridge context, caution is warranted and measures must be taken to eliminate every plausible cause of connection damage. A review of some of these conjectured causes is therefore worthwhile and is presented below. The most important concerns are addressed here, and related issues have been grouped under arbitrarily defined broad categories. A more complete summary of all major and minor concerns expressed following the Northridge earthquake is available elsewhere (SAC 1995a).

8.5.3.1 Workmanship and inspection quality

A percentage of the damage observed following past earthquakes worldwide has been a consequence of substandard workmanship and improper inspection, particularly in countries with poor code enforcement and contractors who hide construction (detailing) mistakes. Hence, as Northridge failures started to appear, many asserted that deficient workmanship and inspection were to blame. Ignorance of standard welding requirements was found to be disconcertingly widespread among structural engineers (SAC 1995a), and some have reported evidence of poor quality welds with defects that escaped detection prior to the earthquake. Nonetheless, although lack of adherence to standard welding procedure generally made matters worse, improved workmanship and inspection quality alone would not have been sufficient to prevent the Northridge failures (specimens constructed under controlled conditions still exhibited erratic behavior in post-Northridge laboratory tests, as described later).

8.5.3.2 Weld design

In the pre-Northridge connection described earlier, the beam web creates an obstacle when one is executing the bottom flange groove weld; deposition of weld metal is interrupted at the beam web at every pass. As a result, there is a high probability of defects in the bottom flange weld at that location. Those defects are particularly difficult to detect through ultrasonic inspection because they are frequently hidden in the portion of the testing signal that is interpreted as interference because of the presence of the beam web.

8.5.3.3 Fracture mechanics

The backup bars used for downhand welding of beam flanges were typically left in place prior to 1994 after completion of the weld. From a strength perspective, these small bars were perceived as additional material that could be left in place without detrimental effects. However, from a fracture mechanics perspective, the small unwelded gap between the edge of the backup bar and the column flange can be considered a notch or crack that acts as a stress raiser, from where new cracks can originate and propagate into the weld or adjacent base metal (see Figure 8.21). This problem is further compounded if the weld metal has low notch-toughness.

Similarly, a large number of defects can exist in the weld runoff tabs installed to allow extension of the weld passes beyond the flange width (as required by the American Welding Society). Runoff tabs collect the defects commonly introduced by the starting and ending of each weld pass in a zone removed from the flange. If left in place, the weld runoff tabs provide an opportunity for these defects, even though located outside the flange, to propagate into the weld proper. This propensity to crack propagation is further accentuated by the very low Charpy-V notch toughness of the E70T-4 electrodes (Figure 8.22) that were commonly used as filler metals in pre-Northridge welds (Kauffman et al. 1996).

8.5.3.4 Base metal elevated yield stress Many engineers have resorted to using A36 steel for beams and A572-Grade 50 for columns to facilitate compliance with the philosophy of strong-column/weak-beam design. The use of Grade 50 steel for columns also increases the panel zone strength, minimizing the need for doubler plates. However, a significant increase in the actual yield and ultimate strengths of the standard A36 steel produced in the United States has been observed over the years, in spite of the absence of changes to the steel grade specification itself. This increase is primarily due to changes in the steel-making process in the 1980s, when integrated mills were replaced by mini-mills that use highly efficient electric arc furnaces to produce steel shapes from scrap steels. SSPC (1994) and SAC (1995b) reports average yield and ultimate strengths of 338 MPa (49 ksi) and 475 MPa (69 ksi) for A36 steel, and maximum yield strengths as high as 496 MPa (72 ksi), as shown in Table 8.1. Some steel producers have also introduced dual-certified steel, which is steel simultaneously in compliance with all the minimum chemical and strength requirements of both A36 and A572-Grade 50 steels. Therefore, engineers who assume A36 steel properties for the design of beams may seriously underestimate the beam flange forces acting on the groove welds, and unintentionally select welds weaker than the base metal, if the contractors supply steel with yield strength in excess of 350 MPa (50 ksi). Furthermore, the intended strong-column/weak-beam design may in practice be a weak-column/strong-beam system if the yield strength of the beam substantially exceeds the nominal value.

Design of Ductile Moment-Resisting Frames

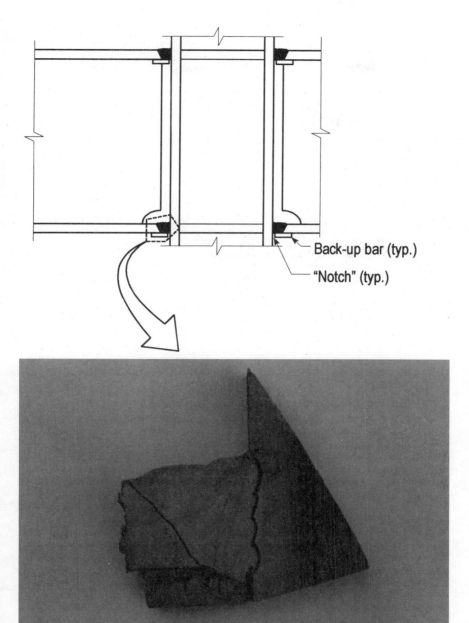

Figure 8.21 Example of fractured beam bottom flange in which a crack originated at the unwelded gap between the edge of the backup bar and the column flange. *(Courtesy of J.E. Patridge, Smith-Emery Co., Los Angeles)*

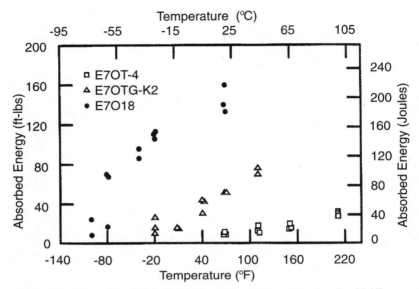

Figure 8.22 Charpy-V notch test results on three different types of weld filler metal. *(From* Modern Steel Construction, *Vol.36, No.1, Achieving ductile behavior of moment connections by E.J. Kaufmann et al., 1996, with permission from the American Institute of Steel Construction)*

8.5.3.5 Welds stress condition The ultimate stress applied to the weld of the beam flange can be estimated if one assumes that the bolted web cannot transfer bending moments. Indeed, researchers have observed that web bolts typically slip during testing, leaving the stiffer welded flanges alone to resist the total applied moment at the connection (Popov et al. 1985, Tsai and Popov 1988). As a result of the incompatible stiffnesses of the bolted web and the welded flanges, the connection resistance is reached when the flanges reach their ultimate tensile stress, F_u (Figure 8.23).

As a result of strain hardening, beams will reach bending moments of 1.2 to 1.3 times the actual plastic moment, M_p^{act} at the required plastic rotations, and flange fracture will develop unless:

$$A_f F_u (d - t_f) = Z_f F_u \geq 1.2 M_p^{act} = 1.2 Z F_y^{act} \qquad (8.17)$$

where A_f and t_f are, respectively, the area and thickness of a beam flange, d is the beam depth, and F_y^{act} is the actual yield stress of the beam. Assuming that the plastic section modulus of the flanges alone, Z_f, is approximately 70 percent of the beam plastic modulus, Z, the ratio of F_y^{act} over F_u needed to develop significant plastic rotations is given by:

TABLE 8.1 Statistical Yield and Tensile Properties for Structural Shapes Based on Data Reported by the Structural Shape Producers Council (SSPC 1994)

Statistics	A36 Steel	Dual Grade	A572 Grade 50
Yield Stress (ksi)*			
Specified	36.0	50	50
Mean	49.2	55.2	57.6
Minimum	36.0	50	50
Maximum	72.4	71.1	79.5
Standard deviation	4.9	3.7	5.1
Mean plus one standard deviation	54.1	58.9	62.7
Tensile Stress (ksi)*			
Specified	58–80**	65 (min.)	65 (min.)
Mean	68.5	73.2	75.6
Minimum	58.0	65.0	65.0
Maximum	88.5	80.0	104.0
Standard deviation	4.6	3.3	6.2
Mean plus one standard deviation	73.1	76.5	81.8
Yield/Tensile Ratio			
Specified	0.62 (max.)	0.77 (max.)	0.77 (max.)
Mean	0.72	0.75	0.76
Minimum	0.51	0.65	0.62
Maximum	0.93	0.92	0.95
Standard deviation	0.06	0.04	0.05
Mean plus one standard deviation	0.78	0.79	0.81

* 1 ksi = 6.895 MPa.
** No maximum for shapes heavier than 426 lb/ft.

$$\frac{F_y^{act}}{F_u} \leq \frac{0.83 Z_f}{Z} \approx 0.6 \qquad (8.18)$$

Given that the mean ratio of F_y^{act}/F_u for currently available steels has been reported to vary between 0.72 and 0.76 (SAC 1995a, SAC 1995b), as shown in Table 8.1, it may not be possible to reliably develop the required plastic deformations in beams, even with perfect groove-welded connections.

8.5.3.6 Stress concentrations The absence of continuity plates opposite the beam flanges in a column produces stress concentrations in the flange near the column web (Figure 8.4). Some engineers have also alleged that this stress concentration cannot be eliminated by the addition of thick continuity plates (Allen et al. 1995). Note that the use of overly thick continuity plates will generally require large welds

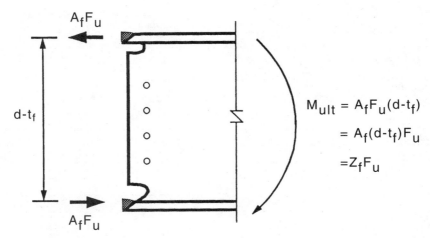

Figure 8.23 Free-body diagram for simplified model of connection strength. *(Courtesy of M.D. Engelhardt, Dept. of Civil Engineering, University of Texas, Austin)*

that will introduce greater residual stresses in the connection: another condition conducive to crack initiation.

8.5.3.7 Effect of triaxial stress conditions Triaxial stress conditions can have an adverse effect on the ductility of steel. This is illustrated in Figure 8.24 in a comparison of the Mohr circles for steel elements with free or constrained lateral deformations when they are subjected to uniaxial yield stress (Blodgett 1995a).

As described in Chapter 2, yielding requires the development of slip-planes. For a steel element unrestrained laterally and subjected to uniaxial stress, ductile behavior develops when the shear stress equivalent to the uniaxial yield stress is exceeded. For a steel with $\sigma_y = \sigma_3 = 350$ MPa (50.8 ksi), the corresponding yield shear stress is 175 MPa (25.4 ksi) from Mohr's circle (Figure 8.24). The corresponding axial strains, obtained from the classical equations of elasticity (Popov 1968), using a value of Poisson's ratio, μ, of 0.3, are $\varepsilon_3 = \sigma_3/E = 0.00175$, and $\varepsilon_2 = \varepsilon_1 = -\mu\sigma_3/E = -0.00053$. However, if the same axial strain $\varepsilon_3 = 0.00175$ is applied when lateral deformations of the steel element are fully restrained (i.e., $\varepsilon_2 = \varepsilon_1 = 0$), the resulting stresses are:

$$\sigma_3 = \frac{E[(1-\mu)\epsilon_3 + \mu\epsilon_2 + \mu\epsilon_1]}{(1+\mu)(1-2\mu)}$$

$$= \frac{(200000)\,[(1.0-0.3)\,(0.00175)]}{(1.3)(0.4)} \qquad (8.19a)$$

$$= 471 \text{ MPa (68.3 ksi)}$$

Design of Ductile Moment-Resisting Frames 333

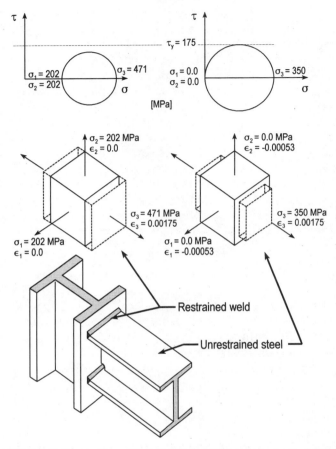

Figure 8.24 Comparison of triaxial stresses in unrestrained and restrained steel elements. *(Adapted from Blodgett 1995a)*

$$\sigma_2 = \frac{E[\mu\epsilon_3 + (1-\mu)\epsilon_2 + \mu\epsilon_1]}{(1+\mu)(1-2\mu)}$$
$$= \frac{(200000)\,[0.3\,(0.00175)]}{(1.3)(0.4)} \qquad (8.19b)$$
$$= 202 \text{ MPa (29.3 ksi)}$$

$$\sigma_1 = \frac{E[\mu\epsilon_3 + \mu\epsilon_2 + (1-\mu)\epsilon_1]}{(1+\mu)(1-2\mu)}$$
$$= \frac{(200000)\,[0.3\,(0.00175)]}{(1.3)(0.4)} \qquad (8.19c)$$
$$= 202 \text{ MPa (29.3 ksi)}$$

As can be seen from the corresponding Mohr circle, even though the axial stress has exceeded the uniaxial yield stress of 350 MPa (50.8 ksi), the maximum shear stress is only 135 MPa (19.6 ksi). The shear stress needed to initiate slip-planes would be reached only at an axial stress of 610 MPa (88.5 ksi), a value most likely in excess of the ultimate yield stress of the material (based on data in Table 8.1). Hence, ductile behavior will not develop, and brittle failure will occur instead. This simplified model also suggests that compression in the column ($\varepsilon_2 < 0$) would enhance the potential for ductile behavior at the weld, while tension ($\varepsilon_2 > 0$) would reduce it. Practically, the above condition of full restraint against lateral deformations is an extreme constraint not encountered in most welds of small to moderate sizes, but may be approached when large welds are executed on very thick steel members. Elasto-plastic studies of the behavior of constrained welds are needed to clarify the relationship between degrees of restraint and ductility.

8.5.3.8 Loading rate Given that all large-scale specimens in past experimental studies prior to the Northridge earthquake had been subjected to quasi-static loading, it was suggested that rate of loading may have had a detrimental effect on the behavior of beam-to-column moment connections. Dynamic testing of pre-Northridge full-size beam-to-column connections with W760x147 beams (W30x99 in U.S. units) revealed that beam flanges experienced strain rates on the order of 10^{-1} mm/mm/sec for moment frames located in buildings having a fundamental period of vibration of approximately 1 second (Uang and Bondad 1996). At such a strain rate, yield stress can be increased by 10 percent (see Figure 2.8), thereby increasing the force demand on the groove-welded joint. It is also known that strain rate will decrease the notch toughness of the material. The combined effects resulted in a poorer cyclic behavior under dynamic loading conditions.

8.5.3.9 Presence of composite floor slab The development of composite action due to the presence of a concrete floor slab may have been responsible for the dominant number of beam bottom flange fractures (compared with top flange fractures). The different neutral axis positions in positive (composite) flexure versus negative (noncomposite) flexure translates into greater axial deformation demands on the beam bottom flange than on the top flange. However, other factors also likely contributed to the greater damage to the beam bottom flanges. For example, the top flange groove weld is easier to accomplish and inspect than the bottom flange weld. Furthermore, the strain demands at the level of the backup bar to the top flange weld are smaller than those on the backup bar to the bottom flange weld, which is farther from the center of the steel section.

Note that in California, engineers have commonly ignored composite action in design of moment-resisting frames, even though 19 mm (3/4 inch) diameter shear studs spaced 300 mm (12 inches) on center are popular to transfer seismic forces from the slab to the steel frame. Welded wire fabric is commonly used as reinforcement in the concrete slab.

8.5.4 Reexamination of pre-Northridge practice

8.5.4.1 Reexamination of past literature The extensive damage to steel moment frames in the Northridge earthquake prompted a reexamination of past experimental data. This review essentially revealed that the Northridge failures should have been expected (Bertero et al. 1994, Roeder and Foutch 1995). Although past experimental studies on standard moment connections generally reported satisfactory performance, sometimes with impressive ductile behavior, most studies reported instances of failures after only a limited amount of inelastic energy dissipation. For example, beyond the numerous sudden failures already reported in Section 8.5.1, Popov and Bertero (1973) reported a number of abrupt specimen failures, sometimes with fractures through welds or flanges, and Popov et al. (1985) noted that most of their specimens failed abruptly after exhibiting more or less satisfactory levels of plastic deformations. That latter test series was conducted to verify the adequacy of the design criteria for beam-to-column joints, using larger specimens than tested to that time and A36 beams framing into A572-Grade 50 columns. The beams' flanges were fully welded, webs were bolted only, and researchers reported hearing the slippage of the web bolts at each load reversal during testing. They also noted that specimens with continuity plates and doubler plates performed better than those without.

The abrupt failures reported in past North American beam-to-column tests were limited to fractures of the welded connections; cracks propagating into columns had not been observed prior to the Northridge earthquake. However, Bertero (1994) reported that Japanese researchers had experienced such column fractures decades earlier (Kato 1969, 1973). In those tests performed on large columns, cracks were observed to propagate from the beam welds through the entire column cross section when the column was subjected to low axial forces; crack propagation stopped after rupture of the column flange when columns were subjected to high axial compression forces.

Thus, beam, column, and weld fractures similar to those documented following the Northridge earthquake have been observed in past studies. Unfortunately, although some of the specimens that exhibited inadequate ductility were brought forth (e.g., Engelhardt and Husain 1993), others instances of erratic behavior received cursory treatment and were attributed to faulty workmanship, even when the test specimens were provided by commercial fabricators.

8.5.4.2 Post-Northridge tests of pre-Northridge details Shortly after the Northridge earthquake, many tests of typical pre-Northridge connections were conducted in an attempt to replicate the observed failures under controlled conditions. A first series of tests involved heavy beam and column specimens (W360x677 A572-Grade 50 columns and W920x233 A36 beams, corresponding to W14x455 and W36x150, respectively, in U.S. units) representative of those that fractured during the earthquake (Engelhardt and Sabol 1994). Special care was taken to ensure superior welding quality and inspection. Backup bars and weld runoff tabs were also removed, and the weld root pass was gouged out and filled with new weld material to locally reinforce the weld. Two specimens had bolted webs (with supplemental welds on the web connector plate), and two specimens had webs fully welded to the column flanges; continuity plates were not used. The four specimens were tested by a standard quasi-static method, which is at strain rates much less than those that typically occur during earthquakes. All specimens failed at a low level of inelastic deformation (attaining plastic rotations of 0.0025 rad to 0.009 rad, depending on the specimen), with brittle fractures observed in both top and bottom flanges. Specimens with fully welded webs did not perform any better than those with bolted webs. These results showed the need for joint reinforcement and/or an alternative welding procedure to be validated through an extensive experimental program.

Tests on eight full-scale specimens of other pre-Northridge connections (W360x262 A572 Grade 50 columns with W760x147 A36 beams, corresponding to W14x176 and W30x99, respectively, in U.S. units) showed similar results (Whittaker et al. 1995, Uang and Bondad 1996a). All eight specimens had the supplemental welds required on the web connector plate. First, three nominally identical specimens (Whittaker et al. 1995) were constructed under close supervision and rigorous inspection and thus were likely of greater than average quality. Tested at low strain rates, these pre-Northridge specimens suffered top flange weld fracture at beam plastic rotations of approximately 0.4 percent, 0.4 percent, and 1.0 percent, respectively (Whittaker et al. 1995). Panel zone yielding, observed in all three specimens, increased the total plastic deformation of the specimens by

0.7 percent, 0.7 percent, and 1.1 percent, respectively. Repairs that consisted of rewelding the failed flanges with toughness-rated filler metal failed in a similar manner at beam plastic rotations of 0.3 percent. Five different specimens tested by Uang and Bondad (1996a) failed in a similar manner. Three specimens, tested quasi-statically, achieved maximum beam plastic rotations ranging between 0.2 percent and 1.6 percent, and total plastic rotations varied from 0.8 percent to 2.3 percent when panel zone plastic deformations were included. Two additional specimens tested at strain rates of 0.1 cm/cm/sec failed without exhibiting any beam plastic rotation; maximum panel zone plastic rotations of 0.15 percent and 1.0 percent were measured respectively in the two tests. The fractures propagated into the column flanges and bolted beam web plates in the dynamically tested specimens, suggesting that loading rate may have contributed to that failure pattern observed in many Northridge-type failures. The propagation of damage in the dynamic tests has been documented on video (Uang 1995).

As soon as the first preliminary test results became available, the prequalified standard moment connection was deleted from most building codes and regulations for applications in moderate to high seismic regions, and it was replaced by general clauses requiring that welded or bolted moment connections be able to sustain inelastic rotations and develop the required strength, as demonstrated by approved cyclic tests or calculations supported by test data. Interpretation of the clauses, particularly regarding what constitutes acceptable levels of inelastic rotations and test procedures, has been left to the regulatory authorities and professional organizations (e.g., SEAOC 1995). As a result, building officials in many jurisdictions have elected to require mandatory testing of any new connection detail not previously proven by cyclic inelastic tests, or any connection with beams and columns larger than tested previously.

8.5.5 Post-Northridge beam-to-column connections design strategies for new buildings

Numerous solutions to the moment frame connection problem have been proposed. The more popular solutions that have been verified by tests are presented below. These solutions will likely evolve further in the coming years, and new solutions will be developed.

Two key strategies have been developed to circumvent the problems associated with the pre-Northridge moment frame connection:

- Strengthening the connection
- Weakening the beam(s) that frame into the connection

Both strategies effectively move the plastic hinge away from the face of the column, thus avoiding the aforementioned problems related to the potential fragility of groove welds subjected to triaxial stress conditions. Other solutions that have been proposed, such as purely metallurgical strategies and friction energy dissipation concepts, as well as outright different design approaches that may emerge, could also win acceptance as confidence is gained through further research.

Target beam plastic rotations of between 0.02 radian and 0.03 radian have been used to date to evaluate the performance of different connections. These limits may be exceeded in buildings located adjacent to a major active fault system, and they may be excessive for buildings located in other seismic zones. However, for the purpose of the following discussion, satisfactory performance requires that a connection can achieve plastic rotations of 0.03 radian without exhibiting strength degradation of more than 20 percent of the plastic moment, per the acceptance criteria suggested in SAC Interim Guidelines (1995b). Note that, as a minimum requirement, it is recommended that experimental validation of proposed connections be done in compliance with the ATC-24 loading protocol (ATC 1992).

It is noteworthy that no proposed detail has achieved the status of prequalified connection at the time of this writing. Explicit design recommendations cannot be formulated from the limited test data collected since the Northridge earthquake, and the reader should interpret the following information with caution. There is no definite answer either as to which of the proposed connections is the most cost effective. Reliable cost comparisons will need to account for cost of connections, royalty fees for proprietary systems, and influence of the connection detail on the weight of the steel frame and the cost of the foundations.

8.5.5.1 Strengthening strategies: cover plates, ribs, haunches, and side plates A number of strategies have been proposed to make the connection stronger than the beams framing into the connection. Cover plates, upstanding ribs, side plates, and haunches (Figure 8.25) have been implemented, and are briefly reviewed below. These should be implemented in conjunction with the use of high toughness weld filler metal, better welding practice, and high-quality inspection. Note that even though removal of the backup bars and weld runoff tabs did not ameliorate performance noticeably in the tests of pre-Northridge connections, the arguments presented earlier regarding the notch effect created by the backup bar are compelling, and their removal is recommended. Alternatively, a fillet weld applied between the backup bar and column flange could be used to seal the cracklike gap described in Section 8.5.3.

Design of Ductile Moment-Resisting Frames 339

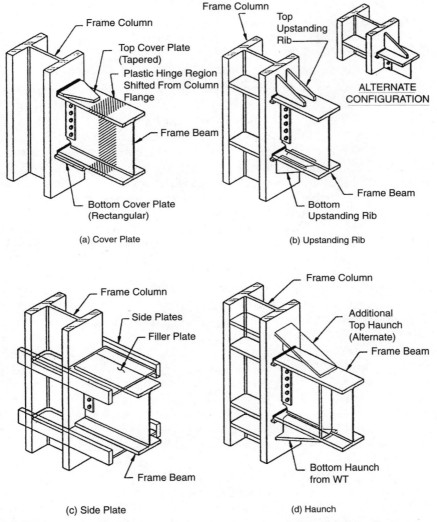

Figure 8.25 Examples of moment connections per strengthening strategies. *(Courtesy of M.D. Engelhardt, Dept. of Civil Engineering, University of Texas, Austin)*

The use of beam strengthening schemes to reinforce beam-to-column connections has the advantage of relocating the plastic hinge away from the column face, and the following disadvantages:

- Increasing the beam moment(s) at the face(s) of the column, thereby increasing the column size to maintain the strong-column/weak-beam system

- Increasing the unbalanced moment on the panel zone
- Increasing the plastic hinge rotation demand (see Section 8.6)

These issues must be considered by the designer.

8.5.5.1.1 Cover Plates and Flange Ribs A number of flange reinforcement strategies have been developed through use of flange cover plates or flange ribs (e.g., Whittaker et al. 1995, Engelhardt and Sabol 1996, Noel and Uang 1996). In nearly all cases (a notable exception is described by Noel and Uang 1996), the top cover plate is tapered and narrower than the beam top flange while the bottom plate is rectangular and wider that the bottom flange (Figure 8.25a). This configuration makes down-hand welding possible for both flanges. Plate tapering is also believed to result in a smoother stress transfer between the flange and cover plate.

Results from a series of cover plate tests by Engelhardt and Sabol (1996) are instructive. Details of 12 specimens considered are summarized in Table 8.2, along with brief description of their performance. Details with bolted web or welded web connections were evaluated, as shown in Figure 8.26a and 8.26b, respectively. Note that for new construction with cover plates, the bottom cover plate can be shop-welded to the column flange and used in the field as an erection seat for the beam. This particular construction sequence also makes it possible to perform ultrasonic testing at various stages of connection assembly and to fully weld the beam web, using the web tab as a backup plate. A welded web can transfer its share of the beam plastic moment, which makes possible the use of smaller cover plates. Smaller plates also minimize residual stresses due to weld shrinkage, and the likelihood of high triaxial tensile stresses at the column face. Separate welds for the flange and cover plate (Figure 8.27) also reduce this likelihood of developing detrimental triaxial stresses in the connection and enable individual ultrasonic inspection of the two welds.

As shown in Table 8.2, two-thirds of the cover-plated specimens developed total plastic rotations of 0.03 rad without brittle fracture. Note that in those specimens, the columns were designed with a very strong panel zone that remained elastic throughout testing, with the exception of specimens SEC-4 and NSF-6 designed with lighter columns and for which panel zone yielding dominated the inelastic response. Results for a specimen with a fully welded web connector plate are shown in Figure 8.28. Yet, cover plates by themselves are not a panacea. As seen in Table 8.2, two of the specimens with bolted webs tested by Engelhardt and Sabol (AISC-3A and AISC-5B) failed in a brittle manner at plastic rotations of less than 0.02 rad, even though the groove welds had passed ultrasonic inspection. Each specimen that failed had a counterpart that exhibited satisfactory behavior.

TABLE 8.2 Summary of Results for Cover-Plated Connection Test Series (Engelhardt and Sabol 1996)

Specimen	Beam size†	Beam flange strength		Beam web strength		Column size† W14x	Top cover plate (tapered) (thickness x width x length) (mm)	Bottom cover plate (rectangular) (thickness x width x length) (mm)	Web connection	Electrode **	Maximum plastic rotation (rad)	Description of Failure
		Fy (MPa)	Fu (MPa)	Fy (MPa)	Fu (MPa)							
AISC-3A	W36x150	294	425	320	435	455	19x300x430	16x355x405	bolted	E70T-4	1.5%	Brittle fracture @ top flange and cover plate groove weld
AISC-3B	W36x150	294	425	320	435	455	19x300x430	16x355x405	bolted	E70T-4	2.5%	Gradual strength deterioration due to local buckling, followed by gradual tearing of bottom flange @ end of cover plate
AISC-5A	W36x150	318	460	375	494	426	25x300x610	25x300x585 *	bolted	E70TG-K2 #	2.5%	Same as AISC-3B
AISC-5B	W36x150	370	492	380	520	426	25x300x610	25x300x585 *	bolted	E70TG-K2 #	0.5%	Brittle fracture @ beam bottom flange connection. Fracture contained within column flange base metal
AISC-7A	W36x150	318	460	375	494	426	19x300x430	16x355x405	bolted	E70T-7	3.5%+	Gradual strength deterioration due to local buckling, and gradual tearing of fillet welds of cover plates to beam flanges
AISC-7B	W36x150	318	460	375	494	426	19x300x430	16x355x405	bolted	E70T-7	5.0%+	Same as AISC-7A
AISC-8A	W36x150	311	444	343	465	426	19x300x430	16x355x405	bolted	E70T-7	3.5%+	Same as AISC-7A
AISC-8B	W36x150	311	444	343	465	426	19x300x430	16x355x405	bolted	E70T-7	3.5%+	Same as AISC-7A
SAC-4	W36x150	292	421	329	437	257	25x300x405	25x355x405	bolted	E70T-8	3.7%+	Same as AISC-7A + significant panel zone yielding
NSF-5	W36x150	296	417	310	415	426	12x300x355	12x380x355	welded	E70T-8	3.3%+	Same as AISC-7A
NSF-6	W30x148	321	445	334	450	257	16x266x355	16x300x355	welded	E70T-8	3.8%+	Gradual tearing of fillet welds of cover plates to beam flanges; panel zone dominated inelastic response
NSF-7	W36x150	340	456	360	467	455	12x300x355	12x380x355	welded	E70T-8	3.8%+	Same as AISC-7A

* Tapered bottom cover plate
† Beam size
** All bolted webs with fully tensioned high strength bolts and supplemental welds on web shear tab to develop 20% of nominal plastic moment of beam web. All welded webs by directly welding to column using complete joint penetration groove weld.
# Except E70T-7 used for bottom cover plate to column flange weld.
† In S.I. units, W36 x 150 is W920 x 223, W30 x 148 is W760 x 221, W14 x 455 is W360 x 677, W14 x 426 is W360 x 634, W14 x 257 is 360 x 382

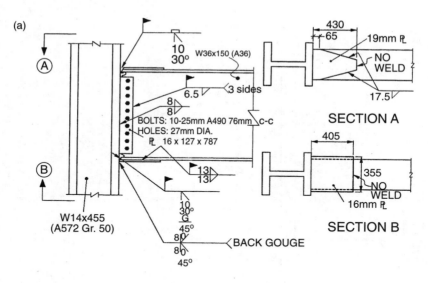
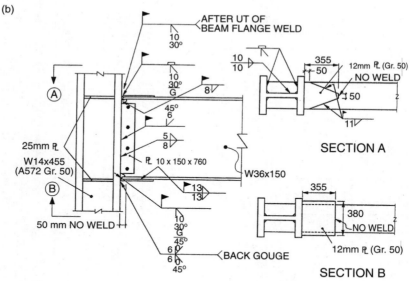

Figure 8.26 Moment connections with cover plates: (a) Bolted web (specimen AISC-3A), (b) Fully welded web (specimen NSF-7). *(Courtesy of M.D. Engelhardt, Dept. of Civil Engineering, University of Texas, Austin)*

Note that for the AISC-#B specimens, a Welding Procedure Specification was written and enforced, whereas for the AISC-#A specimens, the welder was permitted to weld on the basis of his experience.

Design of Ductile Moment-Resisting Frames   343

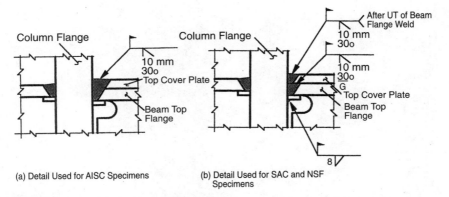

(a) Detail Used for AISC Specimens

(b) Detail Used for SAC and NSF Specimens

Figure 8.27 Typical groove weld details at top flange used for moment connection strengthened by cover plates. *(Courtesy of M.D. Engelhardt, Dept. of Civil Engineering, University of Texas, Austin)*

Two different welders executed the AISC-3A and 3B specimens; both were uncomfortable with the setting recommendations from the electrode manufacturers. The Welding Procedure Specification was enforced for specimen AISC-3B, but the welder of specimen AISC-3A increased the voltage and current of the welding machine to enhance workability. Metallurgical study of the groove welds revealed the

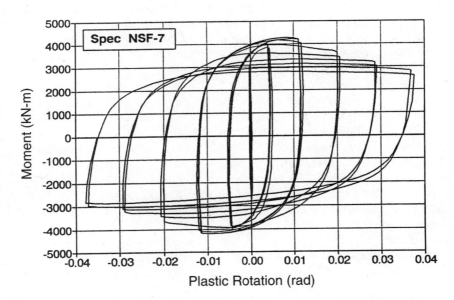

Figure 8.28 Moment connection with cover plates and fully welded web (specimen NSF-7): (a) Hysteretic behavior. *(Part a and b courtesy of M.D. Engelhardt, Dept. of Civil Engineering, University of Texas, Austin)*

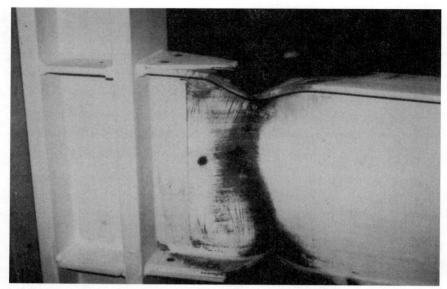

Figure 8.28 Moment connection with cover plates and fully welded web (specimen NSF-7): (b) Specimen state at completion of test.

greater heat input to AISC-3A (that suffered brittle fracture) resulted in a fivefold lower weld toughness than in AISC-3B. As for the other specimen with poor performance, AISC-5B, fracture was attributed to the larger-than-anticipated beam yield strength and the fact that long cover plates were used. These long plates developed the beam plastic moment farther from the column, resulting in larger bending moments at the column face.

Some concerns remain regarding the use of cover plates. First, the panel zones in the very large columns tested by Engelhardt and Sabol (1996) did not yield—poor performance was reported in other tests that developed large panel zone deformations (Obeid 1996, Whittaker and Gilani 1996). Second, overlaid welds should be accomplished only through use of identical electrodes; loss of weld toughness due to the mixing of weld metals has been reported (Wolfe et al. 1996). Note that section J2.7 of AISC (1993) also warns that low notch-toughness welds may result from the mixing of two incompatible weld metals of high notch-toughness. Third, Hamburger (1996) reported that an estimated failure rate of 20 percent has been experienced when laboratory qualification testing of these connections was performed for specific design projects, which suggests that the cover plate detail may not be sufficiently reliable. Fourth, the SAC Interim Guidelines (1995b) indicate that, conceptually, this connection could be exposed to some of the same flaws that plagued the pre-Northridge connections, namely, dependence on weld quality and through-thickness behavior of the column flange, potentially exacerbated by the thicker groove welds made necessary by the addition of cover plates. Finally, SAC (1997) reports that when the bottom cover plate is shop-welded to the column flange to be used as an erection seat for the beam, premature fracture can develop across the column flange as the seam between the bottom flange and cover plate acts as a notch that can trigger crack propagation.

Only a few tests on the use of upstanding beam flange ribs (Figure 8.25b) have been conducted since the Northridge earthquake, although this detail was investigated prior to 1994 (Tsai and Popov 1988). Overall, this type of rib detail appears effective, but additional testing is needed to determine how various design and detailing parameters influence its inelastic performance.

8.5.5.1.2 Top and Bottom Haunches and Bottom Haunches Only

Uang and Bondad (1996b) tested four pre-Northridge specimens repaired with bottom flange triangular T-shaped haunches only, as shown in Figure 8.29. Continuity plates were added to the column at the level of the haunch flange. Two specimens were tested in a quasi-static manner, and two were tested dynamically with a maximum strain rate of 0.1 mm/mm/sec. The repaired specimens performed much better than the pre-Northridge specimens, with beam plastic hinges developing outside the haunch. Plastic deformation of the panel zones was also reduced, and nearly all of the inelastic action was concentrated in the beams. Note that the presence of a haunch increases the depth of the panel zone, thus reducing the extent of panel zone yielding.

Total beam plastic rotations in excess of 3 percent were obtained in the quasi-static tests. Failure was defined by excessive strength degradation due to local buckling of the beam flanges (Figure 8.29), although the specimens could sustain larger plastic rotations and dissipate further hysteretic energy while undergoing further strength degradation. In one of the dynamically tested specimens, in addition to repairing the fractured bottom flange with a haunch, the beam top flange with pre-Northridge type of groove-welded joint was strengthened by the addition of a pair of rib plates on the underside of the flange. This detail, developed for strengthening existing connections, avoids the need to remove the concrete slab around the column, but would still require removal of the building's facade (i.e., cladding panel or other architectural finishes) to provide access to one half of the beam flange for perimeter frames. Although the welded top flange joint fractured during retesting, the two vertical ribs served their intended purpose by maintaining the integrity of the connection.

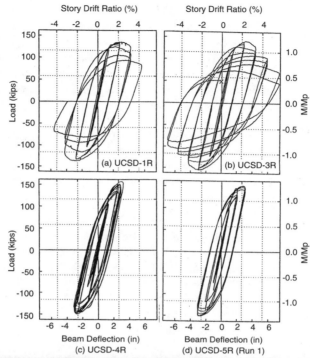

Figure 8.29 Hysteretic behavior of moment connection with bottom flange haunch, in terms of load versus beam tip deflection, and specimen state at first cycle of -7.0 inches tip deflection. Each specimen consists of W360x262 column (W14x176 in U.S. units) of A572 Grade 50 steel, and W760x147 beam (W30x99 in U.S. units) of A36 steel, with point load applied at tip of cantilever beam, 3.6 meters from the center line of the column.

Whittaker et al. (1995) reported adequate performance for pre-Northridge specimens repaired and strengthened by the addition of triangular T-shaped haunches to both top and bottom flanges. Panel zone yielding was substantially eliminated in the strengthened specimen and significant beam plastic rotations were obtained (Figure 8.30). However, with failure defined as the point at which the resistance degraded to 80 percent of the maximum value, beam plastic rotations of 2.7 percent were reached prior to failure.

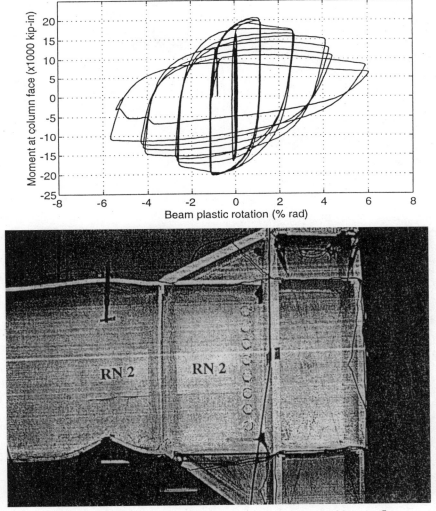

Figure 8.30 Hysteretic behavior of moment connection with top and bottom flange haunches, in terms of moment versus beam plastic rotation, and specimen state upon completion of test.

Hybrid connections with cover plate reinforcement of the top flange and haunch reinforcement of the bottom flange have also been considered. Excellent performance was obtained for a particular configuration and detailing (Figure 8.31). For this particular design, the cover plate was shop-welded to the beam with a fillet weld, and only the cover plate (not the beam top flange) was groove welded to the column; the backup bar was left in position with a closure fillet weld (Noel and Uang 1996).

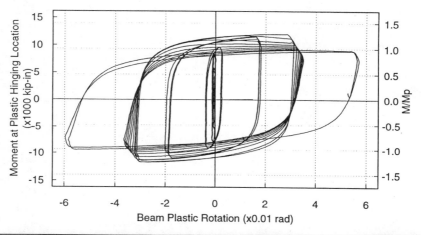

Figure 8.31 Hysteretic behavior of moment connection with top flange cover plate and bottom flange haunch, in terms of moment at hinge location versus beam plastic rotation at hinge location and specimen state at ninth cycle of +3.5 inches tip deflection. *(Courtesy of M.S. Jokerst, Forell/Elsesser Engineers Inc., San Francisco)*

In summary, the available experimental data suggest that using triangular T-shaped haunches is an effective means by which to strengthen a connection. Their high redundancy also contributes to preserve good plastic behavior if one of the full penetration groove welds fails. However, haunches are expensive to construct, and the top haunch, when present, can be an obstruction above the floor level.

Straight haunches have been proposed as a more economical alternative solution (Uang and Bondad 1995). The direct strut action that develops in sloped haunch flanges is not possible in this alternative, and the beam flange force must be transferred to the haunch flange via shear in the haunch web. In the specimen tested, stress concentration at the free end of the haunch fractured the weld between the beam flange and haunch web at that free end (Figure 8.32). Additional stiffeners at the free end of the haunch to tie the beam and haunch together, or the use of a sloped free end to reduce the stress concentration, might be effective in preventing the observed fracture, but the adequacy of such enhancements must be validated by testing.

8.5.5.1.3 Proprietary Side Plates Limited tests of the side-plate strategy illustrated in Figure 8.25c have shown poor inelastic behavior. However, an alternative design using large side plates that extend over the full beam depth has been demonstrated to be effective. More details on this proprietary detail can be found in Nelson (1995).

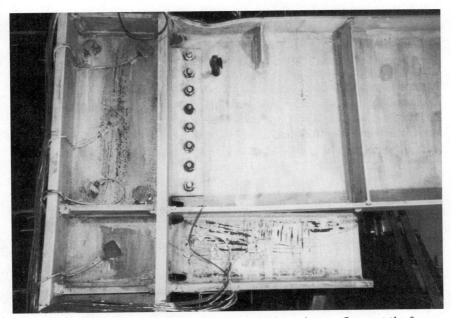

Figure 8.32 Fracture between straight haunches and beam bottom flange at the free end of the haunch and beam top flange local buckling.

8.5.5.2 Weakening strategies: Plastic hinges can be moved away from the face of a column if one reduces the area of the beams' flanges at a selected location. By strategically weakening the beam by a predetermined amount over a small length, at some distance from the welded connection, and by taking into account the shape of the moment diagram to ensure that yielding will occur only at this location of reduced plastic moment capacity, one can effectively protect the more vulnerable beam-to-column connection. One can do this in a number of ways, such as by drilling holes in the flanges and by trimming the flanges. The latter solution has found broad acceptance in a relatively short time.

The idea of shaving beam flanges to improve the seismic performance of steel connections was first proposed and tested by Plumier (1990). Chen and Yeh (1994) confirmed the effectiveness of this approach to enhance the ductility of beam-to-column connections. Although this concept was patented in the United States in 1992, the owner of the patent waived any commercial royalty rights for its public use after the Northridge earthquake.

Two flange shapes have received considerable attention following the Northridge earthquake; see Figure 8.33. The first type (Iwankiw and Carter 1996, Chen et al. 1996) has flanges tapered according to a linear profile intended to approximately follow the varying moment diagram (Figure 8.33a). The second profile (Engelhardt et al. 1996) is shaved along a circular profile as described in Figures 8.33b and 8.33c. Both reduced-beam-section (RBS) profiles (a.k.a. "dogbone" profiles) have achieved plastic rotations in excess of 3 percent, as shown in Figure 8.34. A variant of the linear taper, with additional rib plates welded to the beam flanges to further reduce stresses in the flange groove welds, has also been successfully tested (Uang and Noel 1995).

In all cases, trimming of the flanges delays local buckling, but increases the likelihood of web buckling and lateral-torsional buckling due to the reduction in flange stiffness. The RBS connection usually experiences web local buckling first, followed by flange local buckling and lateral-torsional buckling, resulting in significant strength degradation. The addition of lateral bracing at the reduced beam section delays this strength degradation. High plastic rotation capacities have been achieved when lateral bracing was provided at the end of the dogbone farthest away from the column. Tests indicate a required lateral bracing strength of approximately 4 percent of the actual force developed by the beam flange (Uang and Noel 1996).

8.5.5.3 Extended bolted end plates Extended bolted end plate moment connections are popular where shop-welding and field-bolting is the

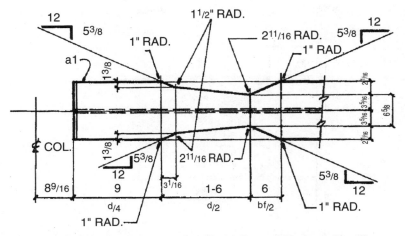

Figure 8.33 Reduced-beam-section designs: (a) Tapered flange profile. *(From Modern Steel Construction, Vol.36, No.4, The Dogbone: A New Idea to Chew On by N.R. Iwankiw and C.J. Carter, 1996, with permission from the American Institute of Steel Construction.)*

preferred assembly method. These connections are known to perform well under monotonic loading applications, but there is limited information on the inelastic response of such connections under severe cyclic loading. The beam-to-plate weld could conceivably be exposed to the same problems that plagued the pre-Northridge connections when the beam plastic moment is developed, particularly if the end plate is very rigid.

Tests on a small number of end-plate connections (Tsai 1988, Tsai and Popov 1990a, Ghobarah et al. 1990) conducted prior to the Northridge earthquake suggest that end-plate connections sized in compliance with the conventional design procedure have limited cyclic plastic deformation capacity. Connections with added plate stiffeners, or with a thicker end plate and stronger bolts, have exhibited superior energy dissipation capacity. End-plate connections also appear to behave well when panel zone yielding develops concurrently (Ghobarah et al. 1992). Nonetheless, studies on the cyclic behavior of such connections are still needed to establish under which conditions they can reliably deliver the required plastic rotations.

8.5.5.4 Bolted connections Nearly all of the past research on seismic beam-to-column connections has focused on fully rigid welded connections. Bolted connections may be viable alternatives to welded connections in some cases, but they have not received much attention from engineers practicing in regions of high seismic risk. Full-scale cyclic tests of rigid or semirigid bolted connections are few (e.g.,

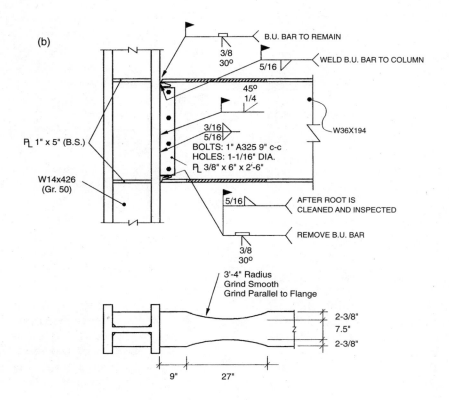
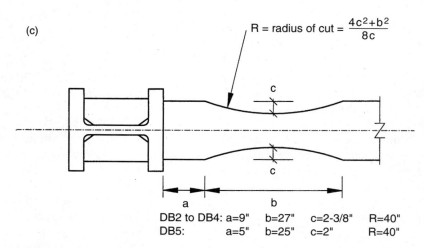

Figure 8.33 Reduced-beam-section designs: (b) Elevation and plan view of radius-cut flange profile, and (c) Information on radius-cut typical flange profile. *(Courtesy of M.D. Engelhardt, Dept. of Civil Engineering, University of Texas, Austin)*

Figure 8.34 Radius cut flange profile moment connection: (a) Specimen state at completion of test. (Parts a-c courtesy of M.D. Engelhardt, Dept. of Civil Engineering, University of Texas, Austin)

Popov and Pinkney 1969, Elnasai and Elghazouli 1994, Radziminsky and Azizinamini 1986, Astaneh et al. 1989, Leon et al. 1996, Sarraf and Bruneau 1996), and much additional work is needed.

Fully rigid bolted connections may be difficult, if not impossible, to implement for deep beams framing into heavy columns using currently available high-strength bolts. Bolting strategies are easier to implement for semirigid connections, whose strength, stiffness and ductility are governed by that of the connecting elements. Semirigid connections are often capable of developing plastic rotations of 0.03 radian, as shown in Figure 8.35 for an existing riveted connection retrofitted to develop a ductile semirigid behavior. Semirigid connections can also be easily repaired if necessary following an earthquake. However, the lower stiffness of semirigid connections with respect to

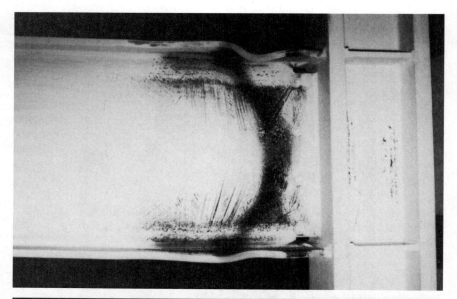

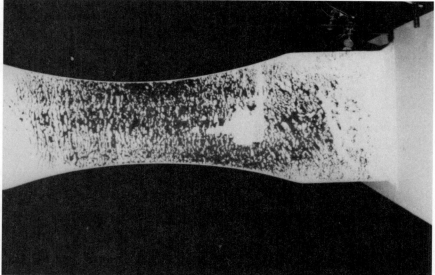

Figure 8.34 Radius cut flange profile moment connection: (b) Specimen state at completion of test: side view and top view. *(Courtesy of M.D. Engelhardt, Dept. of Civil Engineering, University of Texas, Austin)*

fully rigid connections will require stiffer (and heavier) beams and columns to comply with code-specified drift limits and thus larger beam-to-column connections.

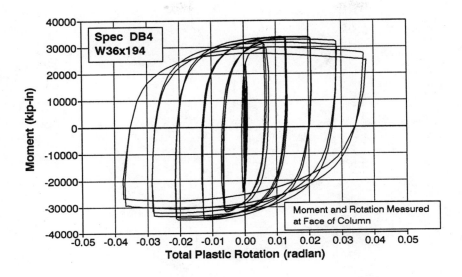

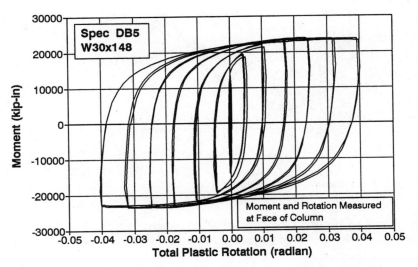

Figure 8.34 Radius cut flange profile moment connection: (c) Hysteretic behavior in terms of moment at column face versus beam plastic rotation.

8.5.5.5 Toughness-rated weld filler metal Some engineers have attributed many of the Northridge failures to the low fracture toughness of the weld metal used in groove welded connections (deposited with E70T-4 electrodes) and weld defects (such as those frequently found at the midwidth of bottom flange, on runoff tabs, etc.). In particular, inadequate weld root penetration frequently occurs near the web weld

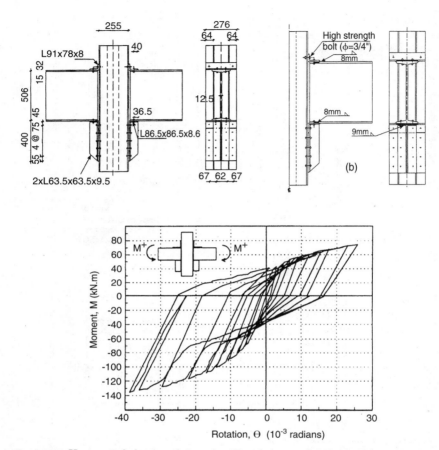

Figure 8.35 Hysteretic behavior of riveted stiffened seat-angle semirigid connection retrofitted using selective welding strategy, in terms of moment versus beam plastic rotation. *(Sarraf and Bruneau 1996)*

access hole in bottom flange connections because downhand welding across the entire width of the bottom flanges requires stopping at mid-width; the beam web creates an obstacle to welding and makes cleaning of slag more difficult there. The beam web is not an impediment to welding of the top flange.

Preliminary dynamic loading tests indicate that the use of weld metals with high notch toughness properties, such as those accomplished with E7018 filler metal, will improve performance when used in conjunction with good detailing practice including removal of backup bars and weld runoff tabs and reinforcement of weld roots and toes and fillet welds around three sides of the web shear tab (Kaufmann et al. 1996, Xue et al. 1996). However, only limited testing of this solution has been conducted at the time of this writing, and more testing is needed. The SAC Interim Guidelines (1995b) do not recommend the use of moment connections details that do not shift the plastic hinge away from the face of the column.

It is worthwhile, however, to report that many researchers agree on the benefits of using filler metal with relatively high notch toughness and better weld quality. SAC (1995b) requires a minimum Chapy-V notch toughness of 20 ft-lb at 0°F for filler metals used in critical joints, such as beam-to-column complete joint penetration welds or other tension applications when cross-thickness loading or triaxial stress states exist. These concepts were integrated into many haunches, cover plates, and other solutions experimentally investigated. E71T-8 electrodes were commonly used for that purpose (e.g., Engelhardt and Sabol 1996, Uang and Bondad 1996b).

8.5.5.6 Column tree configuration Widely used in Japan, column tree construction typically involves the welding of stub-beams to the column prior to shipment to the building site where the remaining beam segments are field-bolted to the stub-beams (Figure 8.36). In principle, all welds of columns, beams, and continuity plates (termed diaphragms in Japan) are done in the shop, with superior welding processes and under tight quality control. For that reason, such connections were thought to be superior to the United States prequalified moment connection described earlier. Unfortunately, the Kobe earthquake, striking exactly one year after the Northridge earthquake, revealed this belief to be partly unfounded.

A investigation by the Steel Committee of the Architectural Institute of Japan covering 988 modern steel buildings following the 1995 Hyogo-ken Nanbu (Kobe) earthquake reported 332 cases of severely damaged buildings, 90 collapses, and 113 buildings for which damage to beam-to-column connections was observed (AIJ 1995). Numerous cases of brittle fractures occurred, and many of the buildings that collapsed were moment frames constructed with the column tree system.

The beam-to-column failures observed during the Kobe earthquake differed somewhat from the Northridge failures in that cracking and fracture were frequently (but not always) accompanied by plastic hinging in the beams. This evidence of plastification was observed mostly in the more modern moment frames having square-tube columns and full penetration welds of the stub-beams to the diaphragms. In the majority of these cases, no sign of plastification was observed in the columns. Most of the fractures occurred in the lower flange of the beams, and the beams exhibited clear signs of plastic hinging accompanied by local buckling of the flanges (Figure 8.37), although, in some cases, the level of plastification was modest.

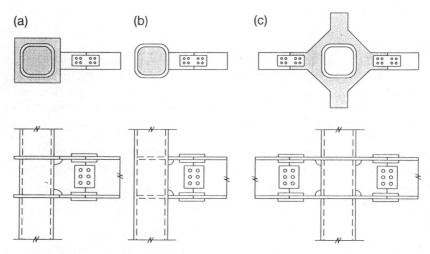

Figure 8.36 HSS columns in Japanese column-tree moment resisting frames, with (a) through-diaphragm, (b) interior diaphragm, (c) exterior diaphragm.

Figure 8.36 HSS column in Japanese column-tree moment resisting frames: (d) typical column-tree construction with through-diaphragm.

Typically, fracture initiated either from the corner of a weld access hole, near a run-off tab or a weld toe, or in the heat-affected zones in the beam flange or diaphragm. In many cases, the fracture progressed into the beam's web (e.g., Figure 8.37b), and, in some cases, propagated into the column flanges (e.g., Figure 8.37d).

Figure 8.37 Damage to Japanese column-tree moment connections in modern moment resisting frames with square-tube columns and full penetration welds at the beams, due to the Hyogo-ken Nanbu (Kobe, Japan) earthquake: (a) Fracture at the lower beam flange. *(Parts a-d from* Performance of Steel Buildings during the 1995 Hyogoken-Nanbu Earthquake *by Architectural Institute of Japan, courtesy of the Committee on Steel Structures of the Kinki Branch of the Architectural Institute of Japan)*

Many beam-to-column connections cracked and fractured without any signs of plastification when fillet welds were used in lieu of full penetration groove welds. These fillet welds were often too small to develop the capacity of the connected members (Figure 8.38a). Many other types of moment connections also suffered serious damage (Figure 8.38b). Notably, when tube columns were used, cracking and fracture frequently occurred in the columns above or below the top or bottom diaphragm (Figure 8.39a). As a result, complete overturning and collapse of the structure occurred (Figure 8.39b). In fact, damage to the beam-to-column connections of at least 59 moment frames having square-tube columns was reported by the AIJ, with about 70 percent of those rated as either collapsed or severely damaged. Although most of those surveyed buildings that collapsed had fillet-welded moment connections, at least three buildings having full-penetration welded moment connections collapsed (AIJ 1995).

8.5.5.7 Proposed repair and rehabilitation schemes There is a fundamental difference between repair and rehabilitation. In essence, repairs are emergency measures that bring a damaged structure back to its pre-earthquake condition. If the exact same earthquake that initially damaged a structure would strike again after completion of

Figure 8.37 Damage to Japanese column-tree moment connections in modern moment resisting frames with square-tube columns and full penetration welds at the beams, due to the Hyogo-ken Nanbu (Kobe, Japan) earthquake: (b) Propagation of fracture in the beam web.

repairs to the structure, one could reasonably expect the same damage to recur (assuming, obviously, that repairs were not accompanied by some measures of strengthening). Rehabilitations (also called retrofits or modifications in some documents) are measures intended to enhance the seismic performance of an existing structure.

Repairs to cracked and fractured steel members usually entail gouging of small cracks and defects, rewelding, grinding, addition of plates or stiffeners, and/or replacement of larger steel pieces or stub sections. This book does not dwell on the topic of repairs following an earthquake. Other documents are available that describe in detail various repair procedures developed following the Northridge earthquake (FEMA 1995).

Design of Ductile Moment-Resisting Frames 361

(c)

(d)

Figure 8.37 Damage to Japanese column-tree moment connections in modern moment resisting frames with square-tube columns and full penetration welds at the beams, due to the Hyogo-ken Nanbu (Kobe, Japan) earthquake: (c) Fracture initiated in the heat-affected zone of the diaphragm, (d) Propagation of fracture in the column.

Figure 8.38 Examples of beam-to-column welded connections damaged by the Hyogo-ken Nanbu (Kobe, Japan) earthquake: (a) Fracture along fillet welds of moment connection at the first story of a multistory residential building.

Seismic rehabilitation is a complex subject whose breadth exceeds the scope of this book. In principle, the connections strategies developed for new construction should be equally effective in existing buildings. Unfortunately, many of those solutions cannot be economically implemented in existing buildings without major modifications. For example, new structural elements added to a connection, such as haunches, will have to work in parallel with the existing flange groove welds recognized as likely to perform poorly in future earthquakes, and additional measures may also be necessary to correct these weld deficiencies. Likewise, moment-resisting frames in existing buildings are frequently located at the edge of buildings (i.e., the optimal location to provide seismic torsional resistance in plan); as a result, access to the outside face of the connection is not possible without removal of the exterior cladding, by itself a practical impediment to the implementation of some seismic rehabilitation strategies.

Beyond such considerations, the concepts and details formulated above for new construction can be applied to seismic rehabilitation.

8.5.6 International significance of the Northridge moment-connection failures

Moment frame connections identical to those that fractured during the Northridge earthquake are also commonly used in other countries

Figure 8.38 Examples of beam-to-column welded connections damaged by the Hyogo-ken Nanbu (Kobe, Japan) earthquake: (b) Large residual interstory drift in a moment frame with damaged connections.

(e.g., Tremblay et al. 1995). Furthermore, irrespective of the types of moment connections used, the Northridge experience reinforces the need for substantial full-scale experimental verification of connection details, for quality workmanship and inspection, and for periodic experimental re-evaluation of accepted practice to assess the significance of accumulated changes in materials properties, welding procedures, and other issues as the steel industry further evolves.

The Northridge (and Kobe) failures also clearly point to the need for a comprehensive reexamination of current beam-to-column connection practice in seismic regions worldwide. This is particularly true because in many of those regions, the beam-to-column details currently used have developed either as extensions of details known to provide a satisfactory behavior under static load conditions or from a few small-scale cyclic tests conducted decades ago.

364 Chapter Eight

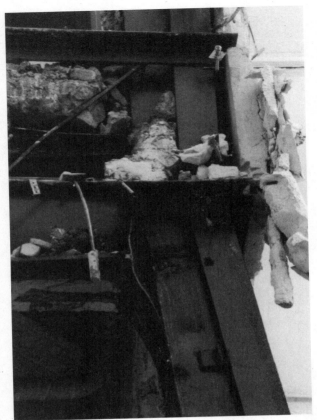

Figure 8.38 Examples of beam-to-column welded connections damaged by the Hyogo-ken Nanbu (Kobe, Japan) earthquake: (c) Close-up view of fractured welds along box-column plate for frame shown in Figure 8-38b.

8.6 Design of a ductile moment frame

8.6.1 Generic design procedure

In light of the above information, a capacity design procedure that relocates the plastic hinges away from the column face is adopted in this book. The various steps that should be followed to implement this approach for the design of ductile moment frames can be summarized as follows:

1. Select a method to relocate the plastic hinges away from the columns (haunches, dogbones, etc.), as shown in Figure 8.40. Select a plan layout with as many moment-frame bays as possible to promote redundancy.

2. Demonstrate the strength and ductility of the chosen connection type by approved cyclic testing, or by calculations if results from past tests on similar details can be used. Demonstration by calculations is

Design of Ductile Moment-Resisting Frames 365

Figure 8.39 Examples of beam-to-column damage due to the Hyogo-ken Nanbu (Kobe, Japan) earthquake: (a) Fracture in column-to-diaphragm welded connections.

Figure 8.39 Examples of beam-to-column damage due to the Hyogo-ken Nanbu (Kobe, Japan) earthquake: (b) Overturning of a building as a consequence of such fractures.

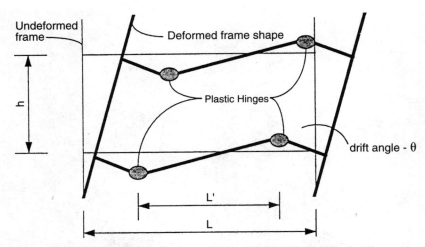

Figure 8.40 Desired plastic collapse mechanism in post-Northridge ductile moment resisting frames. *(From* Interim Guidelines: Evaluation, Repair, Modifications, and Design of Steel Moment Frames, *SAC Joint Venture, 1995, with permission)*

not permitted if it requires extrapolation of test data to connections having bigger structural sections or different material properties, welding materials and processes, construction sequence, relative stiffnesses and strengths of beams, columns, and panel zones, etc., and may require special permission by the building official. For qualification by testing, the connection must be subjected to numerous cycles of inelastic deformations per the ATC-24 testing protocol or equivalent (ATC 1992). A connection is deemed acceptable if it can sustain a minimum beam plastic rotation capacity of 0.03 radians for at least one cycle without strength degradation to less than 80 percent of its plastic moment.

3. For strengthened connections, assume that plastic hinges will form at $d_b/3$ from the toes of haunches or vertical ribs and $d_b/4$ beyond the ends of the cover plates (Figure 8.41), where d_b is the beam depth, unless test data suggests otherwise. For weakened sections, plastic hinges can be assumed to form at midlength of the reduced flanges having circular or straight cut patterns or at $d_b/4$ toward the column from that middle point for tapered reduced flanges, unless test data suggests otherwise. This assumes that total length of a reduced section is approximately $0.75d_b$ to d_b and that flange reduction starts at $d_b/4$ from the column face.

4. Determine the probable plastic moment at the plastic hinge location, as:

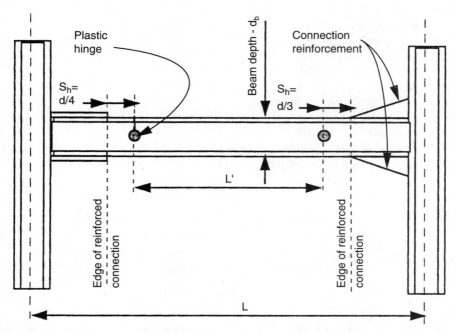

Figure 8.41 Assumed location of plastic hinge locations for the design of post-Northridge ductile moment resisting frames, in lieu of values supported by experimental evidence. *(Adapted from* Interim Guidelines: Evaluation, Repair, Modifications, and Design of Steel Moment Frames, *SAC Joint Venture, 1995.)*

$$M_{pr} = \beta M_p = \beta Z_b F_y = \beta^* Z_b F_y^{act} \qquad (8.20)$$

where Z_b is the beam's plastic modulus, β is a coefficient that accounts for the probable material yield strength, strain hardening effects, and modeling uncertainties, and β^* is a similar coefficient that accounts for only the last two effects. In light of the data presented in Table 8.1, it is recommended that the specified yield strength, F_y, be taken as 50 ksi (345 MPa) for A36 steel, dual-certified steel, and A572 Grade 50 steel. Likewise, the actual steel strength, F_y^{act}, should be taken as 55 ksi (380 MPa) for all those steels. SAC (1997) suggests that β be taken as 1.2. This latter value corresponds to the product of a 1.1 factor to account for yield strengths in excess of the specified values and a 1.1 factor for strain hardening effects. Likewise, β^* is 1.1 to account for strain hardening effects. Note that SAC (1995b) recommended that the β and β^* be magnified by a modeling uncertainty factor approximately equal to 1.16 (to be taken as 1.0 instead when qualification testing was to be performed). This was intended to encourage engineers to require qualification testing, but it had the unfortunate effect of leading some engineers to design excessively large connections having features not conductive to good behavior. This modeling uncertainty factor was therefore dropped by SAC (1997).

5. Calculate the shear, V_p, acting at the plastic hinge locations, as well as the moment at the column face, M_f, and column centerline, M_c, according to the equilibrium equations shown on the respective free-body-diagrams of Figure 8.42.

6. Promote strong-column/weak-beam by satisfying the following equation at each beam-column joint:

$$M_{pcr} = Z_{cr}F_{yc} = \sum Z_c\left(F_{yc} - \frac{P_{uc}}{A_g}\right) \geq \sum M_c \qquad (8.21)$$

where M_c is defined in step 5 above, and all other terms have been defined following Equation 8.1. Recognizing that this approach can be conservative for connections having deep panel zones, SAC (1997) permits that the sum of column moments at top and bottom of the panel zone (but still obtained from the free-body diagrams of Figure 8.42) be substituted for ΣM_c in the above equation. The effect of uneven stiffness distribution between top and bottom columns that can result from the use of a nonsymmetric connection (e.g., cover plate on top flange, haunch on bottom flange) must also be considered.

7. Design the panel zones using Equations 8.7 and 8.15, and $\Delta M = \Sigma M_f$. Although SAC (1995b) permits $\Delta M = \Sigma 0.8 M_f$, in an attempt to develop energy dissipation in both the panel zones and beams, experimental observations indicate that uncertainties in true material strengths make the intended sharing of plastic rotation between the beam and panel zone impossible to control. For example, panel zones in columns with weaker than average yield values would have to provide all of the required hysteretic energy dissipation if coupled with beams having stronger than average strengths, and vice versa. Panel zones can be very reliable energy dissipators, but the kinking of column flanges at large panel shear strain deformations generates complex triaxial stress conditions and possible fracture at the beam flange welds. This kinking can be particularly severe in columns with thinner flanges, with unknown consequences. For those reasons, it is suggested to minimize panel zone yielding, and $\Delta M = \Sigma M_f$ is preferred. Although this step will concentrate plastic hinging in the beams, some panel zone yielding is still possible (up to $4\gamma_y$ in principle) if Equation 8.15 is used.

Beyond those special steps due to the unique choice of plastic hinge locations, the other conventional design issues, such as compliance to drift limits, must be addressed in the usual manner by the designer. Likewise, a proper model of the structure must be formulated for analysis.

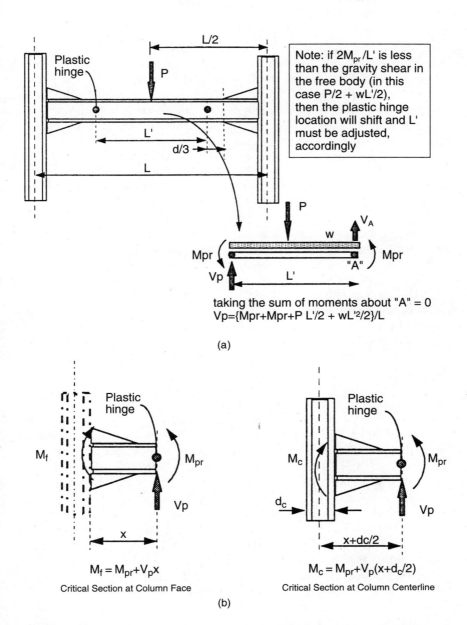

Figure 8.42 Design of post-Northridge ductile moment resisting frames: (a) Free-body-diagram to calculate shear at plastic hinges and (b) Free-body-diagrams to calculate moments at column face and column centerline. *(From* Interim Guidelines: Evaluation, Repair, Modifications, and Design of Steel Moment Frames, *SAC Joint Venture, 1995, with permission)*

8.6.2 RBS design example

For new buildings, moment frames spanning multiple bays are highly recommended to enhance structural redundancy. However, for simplicity, a single bay moment frame is considered for this example (Figure 8.43). That frame is identical in geometry to the concentric braced frame example of Chapter 7, and it is subjected to the same gravity loads. However, in recognition that lower seismic loads are specified for ductile moment-resisting frames than for concentric braced frames, the design lateral loads are taken as 75 percent of those considered in that earlier example. A572 Grade 50 steel is used, with a specified F_y of 350 MPa (50 ksi). Second-order (i.e., P-Δ) analyses of the frame were not conducted in this example (but must be considered in design) to keep the example focused and facilitate verification of the results by the reader.

Figure 8.44 illustrates how beam moment diagrams are used to ensure that plastic hinges develop away from the face of columns. Clearly, for a given beam, reinforced-beam-ends concepts will induce greater moments at the column face than was customary prior to Northridge. On the contrary, smaller column moments are obtained from reduced-beam-section solutions, and one can control their magnitude by varying the magnitude of the flange reduction at the critical section. This can be advantageous given that these moments dictate column and panel zone design. A reduced-beam-section design is considered in this example.

The example frame is first analyzed for the specified loads, without any attention to ductility issues. This phase is called *strength* design, and resulting members are shown in Table 8.3. Note that W760 shapes (W30 in U.S. units) were uniformly selected for all columns, even though some lighter W610 members would have also been satisfactory. The choice of deeper columns was deliberately made in anticipation of the strong-column/weak-beam philosophy to be enforced in the later phases of the design for ductile response.

As is often the case for ductile moment frames in seismic regions, story drifts limits control the selection of the beams and columns in this example. From the *strength* analysis, total story sways of 14.7, 40.4, and 64.8 mm were obtained at the top of the first, second, and third stories respectively. The corresponding interstory drifts of 14.7, 25.7, and 24.4 mm exceeded in two instances the permissible value of 0.005 times the story height (corresponding to 17.5 mm here). Calculations using simplified models revealed that stiffer beams would more effectively control drift than would stiffer columns in this case. Therefore, through use of basic structural analysis concepts assuming inflection points at the middles of beams and columns, the

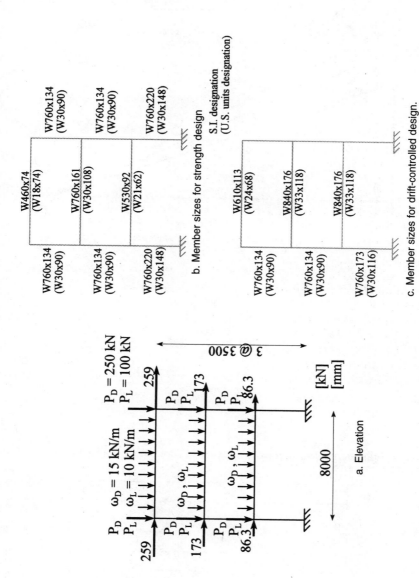

Figure 8.43 Ductile moment resisting frame example.

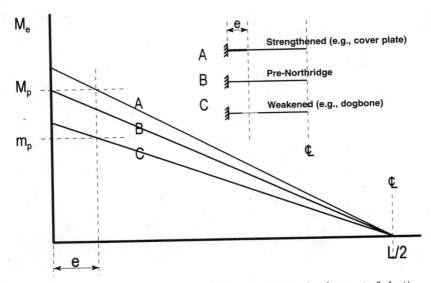

Figure 8.44 Consideration of moment gradient to promote development of plastic hinges at distance e from face of column.

beam moment of inertia needed to limit the drift in the top story was estimated by:

$$\Delta = \Delta_b + \Delta_c = \frac{V_3 h^2}{3E}\left[\frac{L}{2I_b} + \frac{h}{I_c}\right] \leq \frac{17.5\text{mm}}{2} = 8.75\text{mm}$$

$$= \frac{(259{,}000\text{N})(1750\text{mm})^2}{3(200{,}000\text{MPa})}\left[\frac{8{,}000\text{mm}}{2I_b} + \frac{1750\text{mm}}{1500 \times 10^6\text{mm}^4}\right] \leq 8.75\text{mm}$$

(8.22)

TABLE 8.3 Strength and Drift-Controlled Design of Three-Story Ductile Moment-Resisting Frame Example

Element	Story	Strength Design			Drift Control Design		
		S.I. Designations	U.S. units designations	Drift (mm)	S.I. Designations	U.S. units designations	Drift (mm)
Columns	3	W760x134	W30x90		W760x134	W30x90	
	2	W760x134	W30x90		W760x134	W30x90	
	1	W760x220	W30x148		W760x173	W30x116	
Beams	3	W460x74	W18x50	64.8	W610x101	W24x68	44.6
	2	W760x161	W30x108	40.4	W840x176	W33x118	28.3
	1	W530x92	W21x62	14.7	W840x176	W33x118	10.7

where Δ_b and Δ_c are, respectively, the contributions of beam and column flexibility to interstory drift, h is the story height, L is the frame span, I_b and I_c are respectively the beam and column moment of inertia, V_i is the shear force acting in the column at level i in the subassembly of interest, and E is Young's modulus. This equation indicated that a beam moment of inertia greater than 800 x 10⁶ mm⁴ was needed, and a W610x113 roof beam was selected. Likewise, for the second story:

$$\Delta = \Delta_b + \Delta_c = \frac{(V_3 + V_2)h^2}{12E}\left[\frac{L}{I_b} + \frac{h}{2I_c}\right] \le 17.5\text{mm}$$

$$= \frac{(259{,}000 + 432{,}000)(3500)^2}{12(200{,}000)}\left[\frac{8000}{I_b} + \frac{3500}{2(1500 \times 10^6)}\right] \le 17.5\text{mm}$$

(8.23)

and $I_b \ge 2100 \times 10^6$ mm⁴. A W840x176 beam was selected. This procedure was not repeated for the first story because the inflection points at that level were located far from the midheight of the columns. After fine tuning, a satisfactory *drift-controlled* design was achieved. Resulting member sizes and drifts are summarized in Table 8.3.

Tests have shown that radius-cut RBSs are most effective when the distance, e, from their narrowest flange cross section to the column face is equal to 0.75 to 1.0 times the beam depth. A distance $e/d_b = 0.75$ is chosen in this example to minimize the reduction in flange area. Trimmed flanges must also be sized to limit the moment at the column face to M_y of the beam. On this basis,

$$M_c = m_p^{act} + V_p e \le M_y^{act} = SF_y^{act} \tag{8.24}$$

where m_p^{act} is the actual plastic moment at the hinge location, calculated using the actual yield stress, F_y^{act}, and

$$V_p = \frac{2m_p^{act}}{L'} \tag{8.25}$$

where L' is the distance between the plastic hinges. Defining α as the ratio L/L', where L is the clear span between columns, and recalling that $m_p^{act} = \beta^* Z_{RS} F_y^{act}$ where Z_{RS} is the beam's plastic modulus at the reduced section and β^* is defined in the previous section, then:

$$M_c = m_p^{act}\left[1 + \frac{2e}{L'}\right] = m_p^{act}\left[\frac{L}{L'}\right] = m_p^{act}\alpha = \alpha\beta^* Z_{PR} F_y^{act} \le M_y^{act} = SF_y^{act} \tag{8.26}$$

If $\beta^*=1.1$ is used to account for strain hardening, and a shape factor, Z/S, of 1.14 for wide-flange beams (as indicated in Chapter 3) is selected, the maximum plastic modulus of the reduced beam section is given by:

$$Z_{RS} \leq \frac{S}{\alpha\beta^*} \quad \text{and} \quad \frac{Z_{RS}}{Z} \leq \frac{1.0}{\alpha\beta^*\left(\frac{Z}{S}\right)} = \frac{0.80}{\alpha} \qquad (8.27)$$

In principle, this is sufficient to complete design of the reduced beam section. However, the above can be directly expressed in terms of the geometric parameters illustrated in Figure 8.33c. For that purpose, note that $a+(b/2) = e = 3d/4$ in this example and that the beam flange width, b_f, is equal to $b_{RS} + 2c$. Defining $b_{RS} = \rho\, b_f$, it can be established that:

$$\frac{Z_{RS}}{Z} = \frac{Z_w}{Z} + \frac{\rho Z_f}{Z} \quad \text{and} \quad \rho = \frac{\dfrac{Z_{RS}}{Z} - \dfrac{Z_w}{Z}}{\dfrac{Z_f}{Z}} \qquad (8.28)$$

where Z_w and Z_f are respectively the plastic modulus of the beam web and flanges individually. Resulting calculations for the reduced sections are summarized in Table 8.4.

The results presented in Table 8.5 demonstrate that this frame also satisfies the requirements for strong-column/weak-beam design. Had this not been the case, stronger columns would have been necessary.

Finally, the strength of the panel zones is checked using Equation 8.21, with the value of M_c given by Equation 8.26. Calculations show that panel zone strength is sufficient, except for the panel zone at the level of the second floor beam. The horizontal shear acting on that panel zone can be calculated as:

$$V_w = \frac{M_c}{0.95 d_{b1}} - \frac{M_c\left(\dfrac{L_1}{L_{b1}}\right)}{h} \qquad (8.29)$$

$$= \frac{(2238 \times 10^6)}{0.95(835)} - \frac{(2238 \times 10^6)\left(\dfrac{8000}{7250}\right)}{3500} = 2107 \text{kN}$$

The corresponding strength of the panel zone is:

$$V_u = 0.55 F_y d_c t_p \left[1 + \frac{3 b_c t_{cf}^2}{d_b d_c t_p}\right]$$

$$= 0.55(350)(11.9)(750)\left[1 + \frac{3(264)(15.5)^2}{(750)(835)(11.9)}\right] \qquad (8.30)$$

$$= 1718 kN\, [1 + 0.026] = 1762 kN < 2107 kN$$

TABLE 8.4 Design Parameters for Reduced-Beam-Section Connection

Story	S.I. Designations (U.S. units designations)	d mm	e mm	e/d	L mm	L' mm	α	Z_{RS}/Z	Z $\times 10^3$ mm^3	Z_f $\times 10^3$ mm^3	Z_w $\times 10^3$ mm^3	ρ
3	W610x101 (W24x68)	603	450	0.75	7250	6350	1.14	0.69	2900	1998	902	0.564
2	W840x176 (W33x118)	835	625	0.75	7250	6000	1.21	0.65	6890	4480	2340	0.48*
1	W840x176 (W33x118)	835	625	0.75	7238	5988	1.21	0.65	6890	4480	2340	0.48*

* Use the minimum value of 0.5 recommended for ρ, unless lower values shown to be acceptable by tests

TABLE 8.5 Verification of Strong-Column/Weak-Beam Design Requirement

Joint	P_{uc} (kN)	A_g mm^2	$F_{yc} - P_{uc}/A_g$ (MPa)	$\Sigma Z_c (F_{yc} - P_{uc}/A_g)$ (kN-m)	$\Sigma M_c = \alpha\, m_p^{act} = \alpha\, \beta^* Z_{RS} F_y^{act}$ (kN-m)
3	415	13,000	319	1476	968
2	1034	22,400	304	1476+1407 = 2883	2238
1	1656	22,400	332	1407 + 2062 = 3469	2238

Therefore, the column's panel zone has insufficient strength and must be reinforced with doubler plates. Although only a thin plate is needed to provide the required strength, a 5 mm thick doubler plate is used for practical reasons.

The local buckling resistance of the panel zone and doubler plate is then checked through standard structural steel design equations (AISC 1993). Considering the most critical panel zone direction, with $a = d_c - 2t_c$, $h = d_b - 2t_b$, and $a/h = 1.11$:

$$\frac{h}{w} = \frac{797}{11.9} = 67 < 187 \sqrt{\frac{k_v}{F_{yw}}} < 187 \sqrt{\frac{\left(5 + \frac{5}{(a/h)^2}\right)}{F_{yw}}} < 187 \sqrt{\frac{9.06}{50}} = 79.6 \tag{8.31}$$

where k_v is a shear-buckling coefficient. The doubler plate, however, has a width-to-thickness ratio, h/w, of $797/5 = 159$, greater than the limit of 79.6, and plug welds are necessary to prevent its local buckling.

Finally, the limit of Equation 8.16 is checked. Taking advantage of the greater effective panel zone thickness possible now that the doubler plate is plug-welded to the column web:

$$t_z = 11.9 + 5.0 = 16.9 \geq \frac{(d_z + w_z)}{90} = \frac{(719 + 797)}{90} = 16.8 \tag{8.32}$$

Note that plug-welded doubler plates must be added to all panel zones of this ductile moment-resisting frame to satisfy this last requirement.

References

1. AIJ. 1995. *Performance of Steel Buildings during the 1995 Hyogoken-Nanbu Earthquake.* (in Japanese with English summary). Tokyo: Architectural Institute of Japan.
2. AISC. 1993. *Load and Resistance Factor Design Specification for Structural Steel Buildings.* Chicago: American Institute of Steel Construction.
3. AISC. 1992. *Seismic Provisions for Structural Steel Buildings.* Chicago: American Institute of Steel Construction.
4. Allen, J.; Partridge, J.E.; Radau, S.; and Richard, R.M. 1995. Ductile Connection Designs for Welded Steel Moment Frames. Proc. 64th annual convention, Structural Engineers Association of California. 253–269.
5. Astaneh, A.; Nader, M.N.; and Malik, L. 1989. "Cyclic Behavior of Double Angle Connections." *ASCE Journal of Structural Engineering.* Vol. 115, No. 5: 1101–1118.
6. ATC. 1992. *Guidelines for Cyclic Testing of Components of Steel Structures, ATC-24.* Redwood City, CA: Applied Technology Council.
7. Becker, R. 1975. "Panel Zone Effect on the Strength and Stiffness of Steel Rigid Frames." *Engineering Journal.* American Institute of Steel Construction, Vol. 12, No. 1.
8. Bertero, V.V.; Anderson, J.C.; and Krawinkler, H. 1994. *Performance of Steel Building Structures during the Northridge Earthquake.* Report No. UCB/EERC-94/09. Berkeley: Earthquake Engineering Research Center, University of California.
9. Bertero, V.V.; Krawinkler, H.; and Popov, E.P. 1973. *Further Studies on Seismic Behavior of Steel Beam-to-Column Subassemblages.* Report No. UCB/EERC-73/27. Berkeley: Earthquake Engineering Research Center, University of California.

10. Blodgett, O. 1995. *Evaluation of Beam to Column Connections.* SAC Steel moment frame connections, Advisory No. 3, SAC-95-01. Sacramento: SAC Joint Venture.
11. Bruneau, M. and Mahin, S.A. 1991. "Full Scale Tests of Butt Welded Splices in Heavy Rolled Steel Sections Subjected to Primary Tensile Stresses." *Engineering Journal of the American Institute of Steel Construction.* Vol. 28, No. 1: 1–17.
12. Bruneau, M.; Mahin, S.A.; and Popov, E.P. 1987. *Ultimate Behavior of Butt Welded Splices in Heavy Rolled Steel Sections.* Report UBC/EERC-87/10. Berkeley: Earthquake Engineering Research Center, University of California.
13. BSSC. 1994. *NEHRP Recommended Provisions for the Development of Seismic Regulations for New Buildings.* Washington, DC: Building Seismic Safety Council.
14. Chen, S.J., and Yeh, C.H. 1994. *Enhancement of Ductility of Steel Beam-to-Column Connections for Seismic Resistance.* SSRC Technical Session. Lehigh University, Pennsylvania.
15. Chen, S.J.; Yeh, C.H.; and Chu, J.M. 1996. "Ductile Steel Beam-to-Column Connections for Seismic Resistance." *ASCE Structural Journal.* Vol. 122, No. 11: 1292–1299.
16. Chen, W.F., and Liew, J.Y.R. 1992. *Seismic Resistant Design of Steel Moment-Resisting Frames Considering Panel-Zone Deformations.* "Stability and Ductility of Steel Structures under Cyclic Loading." Fukumoto, Y., and Lee, G., editors. 323–334.
17. EERI. 1990. Loma Prieta Earthquake Reconnaissance Report. *Earthquake Spectra.* Supplement to Vol. 6. Oakland, California.
18. EERI. 1995. Northridge Earthquake Reconnaissance Report. Vol. 1. *Earthquake Spectra.* Supplement C to Vol. 11. Oakland, California.
19. EERI. 1996. Northridge Earthquake Reconnaissance Report. Vol. 2. *Earthquake Spectra.* Supplement C to Vol. 11. Oakland, California.
20. Elnashai, A.S., and Elghazouli, A.Y. 1994. "Seismic Behavior of Semi-Rigid Steel Frames." *Journal of Constructional Steel Research.* Vol. 29, No. 1-3: 149–174.
21. Engelhardt, M.D., and Husain, A.S. 1993. Cyclic Loading Performance of Welded Flange-Bolted Web Connections. *Journal of Structural Engineering.* American Society of Civil Engineers. Vol. 119, No. 12: 3537–3550.
22. Engelhardt, M.D., and Sabol, T.A. 1995. *Lessons Learned from the Northridge Earthquake: Steel Moment Frame Performance.* Proceedings, New Directions in Seismic Design. Tokyo. October. 1–12.
23. Engelhardt, M.D., and Sabol, T.A. 1996. *Reinforcing of Steel Moment Connections with Cover Plates: Benefits and Limitations.* Proceedings, U.S.-Japan Seminar on Innovations in Stability Concepts and Methods for Seismic Design in Structural Steel. Honolulu, Hawaii.
24. Engelhardt, M.D., and Sabol, T.A. 1994. *Testing of Welded Steel Moment Connections in Response to the Northridge Earthquake.* Progress report to the AISC Advisory Committee on special moment-resisting frame research. October.
25. Engelhardt, M.D.; Winneberger, T.; Zekany, A.J.; and Potyraj, T.J. 1996. "The Dogbone Connection: Part II." *Modern Steel Construction.* Vol. 36, No. 8: 46–55.
26. Englekirk, R. 1994. *Steel Structures—Controlling Behavior through Design.* New York: John Wiley & Sons.
27. Fielding, D.J., and Chen, W.F. 1973. "Steel Frame Analysis and Connection Shear Deformation" *ASCE Journal of the Structural Division.* Vol. 99, ST1: 1–18.
28. Fielding, D.J., and Huang, J.S. 1971. "Shear in Steel Beam-to-Column Connections." *Welding Journal.* July.
29. Ghobarah, A.; Korol, R.M.; and Osman, A. 1992. "Cyclic Behavior of Extended End-Plate Joints." *ASCE Structural Journal.* Vol. 118, No. 5: 1333–1353.
30. Graham, J.D.; Sherbourne, A.N.; and Khabbaz, R.N. 1959. *Welded Interior Beam-to-Column Connections.* American Institute of Steel Construction.
32. Hamburger, R. 1996. "More on Welded Moment Connections." *NEWS.* April. San Francisco: Structural Engineers Association of California (also available at http://www.seaoc.org/seaoc/seaonc/nl496/ncnl496.htm).
33. ICBO. 1988, 1997. *Uniform Building Code.* International Conference of Building Officials. Whittier, California.
34. Ivankiw, R.N., and Carter, C.J. 1996. "The Dogbone: A New Idea to Chew On." *Modern Steel Construction.* Vol. 36, No. 4: 18–23.

35. Kato, B. 1973. *Brittle Fracture of Heavy Steel Members.* Proc. of the National Conference on Tall Buildings. Architectural Institute of Japan. August. Tokyo, Japan. 99–100.
36. Kato, B., and Morita, K. 1968. "Brittle Fracture of Heavy Steel Structural Members "(in Japanese). *Transactions of the Architectural Institute of Japan.* No. 156, February. Tokyo, Japan.
37. Kaufmann, E.J.; Xue, M.; Lu, L.W.; and Fisher, J.W. 1996. "Achieving Ductile Behavior of Moment Connections." *Modern Steel Construction.* Vol. 36, No. 1.
38. Krawinkler, H. 1978. "Shear in Beam-Column Joints in Seismic Design of Steel Frames." *Engineering Journal, American Institute of Steel Construction.* Vol. 5, No. 3: 82–91.
39. Krawinkler, H.; Bertero, V.V. and Popov, E.P. 1971. *Inelastic Behavior of Steel Beam-to-Column Subassemblages.* Report No. UCB/EERC-71/7. Berkeley: Earthquake Engineering Research Center, University of California.
40. Krawinkler, H.; Bertero, V.V.; and Popov, E.P. 1975. "Shear Behavior of Steel Frame Joints." *Journal of the Structural Division, ASCE.* Vol. 101, ST11: 2317–2336.
41. Nelson, R.F. 1995. Proprietary Solution. *Modern Steel Construction.* Vol. 36, No. 1: 40–44.
42. Noel, S., and Uang, C.M. 1996. *Cyclic Testing of Steel Moment Connections for the San Francisco Civic Center Complex.* Test report to HSH Design/Build, Structural Systems Research Project. Division of Structural Engineering Report No. TR-96/07 University of California, San Diego.
43. Obeid, K. 1996. "Steel Moment Frame Connections: Shear in the Panel Zone." *NEWS.* April. Structural Engineers Association of California. San Francisco.
44. Osman, A.; Korol, R.M.; and Ghobarah, A. 1990. *Seismic Performance of Extended End-Plate Connections.* Proc. 4th National U.S. Conference on Earthquake Engineering. Earthquake Engineering Research Institute. Oakland, California.
45. Plumier, A. 1990. *New Idea for Safe Structures in Seismic Zones.* University of Liege. IABSE Symposium. Brussels, Belgium.
46. Popov, E.P. 1968. *Introduction to Mechanics of Solids*, New Jersey: Prentice Hall.
47. Popov, E.P. 1987. "Panel Zone Flexibility in Seismic Moment Joints." *Journal of Constructional Steel Research.* Vol. 8, No. 1: 91–118.
48. Popov, E.P.; Amin, N.R.; Louie, J.J.; and Stephen, R.M. 1985. "Cyclic Behavior of Large Beam-Column Assemblies." *Earthquake Spectra.* Earthquake Engineering Research Institute, Vol. 1, No. 2: 203–238.
49. Popov, E.P., and Bertero, V.V. 1973. "Cyclic Loading of Steel Beams and Connections." *Journal of the Structural Division, ASCE.* Vol. 99, ST6.
50. Popov, E.P.; Bertero, V.V.; and Chandramouli, S. 1975. *Hysteretic Behavior of Steel Columns.* Earthquake Engineering Research Center Report UCB/EERC-75-11. University of California, Berkeley.
51. Popov, E.P., and Pinkney, R.B.1969. "Cyclic Yield Reversals in Steel Building Connections." *Journal of the Structural Division, ASCE.* Vol. 95, ST3: 327–353.
52. Popov, E.P., and Stephen, R.M. 1972. *Cyclic Loading of Full-Size Steel Connections.* American Iron and Steel Institute. Bulletin No. 21. New York.
53. Popov, E.P., and Stephen, R.M. 1970. *Cyclic Loading of Full-Size Steel Connections.* Earthquake Engineering Research Center Report UCB/EERC-70-3. University of California, Berkeley.
54. Popov, E.P., and Stephen, R.M. 1977. "Tensile Capacity of Partial Penetration Welds." *ASCE Journal of the Structural Division.* Vol. 103, No.ST9.
55. Popov, E.P., and Tsai, K.C. 1989. "Performance of Large Seismic Steel Moment Connections under Cyclic Loads." *Engineering Journal.* American Institute of Steel Construction. Vol. 26, No. 2: 51–60.
56. Popov, E.P.; Tsai, K.C.; and Engelhardt, M.D.. 1989. "On Seismic Steel Joints and Connections." *Engineering Structures*, Vol. 11, No. 4: 193–209.
57. Radziminski, J.B., and Azizinamini, A. 1986. *Low Cyclic Fatigue of Semi-Rigid Steel Beam-to-Column Connections.* Proc. 3rd U.S. National Conference on Earthquake Engineering. Vol. 2: 1285–1296. Earthquake Engineering Research Institute. Oakland, California.

58. Roeder, C.W.; Carpenter, J.E.; and Taniguchi, H. 1989. "Predicted Ductility Demands for Steel Moment-Resisting Frames." *Earthquake Spectra*. Vol. 5, No. 2: 409–427.
59. Roeder, C.W., and Foutch, D.A.. 1995.Experimental Results for Seismic Resistant Steel Moment Frame Connections. *Journal of Structural Engineering*. American Society of Civil Engineers. Vol. 122, No. 6: 581–588.
60. Rosenbaum, D.B. 1996. "Welds in Bay Area Hit by Quake, Too." *Engineering News Record*. Vol. 237, No. 11: 10.
61. SAC, 1995a. *Steel Moment Frame Connections—Advisory No. 3*. SAC-95-01. SAC Joint Venture. Sacramento, California.
62. SAC, 1995b. *Interim Guidelines: Evaluation, Repair, Modification and Design of Welded Steel Moment Frame Structures*. Program to Reduce the Earthquake Hazards of Steel Moment Frame Structures. Federal Emergency Management Agency. Report FEMA 267/SAC-95-02. SAC Joint Venture. Sacramento, California.
63. SAC, 1997. *Interim Guidelines Advisory No. 1: Supplement to FEMA 267*. Program to Reduce the Earthquake Hazards of Steel Moment Frame Structures. Federal Emergency Management Agency. Report FEMA 267A/SAC-96-03. SAC Joint Venture. Sacramento, California.
64. Sarraf, M., and Bruneau, M. 1996. "Cyclic Testing of Existing and Retrofitted Riveted Stiffened-Seat Angle Connections." *ASCE Structural Journal*. Vol. 122, No. 7: 762–775.
65. Schneider, S.P.; Roeder, C.W.; and Carpenter, J.E. 1992. "Seismic Behavior of Moment-Resisting Steel Frames: Experimental Study." *ASCE Structural Journal*. Vol. 119, No. 6: 1885–1902.
66. SEAOC. 1995. *Interim Recommendations for Design of Steel Moment-Resisting Connection*. Structural Engineers Association of California. Sacramento, California.
67. SEAOC. 1980, 1988, 1990. *Recommended Lateral Force Requirements and Commentary*. Seismology Committee. Structural Engineers Association of California. Sacramento, California.
68. SSPC. 1994. *Statistical Analysis of Tensile Data for Wide Flange Structural Shapes*. Structural Shape Producers Council.
69. Tremblay, R.; Bruneau, M.; Nakashima, M.; Prion, H.G.L.; Filiatrault, A.; and DeVall, R. 1996. "Seismic Design of Steel Buildings: Lessons from the 1995 Hyogo-ken Nanbu Earthquake." *Canadian Journal of Civil Engineering*. Vol. 23, No. 3: 757–770.
70. Tremblay, R.; Timler, P.; Bruneau, M.; and Filiatrault, A. 1995. "Performance of Steel Structures during the January 17, 1994, Northridge Earthquake." *Canadian Journal of Civil Engineering*. Vol. 22, No. 2: 338–360.
71. Tsai, K.C., and Popov, E.P. 1990a. "Cyclic Behavior of End-Plate Moment Connections." *ASCE Structural Journal*. Vol. 116, No. 11: 2917–2930.
72. Tsai, K.C., and Popov, E.P. 1990b. "Seismic Panel Zone Design Effect on Elastic Story Drift in Steel Frames." *ASCE Structural Journal*. Vol. 116, No. 12: 3285–3301.
73. Tsai, K.C., and Popov, E.P. 1988. *Steel Beam-Column Joints in Seismic Moment-Resisting Frames*. Earthquake Engineering Research Center Report UCB/EERC-88/19. University of California, Berkeley.
74. Tsai, K.C.; Wu, S.; and Popov, E.P. 1995. "Experimental Performance of Seismic Steel Beam-Column Moment Joints." *ASCE Structural Journal*. Vol. 121, No. 6: 925–931
75. Uang, C.M. 1995. *Dynamic Testing of Large-Size Steel Moment Connections*. VHS-video of tests. University of California, San Diego.
76. Uang, C.M., and Bondad, D.M. 1996. *Dynamic Testing of Full-Scale Steel Moment Connections*. Proceedings of 11th World Conference on Earthquake Engineering. Acapulco. CD-ROM, Paper 407, Permagon Press: New York.
77. Uang, C.M., and Noel, S. 1995. *Cyclic Testing of Rib-Reinforced Steel Moment Connection with Reduced Beam Flanges*. Test report to Ove Arup & Partners. Structural systems research project. Division of Structural Engineering Report No. TR-95/04. University of California, San Diego.
78. Uang, C.M., and Noel, S. 1996. *Cyclic Testing of Strong- and Weak-Axis Steel Moment Connection with Reduced Beam Flanges*. Final report to the City of Hope. Division of Structural Engineering Report No. TR-96/01. University of California, San Diego.

79. Wolfe, J.; Nienberg, M.; Manmohan, D.; Halle, J.; and Quintana, M. 1996. "Welding Alert, Cover Plated Moment Frame Connections." *NEWS*. April. Structural Engineers Association of California. San Francisco.
80. Wakabayashi, M. 1986. *Design of Earthquake-Resistant Buildings*. New York: McGraw-Hill.
81. Whittaker, A.; Bertero., V.; and Gilani, A.. 1995. *Testing of Full-Scale Steel Beam-Column Assemblies*. SAC Phase I Report. SAC Joint Venture. Sacramento, California.
82. Whittaker, A., and Gilani, A. 1996. *Cyclic Testing of Steel Beam-Column Connections*. Report No. EERC-STI/96-04. Berkeley: Earthquake Engineering Research Center, University of California.
83. Xue, M.; Kaufmann, E.J.; Lu, L.W.; and Fisher, J.W. 1996. "Achieving Ductile Behavior of Moment Connections—Part II." *Modern Steel Construction*. Vol. 36, No. 6: 38–42.
84. Youssef, N.F.G.; Bonowitz, D.; and Gross, J.L. 1995. *A Survey of Steel Moment-Resisting Frame Buildings Affected by the 1994 Northridge Earthquake*. Report NISTIR 5625. National Institute of Standards and Technology. Technology Administration, United States Department of Commerce. Gaithersburg, Maryland.

Chapter 9

Limit State Philosophy in Seismic Design Provisions

9.1 Introduction

The widely accepted limit state philosophy for seismic design considers at least two limit states associated with different levels of earthquake excitation: the service limit state for moderate earthquake shaking and the ultimate limit state for severe earthquake shaking. The building seismic design provisions in North America and Japan are used to demonstrate how these two limit states are implemented.

After reviewing the seismic provisions adopted in several countries, this chapter presents simple formulations of the seismic response factors that are used for ultimate limit state design; such formulations form the basis for a direct comparison between different seismic provisions. A comparison for service limit state requirements is also presented. The chapter concludes with a historical overview of code development in the United States.

9.2 Seismic limit state philosophy

The limit state philosophy of modern seismic provisions is best described by the performance criteria set forth by the Structural Engineers Association of California (SEAOC) Recommended Lateral Force Requirements (SEAOC 1990). Structures designed in conformance with these recommendations should, in general, be able to:

- Resist minor levels of earthquake ground motion without damage.
- Resist moderate levels of earthquake ground motion without structural damage, but possibly some nonstructural damage.
- Resist major levels of earthquake ground motion without collapse, but likely with some structural and nonstructural damage.

One can achieve the first two levels of performance, called the service limit state, by:

- Defining the level of moderate earthquake shaking.
- Limiting stresses or internal forces in structural members.
- Limiting the story drift ratio, defined as the ratio between interstory drift and story height.

The third performance level, which is often termed the ultimate limit state, can be achieved by:

- Defining the level of severe earthquake shaking.
- Providing sufficient strength, ductility, and deformation capacity to elements of the seismic framing system and providing a deformation-capable gravity load-resisting frame.
- Limiting the maximum story drift to ensure that structural integrity and stability are maintained.

9.3 Seismic design procedures in modern codes

9.3.1 General

Design procedures for the following four seismic provisions are summarized below: the Building Standard Law of Japan (1981), the NEHRP Recommended Provisions for the Development of Seismic Regulations for New Buildings (1994), the Uniform Building Code (1994) of the United States, and the National Building Code of Canada (1990). The Japanese code is selected for comparison because its approach is very different from that of the codes and regulations adopted in North America. Further, design concepts and notations vary from one code to another. In the presentation that follows, consistent notation for describing design earthquake ground motions will be used.

9.3.2 Japanese Building Standard Law (BSL)

The Building Standard Law (BSL) of Japan presents a two-level design procedure. Designers must consider service limit state requirements for the moderate earthquake associated with the Level 1 design and ultimate limit state requirements for the severe earthquake associated with the Level 2 design.

- **Level 1 Design.** In Japan, moderate earthquake shaking corresponds to a peak ground acceleration of between 0.07g and 0.10g (Kato 1986). The BSL service limit state requires that a regular building remain in the elastic range when subjected to lateral seismic forces associated with a base shear ratio, C_w:

$$C_w = 0.2ZR_t \tag{9.1}$$

where ZR_t ($\equiv C_{eu}$) represents the linear elastic design response spectrum for *severe* earthquake shaking. The intensity of the moderate earthquake is 20 percent of that of the severe earthquake. Figure 9.1 shows an elastic design spectrum, ZR_t, for soil type 2, which approximately corresponds to soil type 3 in the UBC. To control nonstructural damage, the maximum story drift is limited to 0.5 percent of the story height. To avoid structural damage, the maximum allowable stress for steel design is limited to approximately 90 percent of the yield stress. One establishes this stress level by increasing the basic allowable stress for gravity-load design, which is about 60 percent of the yield stress, by 50 percent. Because the structure is expected to respond in the elastic range, ductility is not considered for the service limit state check.

- **Level 2 Design.** A severe earthquake is assumed to have a peak ground acceleration ranging from 0.34g to 0.4g (Kato 1986). The base shear ratio, C_y, is computed as:

$$C_y = D_s ZR_t \tag{9.2}$$

where D_s is a structural characteristics factor that accounts for the energy dissipation capacity (ductility) of the structure (Kato and Akiyama 1982). For steel building structures, the value of D_s ranges from 0.25 for a ductile system to 0.50 for a nonductile system (see Table 9.1).

In summary, the C_w force level is used for the service limit state design, and the C_y force level is used to check the ultimate limit state design. For buildings that satisfy certain height limitations and regularity requirements, a simplified, yet conservative, one-level design procedure can be used.

TABLE 9.1 Typical D_s Values for Steel Structures (1981 BSL of Japan)

Types of Moment-Resisting Portion of Braced Frames*	Types of Braces	
	Moment Frames or Braced Frames with $\beta_u^\dagger \leq 0.3$	Braced Frames with $\beta_u^\ddagger > 0.7$
FA	0.25	0.35
FB	0.30	0.35
FC	0.35	0.40
FD	0.40	0.50

*Classification of moment-resisting frames is based on the compactness ratio of beams and columns. See IAEE (1988) for details.

†β_u = ratio of ultimate shear carried by braces to total ultimate shear.

‡For braces with effective slenderness ratio between $50/\sqrt{F_y}$ and $90/\sqrt{F_y}$ only, where yield stress F_y has a unit value of t/cm². See IAEE (1988) for other ranges of effective slenderness ratio.

9.3.3 NEHRP Recommended Provisions

The 1994 NEHRP Recommended Provisions are presented for strength design and not working stress design. The main requirements for the ultimate limit state check are summarized as follows:

- The severe design earthquake is characterized by an elastic design response spectrum:

$$C_{eu} = \frac{1.2C_v}{T^{2/3}} \leq 2.5C_a \tag{9.3}$$

where C_v and C_a are seismic coefficients based on the soil profile and the effective peak velocity or the effective peak acceleration, respectively, and T is the fundamental period of vibration. Figure 9.1 shows the elastic design spectrum for soil type C.

- The inelastic seismic design base shear coefficient, C_s, is reduced from the elastic level, C_{eu}, by a seismic force reduction (or response modification) factor, R, to account for ductility and structural overstrength (or reserve strength):

$$C_s = \frac{C_{eu}}{R} = \frac{1.2C_v}{RT^{2/3}} \leq \frac{2.5C_a}{R} \tag{9.4}$$

Design actions are calculated by imposing lateral forces corresponding to C_s on a linearly elastic mathematical model of the building.

- Maximum inelastic displacements are computed as follows:

$$\Delta_{max} = C_d \Delta_s \tag{9.5}$$

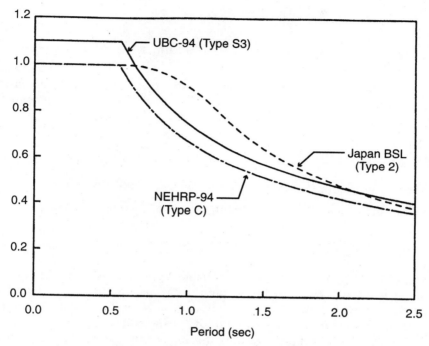

Figure 9.1 Elastic design spectra for comparable soil conditions.

where Δ_s is the elastic displacement due to the application of lateral forces corresponding to C_s, and C_d is a displacement amplification factor. The value of Δ_{max} is limited to $0.02h$ for typical multistory buildings (or $0.01h$ for essential facilities that are required for postearthquake recovery), where h is the story height.

Values of R and C_d for different steel framing systems are listed in Table 9.2. The objective of the NEHRP Recommended Provisions is to provide life safety during severe earthquake shaking. No service requirements are specified to minimize structural and nonstructural damage in minor or moderate earthquakes.

9.3.4 Uniform Building Code (UBC)

The seismic provisions in the 1994 Uniform Building Code (UBC) are written for working stress design. The key requirements for the ultimate limit state are as follows:

- The severe design earthquake is characterized by an elastic design spectrum, C_{eu}:

$$C_{eu} = \frac{1.25ZIS}{T^{2/3}} \leq 2.75ZI \qquad (9.6)$$

where Z is a seismic zone factor, I is an importance factor, and S is the site soil coefficient.

- Design actions in components of the seismic framing system are calculated through use of lateral forces reduced from C_{eu} by a seismic force reduction (or structural performance) factor, R_w:

$$C_w = \frac{C_{eu}}{R_w} = \frac{1.25ZIS}{R_w T^{2/3}} \leq \frac{2.75ZI}{R_w} \qquad (9.7)$$

See Table 9.2 for the values of R_w applicable to steel buildings.

- Maximum inelastic displacements are computed as follows:

$$\Delta_{max} = \frac{3R_w}{8}\Delta_w \qquad (9.8)$$

where Δ_w represents the elastic displacements due to the application of lateral forces corresponding to C_w, and $3R_w/8$ is a displacement amplification factor. Unlike the NEHRP Recommended Provisions, no limit is specified for Δ_{max}. Instead, Δ_{max} is used in the UBC to calculate the minimum building separation to avoid pounding, to estimate maximum story drifts, to check deformation capacity of critical structural members (e.g., shear links in eccentrically braced frames), to check P-delta effects, and to detail connections for nonstructural components.

For structural steel design using the AISC ASD Specification (AISC 1989), the stresses produced by the prescribed design seismic forces (Equation 9.7) and gravity loads must not exceed the allowable stress. The allowable stress for seismic design can be increased by one-third above that for gravity-load design. For example, whereas an allowable stress of $0.66F_y$ is considered for gravity-load design, where F_y is the specified yield stress, the corresponding allowable stress for seismic design is $0.88F_y$. When the AISC LRFD Specification (AISC 1993) is used, the UBC-prescribed seismic forces must be increased by a load factor of 1.4 for strength design, and the maximum forces in the members shall not exceed the design strength (ϕR_n).

The UBC seismic provisions are traditionally adopted with minor modifications from the SEAOC Recommended Lateral Force Requirements (1990). Although the Commentary of the SEAOC Requirements states

TABLE 9.2 R and C_d Factors specified in the NEHRP and R_w Factors specified in the UBC

Structural Steel System	R	R_w	C_d	C_d/R
Moment-resisting frame system SMRF OMRF	8 4.5	12 6	5.5 4	0.73 0.89
Dual system EBF + SMRF CBF + SMRF	8 6	12 10	4 5	0.50 0.83
Building frame system EBF CBF	8 5	10 8	4 4.5	0.50 0.90
Bearing wall system CBF Framed walls with shear panels	4 6.5	6 8	3.5 4	0.88 0.62
Inverted pendulum structures SMRF OMRF	2.5 1.25	3 3	2.5 1.25	1.00 1.00

SMRF = Special moment-resisting frame
OMRF = Ordinary moment-resisting frame
EBF = Eccentrically braced frame
CBF = Concentrically braced frame

that the service limit state (see Section 9.2) should be checked, it does not specify how this is to be achieved. Because a design force level corresponding to the moderate design earthquake is not specified in these provisions, it is generally considered that the UBC adopts a one-level seismic design procedure only.

In the UBC, the story drift ratio produced by the $C_w W$ seismic forces has to be limited as follows:

- For buildings with a fundamental period larger than 0.7 second,

$$\frac{\Delta_w}{h} \leq \text{maximum} \left\{ \frac{0.03}{R_w}, 0.004 \right\} \quad (9.9a)$$

- For buildings with a fundamental period equal to or less than 0.7 second,

$$\frac{\Delta_w}{h} \leq \text{maximum} \left\{ \frac{0.04}{R_w}, 0.005 \right\} \quad (9.9b)$$

The drift limit of $0.04/R_w$ in Equation 9.9b is a direct conversion of the requirement of the 1985 UBC, which stipulated a drift limit of $0.005K$ of the story height where K is a horizontal force factor (the predecessor of R). It will be shown in Section 9.5 that the R_w-dependent (or K-dependent) UBC drift limit can be interpreted as an indirect way of achieving the service limit state in this one-level design procedure.

9.3.5 National Building Code of Canada

The seismic provisions in the 1990 National Building Code of Canada (NBCC) are written for strength design. The main requirements for the ultimate limit state are the following:

- The severe design earthquake is characterized by an elastic design spectrum, C_{eu},

$$C_{eu} = vSIF \tag{9.10}$$

where v is the specified horizontal ground velocity expressed as a ratio of 1m/s, S is a period-dependent seismic response factor, I is a seismic importance factor, and F is a foundation factor.

- The prescribed design seismic force is reduced from C_{eu} by two factors:

$$C_s = \frac{C_{eu}}{(R/U)} \tag{9.11}$$

where R is the force modification factor that reflects the capacity of a structure to dissipate energy through inelastic behavior, and U (= 0.6) is a calibration factor "...representing level of protection based on experience..." (NBC 1990). See Table 9.3 for selected values of R specified in the NBCC.

TABLE 9.3 Typical Values of R for Steel Structures (1990 NBCC)

Steel Structural System (U.S. Description)	R
1. Moment-resisting frame system	
a. Ductile moment-resisting frame (SMRF)	4.0
b. Moment-resisting frame with nominal ductility (OMRF)	3.0
2. Braced frame	
a. Ductile braced frame—EBF (EBF)	3.5
b. Ductile braced frame—CBF (SCBF)	3.0
c. Braced frames with nominal ductility (CBF)	2.0
3. Others	1.5

- Maximum displacements are calculated as follows:

$$\Delta_{max} = R\Delta_s \qquad (9.12)$$

where Δ_s are the displacements produced by the design seismic forces, $C_s W$, acting on a linearly elastic model of the building. The value of Δ_{max} is limited to $0.02h$ (or $0.01h$ for postdisaster buildings).

9.4 Seismic force reduction and displacement amplification factors

9.4.1 General

All four seismic provisions described above use a seismic force reduction factor to reduce elastic spectral demands to the design level and a displacement amplification factor to calculate the likely inelastic displacements from the elastic displacements corresponding to the prescribed (or reduced) seismic forces. The physical meaning of these two factors, which are defined as R and C_d in the NEHRP Seismic Provisions, are formulated first (Uang 1991b). Based on such a formulation, it is then possible to compare the different approaches used in the seismic provisions described in the previous section.

9.4.2 Structural response and design seismic forces

Consider the typical global structural response envelope of Figure 9.2 that shows the relationship between story drift and required elastic strength, which is expressed in terms of the base shear ratio, C_{eu}. A framing system that is designed and detailed per modern codes for seismic resistance can deform well beyond the yield limit (Δ_y) and exhibit substantial reserve strength beyond the nominal design strength. For the purpose of the following discussion it is assumed that the framing system can sustain a maximum displacement of Δ_{max} and attain a maximum strength of C_y; see Figure 9.2.

For design, the NEHRP Recommended Provisions use C_s forces that correspond to the first significant yield level—the level beyond which the global structural response starts to deviate significantly from the elastic response. This design force level is consistent with material codes that use strength design such as the AISC LRFD Specification (AISC 1993). To be consistent with material codes that use allowable (or working) stress design methods, such as the AISC ASD Specification (AISC 1989), the UBC uses working stress forces C_w (see

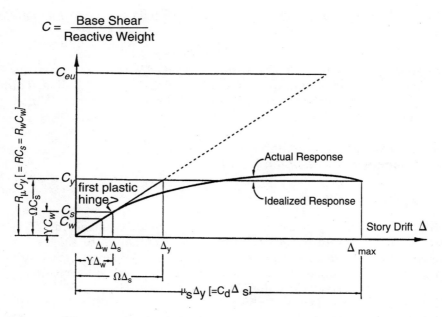

Figure 9.2 General structural response (Uang 1991b).

Figure 9.2). The advantage of using C_s or C_w forces for design is that these forces are suitable for use with elastic structural analysis.

One problem with the design procedures is that the true strength ($=C_y W$) of the building is unknown. If a structure's reserve strength (or overstrength) beyond the design force level (C_s or C_w) is significantly less than that implicitly assumed, the performance of the structure is not likely to be satisfactory during design earthquake shaking. Further, the maximum inelastic displacements cannot be calculated from the results of an elastic analysis. As an approximation, the NEHRP Recommended Provisions uses the displacement amplification factor, C_d, and the UBC uses a factor of $3R_w/8$ to predict maximum inelastic displacements from the elastic displacements (Δ_s or Δ_w) resulting from the application of the design seismic forces.

9.4.3 Definition of relevant terms

The terms needed to establish the seismic force reduction factors (R and R_w) and the displacement amplification factor (C_d) are bulleted below:

- **Structural ductility factor** (μ_s). If the actual structural response curve can be idealized by the linearly elastic-perfectly plastic curve of Figure 9.2, the structural ductility factor, μ_s, can be defined as

$$\mu_s = \frac{\Delta_{max}}{\Delta_y} \tag{9.13}$$

where deformation is expressed in terms of story drift, Δ.

- **Ductility reduction factor** (R_μ). A building is capable of dissipating hysteretic energy if deformations in excess of the yield deformation can be sustained. The elastic design force level (C_{eu}) can be reduced to a yield strength level (C_y) by the factor R_μ:

$$R_\mu = \frac{C_{eu}}{C_y} \tag{9.14}$$

The nominal yield strength of a building is greater than its strength at first significant yielding. For a single-degree-of-freedom system, the relationship between μ_s and R_μ is well established (Newmark and Hall 1982) and presented in Section 9.4.6. Other relationships between μ_s and R_μ have been summarized by Miranda and Bertero (1994).

- **Structural overstrength factor** (Ω). The reserve strength that exists between the yield level (C_y) and the first significant yield level (C_s) is defined as the structural overstrength factor, Ω:

$$\Omega = \frac{C_y}{C_s} \tag{9.15}$$

Structural overstrength results from a number of factors including internal force redistribution, code requirements for multiple loading combinations, code minimum requirements regarding proportioning and detailing, material strength higher than that specified in the design, strain hardening, deflection constraints on system performance, member oversize, effect of nonstructural elements, and strain rate effects.

- **Allowable stress factor** (Y). This factor is used to account for differences in the format of the codes. For allowable (or working) stress design, the design force level $(C_w$ in Figure 9.2) is reduced from C_s by a factor Y, that is:

$$C_w = \frac{C_s}{Y} \tag{9.16}$$

The value of Y calculated by analysis of U.S. seismic provisions is approximately 1.5.

9.4.4 Formulation of R (or R_w) and C_d factors

The seismic force reduction factor (or the NEHRP response modification factor R) for use with strength design can be derived as follows (see Figure 9.2):

$$R = \frac{C_{eu}}{C_s} = \frac{C_{eu}}{C_y} \frac{C_y}{C_s} = R_\mu \Omega \qquad (9.17)$$

Similarly, the force reduction factor for allowable stress design is

$$R_w = \frac{C_{eu}}{C_w} = \frac{C_{eu}}{C_y} \frac{C_y}{C_s} \frac{C_s}{C_w} = R_\mu \Omega Y \qquad (9.18)$$

The displacement amplification factor of the NEHRP Recommended Provisions, C_d, is the ratio of Δ_{max} and Δ_s, (see Figure 9.2):

$$C_d = \frac{\Delta_{max}}{\Delta_s} = \frac{\Delta_{max}}{\Delta_y} \frac{\Delta_y}{\Delta_s} \qquad (9.19a)$$

where Δ_{max}/Δ_y is the structural ductility factor (see Equation 9.13), and Δ_y/Δ_s from Figure 9.2 is equal to:

$$\frac{\Delta_y}{\Delta_s} = \frac{C_y}{C_s} = \Omega \qquad (9.19b)$$

Therefore Equation 9.19a can be rewritten as

$$C_d = \mu_s \Omega \qquad (9.20)$$

Both R (or R_w) and C_d are functions of the structural overstrength factor, structural ductility factor, and damping ratio. The effect of damping is generally included in the ductility reduction factor, R_μ. Furthermore, Equations 9.17 and 9.18 show that it is misleading to refer to R (or R_w) as a ductility reduction factor because overstrength and ductility may contribute equally to R. Similarly, the displacement amplification factor is generally not equal to the structural ductility factor.

9.4.5 Seismic force reduction factor—a comparison of seismic provisions

The seismic force reduction factors incorporated in the four design provisions reviewed in Section 9.3 cannot be compared directly. Numerical values assigned to identical framing systems vary significantly. For example, for ductile steel moment-resisting frames, the $1/D_s$ value in the BSL is 4, the value of R in NEHRP is 8, and the value of R in NBCC is 4. Nevertheless, a direct comparison can be made on the basis of Equation 9.17. See Table 9.4 for a summary of the approaches to design in the U.S.A., Canada, and Japan.

In the United States, the ductility reduction factor (R_μ) and structural overstrength factor (Ω) are not specified; rather, empirical values of R are used to reduce the elastic seismic force demand to the design level. The BSL of Japan specifies a factor D_s related to R_μ (=$1/D_s$) and engineers must compute the structural overstrength factor, Ω. (An evaluation of structural overstrength, which requires nonlinear analysis, can be waived for buildings less that 31 meters in height, whose frames are symmetric or nearly symmetric, and whose lateral story stiffness does not change sharply with height. In such cases, the Level 1 seismic design forces for service considerations must be increased, which is equivalent to using a smaller value of R (Uang 1991a). The NBCC specifies values for R_μ (which is defined as R in NBCC; see Table 9.3 for values) and a single value of Ω (=$1/U$ = 1.67) for all lateral load-resisting systems.

9.4.6 Displacement amplification factor—a comparison of seismic provisions

Based on Equations 9.17 and 9.20, the ratio between the seismic displacement amplification factor (DAF) and the force reduction factor (FRF) can be computed as follows:

$$\frac{DAF}{FRF} = \frac{C_d}{R} = \frac{\mu_s \Omega}{R_\mu \Omega} = \frac{\mu_s}{R_\mu} \qquad (9.21)$$

This equation indicates that the ratio is independent of the overstrength factor. For single-degree-of-freedom (SDOF) systems, the ratio between μ_s and R_μ in Equation 9.21 can be expressed as (Newmark and Hall 1982):

- For moderate and long period structures (i.e., in the velocity and displacement amplification regions of the response spectrum):

$$\frac{\mu_s}{R_\mu} = \frac{\mu_s}{\mu_s} = 1 \qquad (9.22)$$

TABLE 9.4 Comparison of Seismic Design Approaches

Seismic Provisions	R	R_μ	Ω
NEHRP (U.S.A.)	R	*[†]	*
NBC (Canada)	*	R	1.67
BSL (Japan)	*	$1/D_s$	Ω[‡]

[†] Value not given
[‡] Value to be calculated by analysis

- For short period structures (i.e., in the acceleration amplification region of the response spectrum):

$$\frac{\mu_s}{R_\mu} = \frac{\mu_s}{\sqrt{2\mu_s - 1}} \geq 1 \qquad (9.23)$$

If a building frame responds as an SDOF system, substitution of Equations 9.22 and 9.23 into Equation 9.21 suggests that the *DAF* should not be less than the *FRF*.

The *DAF*/*FRF* ratios for several steel framing systems designed in accordance with the NEHRP Recommended Provisions are listed in the fifth column ($=C_d/R$) in Table 9.2. Note that the *DAF*/*FRF* ratios are no larger than one.

Table 9.5 compares *DAF*/*FRF* ratios for several codes and provisions: the three seismic provisions in North America, the Mexico Building Code (Seismic 1988), and the 1988 Eurocode (CEC 1988). Because the magnitude of the *DAF* also depends on how much the elastic seismic force has been reduced, and because different seismic codes use different values of *FRF*, a rational comparison should be based on the ratio of *DAF* and *FRF*, not *DAF* alone. Table 9.5 indicates that the *DAF*/*FRF* ratios vary considerably. At one extreme, both the Mexico Building Code and the Eurocode use a *DAF* that is no smaller than the *FRF*. At the other extreme, the 1994 UBC uses a single value of *DAF* that is only 38 percent that of the *FRF*.

Nonlinear analysis of two instrumented steel buildings in California was conducted to evaluate sample *DAF*/*FRF* ratios (Uang and Maarouf 1994). The first was a 13-story steel moment-resisting frame with a measured fundamental period of 2.2 seconds. The second building was a two-story eccentrically braced steel frame with a period of 0.3 second (see Figure 9.3). These two buildings were analyzed using eight historical ground motions. Figure 9.2 shows that the ratio of *DAF* and *FRF* is equivalent to the ratio of the inelastic drift (Δ_{max}) to the elastic drift (Δ_{eu}) because:

$$\frac{DAF}{FRF} = \frac{\mu_s}{R_\mu} = \frac{\frac{\Delta_{max}}{\Delta_y}}{\frac{C_{eu}}{C_y}} = \frac{\frac{\Delta_{max}}{\Delta_y}}{\frac{\Delta_{eu}}{\Delta_y}} = \frac{\Delta_{max}}{\Delta_{eu}} \qquad (9.24)$$

Drifts Δ_{eu} and Δ_{max} were obtained directly from elastic and inelastic dynamic analysis. When the intensity of earthquake input motions was varied, different levels of ductility reduction were realized.

Assuming Δ is the roof drift, Figure 9.4a shows the variations of the average *DAF*/*FRF* ratio. The ratio is significantly higher than the value of 0.375 used in the UBC. The *DAF* will be larger than *FRF* when the ductility reduction factor is large, and when the fundamental

TABLE 9.5 DAF/FRF Ratios of Several Seismic Design Provisions

Building Code	FRF	DAF	DAF/FRF
Uniform Building Code (1994)	R_w	$3R_w/8$	0.375
NEHRP Seismic Provisions (1994)	R	C_d	0.5–1.0
National Building Code of Canada (1990)	$R/0.6$	R	0.6
Mexico Building Code (1987)	$Q^\dagger$	Q	1.0
Eurocode No. 8 (1988)	q	q	1.0

† Less than Q in short period range

period (T) of the structure is short relative to the predominant period (T_g) of the earthquake ground motion, the threshold value of T/T_g is about 0.5.

Assuming Δ is the story drift, the DAF/FRF ratio is no less than 1.0 (see Figure 9.4b). This observation is consistent with the expectation that structural damage in multistory frames tends to concentrate in a limited number of stories.

It appears from the case study that the DAF used in the American and Canadian codes is too small.

9.5 Comparison of service limit state requirements

9.5.1 Introduction

Neither the NEHRP Recommended Provisions nor the NBCC considers the service limit state for moderate earthquakes. The BSL considers the service limit state in its Level 1 design; both the design seismic force level and the limiting story drift ratio (= 0.005) are independent of ductility-related factors. Although the UBC considers only single-level design for severe earthquake shaking, it is interesting to note that the UBC drift limits in Equations 9.9a and 9.9b are significantly lower than those specified in the NEHRP Recommended

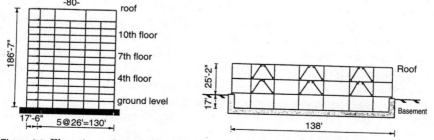

Figure 9.3 Elevations of two instrumented steel building frames. (Maarouf and Uang 1994)

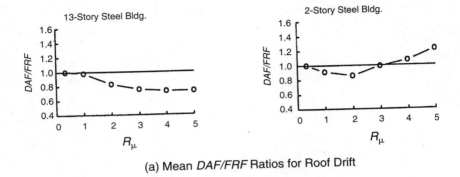

(a) Mean *DAF/FRF* Ratios for Roof Drift

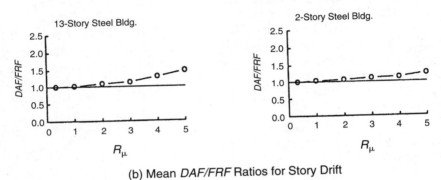

(b) Mean *DAF/FRF* Ratios for Story Drift

Figure 9.4 Relationships between R_μ and mean *DAF/FRF* ratios. *(Maarouf and Uang 1994)*

Provisions and the NBCC, and they are closer in value to the BSL drift limit for Level 1 design. The major difference between the UBC and BSL drift limits is that the UBC drift limits are dependent on R_w. Although the UBC does not contain explicit design provisions for the service limit state, the UBC drift limits have been interpreted by some (e.g. Uang and Bertero 1991) as a means to achieve the service limit state in one-level design procedure. Such an interpretation also provides a basis for the ductility- (or R_w) dependent drift limits that are not found in other codes and provisions.

Because the UBC does not specify an elastic design spectrum for the service limit state, it has been shown that the single-valued drift limit of 0.005 and 0.004 in Equations 9.9a and 9.9b should not apply (Uang and Bertero 1991). Although the UBC drift limit of $0.04/R_w$ or $0.03/R_w$ involves R_w (a factor developed for the ultimate limit state) one simple way to show that these drift limits can be interpreted as a requirement for the service limit state is shown in Figure 9.5.

Limit State Philosophy in Seismic Design Provisions

For example, consider a building with $T < 0.7$ second. Point A in Figure 9.5 represents an optimal UBC design. Note that both the design base shear ratio and story drift limit are functions of R_w. Therefore, the slope of line OA, which is a measure of the building lateral stiffness, is independent of R_w. Accepting a drift limit of 0.005, which is used in the BSL and the 1985 UBC seismic provisions, to control nonstructural damage for moderate earthquakes, it is observed that the corresponding design base shear ratio, C_{es}, at point B should satisfy the following relationship:

$$\frac{C_{es}}{0.005} = \frac{\left(\frac{C_{eu}}{R_w}\right)}{\left(\frac{0.04}{R_w}\right)} = \frac{C_{eu}}{0.04} \qquad (9.25)$$

Solving for C_{es} gives the following:

$$C_{es} = \frac{C_{eu}}{8} \qquad (9.26a)$$

Similarly, it can be shown that for buildings with $T > 0.7$ second, the UBC drift limit of $0.03/R_w$ in Equation 9.9b suggests the following:

$$C_{es} = \frac{C_{eu}}{6} \qquad (9.26b)$$

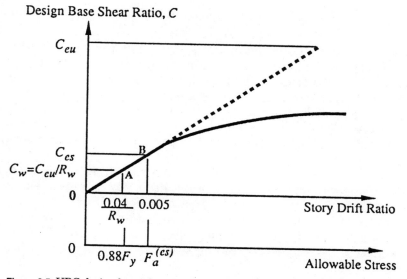

Figure 9.5 UBC design base shear and story drift limit requirement.

Despite its empirical nature, the UBC drift limit ($0.03/R_w$ or $0.04/R_w$) can be interpreted as a means by which to satisfy the service limit state drift requirement for moderate earthquakes. Nevertheless, this approach will not always satisfy the second requirement of the service limit state, which is to minimize structural damage in moderate earthquakes by controlling member forces or stresses. This is explained as follows.

As an example, consider steel design using the AISC ASD Specification (AISC 1989). When the stresses produced by gravity loads are insignificant, the actual allowable stress, $F_a^{(es)}$, at the moderate earthquake level (Equation 9.8a) is:

$$\frac{F_a^{(es)}}{0.88 F_y} = \frac{C_{es}}{C_w} = \frac{\left(\frac{C_{eu}}{8}\right)}{\left(\frac{C_{eu}}{R_w}\right)} = \frac{R_w}{8} \tag{9.27}$$

where $0.88 F_y$ ($= 0.66 F_y$ increased by one-third) is the allowable stress for seismic design. Solving for $F_a^{(es)}$ gives the following:

$$F_a^{(es)} = \frac{R_w}{8}(0.88 F_y) = (0.11 F_y) R_w \tag{9.28a}$$

A similar derivation for buildings with T > 0.7 second gives the following:

$$F_a^{(es)} = \frac{R_w}{6}(0.88 F_y) = (0.15 F_y) R_w \tag{9.28b}$$

Figure 9.6 shows the variations of $F_a^{(es)}$ for different lateral-load-resisting systems (i.e., different R_w values.) It shows that the design approach of the UBC does not provide uniform protection against structural yielding or damage in moderate earthquakes. Under such a low level of ground shaking, about one-sixth to one-eighth as intense as the severe (design) earthquake, excessive yielding might develop in structural members in ductile systems designed using large values of R_w.

The above observations also hold true if strength design is used. Unless reserve strength beyond the prescribed design strength exists because of other design criteria (e.g., gravity-load design criterion, drift limitations, strong-column/weak-beam requirement), it appears that the single-level design procedure used in the UBC does not prevent member stresses from exceeding the yield stress at the service limit state.

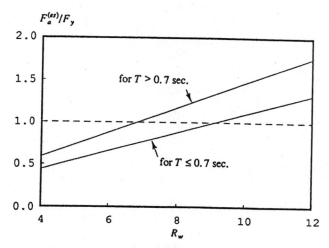

Figure 9.6 Variations of $F_a^{(es)}$ with R_w.

9.6 Future directions

It is well known that a properly proportioned and detailed structure designed for significantly lower seismic forces than those necessary to ensure elastic response can survive strong earthquake shaking without collapse. Ductility and structural overstrength permit design to be based on forces smaller than the elastic forces. Methods to improve the ductility of structural elements of different materials are well developed, and seismic force reduction rules of single-degree-of-freedom systems accounting for ductility have been developed.

Although the contribution of structural overstrength was acknowledged by the writers of the first SEAOC Recommended Lateral Force Requirements, it was not until the early 1980s that the subject of structural overstrength was studied in detail. Through a series of earthquake simulation tests by Bertero (1986) and others of both reinforced concrete and steel frames, it was concluded that structural overstrength was necessary for modern buildings to survive severe earthquake shaking and that the response modification factor (R) could be expressed as the product of a structural overstrength factor ($R_s = \Omega$) and a system ductility factor (R_μ) (Uang and Bertero 1986).

Fischinger and Fajfar (1990) reported values of R_s equal 1.4 to 2.8 for reinforced concrete moment-resisting frames and dual shear wall moment-resisting frame systems designed in accordance with the 1988 NEHRP Recommended Provisions, but the structural overstrength factor remained very close to 1.0 for reinforced concrete shear wall buildings. Figure 9.7 shows the general trend of the structural overstrength and ductility reduction (R_μ) factors. Higher values of R_s in the short period range somewhat compensate for smaller values of R_μ in that range.

400　Chapter Nine

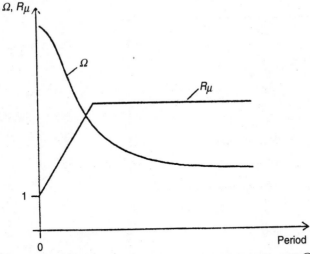

Figure 9.7 Typical qualitative relationships between R_μ and Ω. *(Fischinger and Fajfar 1994)*

Osteraas and Krawinkler (1990) studied structural overstrength (ΩY) of steel framing systems designed in compliance with the UBC working-stress-design provisions. They observed that, as shown in Figure 9.8, perimeter moment-resisting frames have a smaller structural overstrength than do moment-resisting space frames because

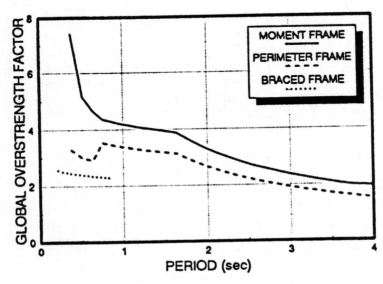

Figure 9.8 Variations of structural overstrength (ΩY) of steel frames. *(Osteraas and Krawinkler 1990)*

gravity load effect actions do not substantially impact the design of perimeter moment-resisting frames. Uang (1991b) reported structural overstrength of 2 to 3 for 4- to 12-story special steel moment-resisting frames. Jain and Navin (1995) reported structural overstrength for reinforced concrete frames designed per the working-stress design procedures of the 1984 Indian code. Figure 9.9 shows that the dependence of structural overstrength on seismic zone is considerable, a finding that is valid for the American practice.

Recently, the ATC-19 report (1995) proposed that R be computed as $R = R_s R_\mu R_R$, where R_R is defined as the redundancy factor. Independently, an ad hoc subcommittee of the SEAOC Seismology Committee was established in late 1993 to develop a new strength design code. Several major changes and new concepts were proposed by the committee, including the introduction of R_o for structural overstrength and R_d for ductility reduction; a structural overstrength factor (Ω_o) to replace $3R_w/8$; a redundancy or reliability factor, ρ; and a new displacement amplification factor to replace $3R_w/8$. These provisions are currently published as an appendix to the 1996 SEAOC Recommended Lateral Force Requirements and Commentary (SEAOC 1996).

For strength design, the design earthquake force, E, is computed as follows:

$$E = \rho E_h + E_v \qquad (9.29)$$

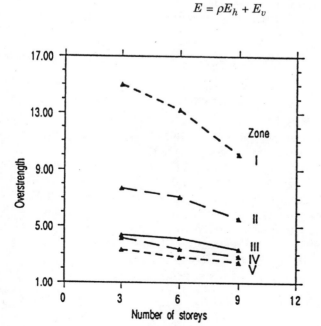

Figure 9.9 Variations of structural overstrength (ΩY) of R.C. frames. *(Jain and Navin 1995)*

where

$$\rho = 2 - \frac{20}{r_{max}\sqrt{A_B}} \quad (9.30a)$$

$$E_v = 0.5 C_a I D \quad (9.30b)$$

$$E_h = \frac{C_v}{RT} W \leq \frac{2.5 C_a}{R} W \quad (9.30c)$$

and

$$R = \frac{R_d R_o}{I} \quad (9.30d)$$

In Equation 9.30, r_{max} is the maximum element-story shear ratio (i.e., the largest value of the ratio of the design shear in an individual element to the design story shear, for all elements in all stories in the lower half of the building). A_B is the ground floor area (in ft^2), C_a and C_v are seismic coefficients dependent on the seismic zone and the soil profile type, I is an importance factor, and D is the dead load. R_d and R_o in Equation 9.30d are numerical coefficients representing the global ductility capacity and the structural overstrength of the lateral-force-resisting system, respectively.

To estimate the maximum story drift in the design earthquake (Δ_m), the elastic story drift (Δ_s) at the design force level is amplified as follows:

$$\Delta_m = 0.7 R_d R_o \Delta_s \quad (9.31)$$

The maximum value of Δ_m is limited to 0.025 of the story height for $T < 0.7$ second and 0.02 otherwise. The importance factor is introduced into Equation 9.30d and not Equation 9.30c to reduce R (i.e., reduce damage) for important structures, not to increase the intensity of the design earthquake. The product of $0.7 R_d R_o$, and not $0.7R$ is used to estimate Δ_m to avoid the cancellation of I when more stringent drift limits for important structures are imposed.

The continued use of $3R_w/8$ in the UBC (or $2R/5$ in the NEHRP Recommended Provisions) as a force multiplier for elements that may fail in a brittle manner (e.g., columns, column splices) has been questioned (e.g., Uang 1993). The following amplified earthquake load was adopted in the 1996 SEAOC Recommended Lateral Force Requirements in an attempt to protect such elements:

$$E_m = \Omega_o E_h \quad (9.32)$$

where $\Omega_o = 1.1R_o$. Although the role of structural overstrength in the formulation of R in Equation 9.17 is clear, it should be noted that a lower bound value for the overstrength ($=R_o$) is used in Equation 9.30d and an upper bound value for the overstrength ($=\Omega_o$) is used in Equation 9.32. The SEAOC recommendations were recently adopted with some changes for the 1997 UBC.

Similar changes to the NEHRP Recommended Provisions were proposed by the NEHRP Provisions Update Committee; this activity started in 1995. It was decided not to split R in the 1997 Recommended Provisions because reliable values for R_d and R_o for different framing systems were unavailable. Although the split-R approach was not adopted, the force multiplier $2R/5$ was replaced by a system overstrength factor, Ω_o. The values of C_d/R in the 1994 NEHRP Recommended Provisions are significantly higher than the value of 3/8 used in the 1994 UBC, yet these values can be smaller than 0.7 (see Equation 9.31). Nonetheless, the Committee decided not to change the values of C_d for the 1997 Recommended Provisions.

9.7 Historical perspective on force reduction factors

Although considerable changes to the practice of seismic design are likely in the next 10 years, it is proper and instructive, in closing this chapter to provide a historical basis for the force reduction factors, R or R_w. The numerical values assigned to those factors by codes for various types of structural systems (e.g., Table 9.2) were not obtained by rigorous analysis and experimentation, but rather by consensus of expert engineers.

The first North American design requirements intended to prevent building collapse during earthquakes originated in California. Interestingly, after a major earthquake struck San Francisco in 1906, reconstruction of the devastated city proceeded with an updated building code that required the consideration of a wind force of 30 pounds per square foot (1.44 kPa) for the design of new buildings (Bronson 1986). No specific earthquake-resistant design clauses were introduced. Given that many building codes of that time did not even have requirements for wind resistance (such as the Los Angeles building code in which wind pressure was not considered in design until 1924), it was hoped that the new "stringent" wind pressure requirement would simultaneously address both wind and earthquake effects.

The 1927 Uniform Building Code introduced the first seismic design requirements in North America in 1927 partly in response to the Santa Barbara earthquake of 1925. This model code proposed clauses for consideration for possible inclusion in the building codes of various cities, at their discretion, and was not binding. The 1927 UBSC proposed that a single horizontal point load, F, equal to 7.5 percent or 10 percent (depending on the soil condition) of the sum of the building's total dead and live load, W, be considered to account for the effect of earthquakes.

Hard soils/rock: $\qquad F = CW = 0.075W \qquad$ (9.33a)

Soft soil: $\qquad F = CW = 0.10W \qquad$ (9.33b)

where C is a seismic coefficient. No justification can be found for these values of C but they likely reflected the consensus of the engineering community. Interestingly, in 1932, Dr. Kyoji Suyehiro of Japan visited California and reported in a series of lectures that buildings designed using a value of C equal to 0.10 in Japan survived the tragic Kanto (Tokyo) earthquake of Richter Magnitude 8.2 in which 140,000 died (Suyehiro 1932).

Enforceable earthquake-resistant design code provisions in North America were implemented following the 1933 Long Beach earthquake of Richter Magnitude 6.3. This earthquake produced damage in Long Beach and surrounding communities in excess of $42 million in 1933 dollars (more than $400 million in 1995 dollars), and the death toll exceeded 120 (Alesch and Petak 1986, Iacopi 1981). It was significant that a large number of the buildings that suffered damage were schools, and that the total number of casualties and injuries would have undoubtedly been considerably larger had this earthquake not occurred at 5:54 P.M., when the schools were fortunately empty. Nonetheless, this economic and physical loss provided the necessary political incentive to implement the first mandatory earthquake-resistant design regulations. The California State Legislature passed the Riley Act and the Field Act, the former requiring that all buildings in California be designed to resist a lateral force equal to 2 percent of their total vertical weight, the latter mandating that all public schools be designed to resist a similar force equal to between 2 percent and 10 percent of the dead load plus a fraction of the live load; the magnitude of the design lateral force depended on the building type and the soil condition. At the same time, a Los Angeles building ordinance was issued, calling for 8 percent of the sum of the dead load plus half of the live load to be used as a design lateral force.

Once researchers brought forth the difference between the dynamic and static response of structures, showing that the seismically induced forces in a flexible (high-rise) building are typically smaller than those in stiff (low-rise) ones, simplified empirical equations to attempt to capture this observed dynamic behavior, and suitable for hand calculations, were developed. The 1943 Los Angeles Building Code was the first to introduce a seismic coefficient and a lateral force distribution that indirectly reflected building flexibility. The lateral forces were calculated as $V = CW$, where V and W were the story shear and total weight of the building above the story under consideration, respectively. The seismic coefficient was calculated as:

$$C = \frac{0.60}{N + 4.5} \tag{9.34}$$

where N is the number of stories above the story under consideration. This formula was slightly modified (SEAOC 1980) when the building height restriction of 13 stories, in effect in Los Angeles in 1943, was removed in 1959.

The 1950s saw the introduction into the lateral force equation of a numerical coefficient, K, intended to reflect the relative seismic performance of various types of structural systems and a more refined consideration of building flexibility through calculation of the fundamental period of vibration, T, of the building in the direction under consideration (Anderson et al. 1952, Green 1981). The generic expression for the base shear became:

$$V = KCW \tag{9.35}$$

where

$$C = \frac{0.05}{T^{1/3}} \tag{9.36a}$$

and

$$T = \frac{0.05H}{\sqrt{D}} \tag{9.36b}$$

where V is the base shear, W is the total dead load and H and D are, respectively, the height of the building and its dimension (in feet) in the direction parallel to the applied forces. The distribution of the base shear along the building height was specified to be inverted-triangular. Types of construction that had been observed to perform better in past earthquakes were assigned low values of K, whereas those that had not performed as well were assigned high values of K. Buildings relying on ductile moment-resisting space frames to resist seismic forces were designed with $K = 0.67$. Buildings with dual structural systems were assigned a value of K equal to 0.8; K for bearing wall systems was set equal to 1.33, and buildings with types of framing systems other than those specified above were assigned a value of K equal to 1.00 (SEAOC 1959). Over time, the equation evolved slightly to include an importance factor, I (equal to 1.0 for normal buildings), a seismic zone factor, Z (equal to 1.0 in the more severe seismic zones), and a soil condition factor varying between 1.0 and 1.5, depending on site conditions. The magnitude of the specified base shear was also increased in 1974, following the San Fernando earthquake of 1971, because many felt that it was too low. This was accomplished by changing the seismic coefficient to the following:

$$C = \frac{1}{15\sqrt{T}} \tag{9.37}$$

Detailed descriptions of the significance of each of the above factors and the way to calculate them, as well as descriptions of the various changes that occurred in seismic codes in the 1960s, 1970s, and some of the 1980s, are available elsewhere (SEAOC 1980, Green 1981, ATC 1995b). However, it is of utmost importance to appreciate that numerical values for K that were introduced into the SEAOC Recommended Lateral Force Requirements in 1959, (and that eventually made their way into other codes worldwide) were based largely on judgment, reflecting the consensus of the SEAOC code committee membership (consisting of expert design professionals and academicians).

A fundamental change in the format of the base shear equation was proposed in 1978 with publication of the ATC-3-06 (ATC 1978) report "Tentative Provisions for the Development of Seismic Regulations for Buildings." That document, prepared by multidisciplinary task groups of experts, proposed new comprehensive seismic provisions that introduced many innovative concepts, among which were the new equations for the seismic coefficients presented in Equations 9.3 and 9.4. The authors of the ATC-3-06 elected not to substantially change the required force levels but rather to concentrate on providing ductile detailing (ATC 1995a). This was a paradigm shift that essentially promoted ductile detailing as a top consideration for design.

Numerical values for R were determined largely by calibration to past practice (ATC 1995a). For example, for ductile steel moment-resisting space frames, equating the proposed ATC equation for base shear at the strength level V_{ATC}, to that in effect at the time, V_{SEAOC}: (SEAOC 1974).

$$V_{SEAOC}\left(\frac{1.67}{1.33}\right) = \frac{V_{ATC}}{0.9} \qquad (9.38)$$

where 1.67 was the typical margin of safety between allowable-stress and ultimate-strength design values, 1.33 accounted for the 33 percent increase in allowable stresses that was permitted by these codes for load combinations involving earthquakes or wind, and 0.9 was the capacity reduction factor for flexure in the context of ultimate strength design. Substituting the respective base shear equations in this expression:

$$(ZIKCS)_{SEAOC} W \left(\frac{1.67}{1.33}\right) = \frac{V_{ATC}}{0.9} \qquad (9.39)$$

Assuming a site in California, a fundamental period of 1.0 second, identical soil conditions for which $S_{SEAOC} = 1.5$ and $S_{ATC} = 1.2$, and using $Z = 1.0, A = 0.4, I = 1.0, T = 1.0$:

$$(1.0(1.0)K\left(\frac{1}{15\sqrt{(1.0)}}\right)(1.5)W\left(\frac{1.67}{1.33}\right) = \frac{1.2(0.4)(1.2)}{0.9R(1.0)^{2/3}} W \qquad (9.40)$$

and:

$$R = \frac{5.1}{K} \qquad (9.41)$$

For a ductile steel moment-resisting frame, K per SEAOC (1960) was 0.67 giving a value of R equal to approximately 8.0 (rounded up from the calculated value of 7.61). Values of R for other types of structural systems (see Table 9.2), were also calculated using Equation 9.38 and adjusted to reflect the consensus of the ATC-3-06 committee members. The ATC-3-06 equations have also been implemented in the Uniform Building Code, with some minor modifications, as Equations 9.6 and 9.7. Hence, through use of a procedure similar to that described above (ATC 1995a), the following relationship is obtained:

$$R_w = \frac{7.86}{K} = \frac{7.86R}{5.1} = 1.54R \approx \frac{8}{K} \qquad (9.42)$$

Reference

1. AISC. 1989. *Allowable Stress Design (ASD) Manual of Steel Construction*. Chicago: American Institute of Steel Construction.
2. AISC. 1993. *Load and Resistance Factor Design Specification for Structural Steel Buildings (LRFC)*. Chicago: American Institute of Steel Construction.
3. Alesch, D.J., and Petak, W.J. 1986. *The Politics and Economics of Earthquake Hazard Mitigation*. Boulder: Institute of Behavioral Science. University of Colorado.

4. Anderson, A.W., and Blume, J.A., et al. 1952. "Lateral Forces of Earthquake and Wind." *Transaction*. Vol. 117. ASCE.
5. ATC. 1995a. *Structural Response Modification Factors*. Report No. ATC-19, Applied Technology Council. Redwood City, CA.
6. ATC. 1995b. *A Critical Review of Current Approaches to Earthquake Resistant Design*. Report No. ATC-34, Applied Technology Council. Redwood City, CA.
7. Bertero, V.V. 1986. *Implications of Recent Earthquakes and Research on Earthquake-Resistant Design and Construction of Buildings*. Report No. UCB/EERC-86/03. Berkeley: Earthquake Engineering. Research Center, University of California.
8. Bronson, W. 1986. *The Earth Shook, the Sky Burned—A Moving Record of America's Great Earthquake and Fire*. San Francisco: Chronicle Book.
9. BSSC. 1994. *NEHRP Recommended Provisions for the Development of Seismic Regulations for New Buildings*. Washington, DC: Building Seismic Safety Council.
10. CEC. 1988. *Eurocode No. 8: Structures in Seismic Regions*. EUR 12266 EN. Luxembourg: Commission of the European Communities.
11. Fischinger, M., and Fajfar, P. 1990. *On the Response Modification Factors for Reinforced Concrete Buildings*. Proc. 4th U.S. National Conference on Earthquake Engineering., Vol. 2: 249–258. Palm Springs, CA: Earthquake Engineering Research Institute.
12. Fischinger, M., and Fajfar, P. 1994. "Seismic Force Reduction Factors," *in Earthquake Engineering*, A. Rutenberg (editor), Balkema, pp. 279–296.
13. Freeman, S.A. 1985. "Drift Limits: Are They Realistic?" *Earthquake Spectra*. Vol. 1, No. 2: 355–362. Earthquake Engineering Research Institute.
14. Green, N.B. 1981. *Earthquake Resistant Building Design and Construction*. New York: Van Nostrand Reinhold.
15. Iacopi, R. 1981. *Earthquake Country—How, Why, and Where Earthquakes Strike in California*. Menlo Park, CA: Sunset Books, Lane Publishing.
16. IAEE. 1988. *Earthquake Resistant Regulations—A World List*. Tsukuba, Japan: IAEE.
17. ICBO. 1994. *Uniform Building Code (UBC)*. Whittier, CA: International Conference of Building Officials.
18. Jain, S.K., and Navin, R. 1995. "Seismic Overstrength in Reinforced Concrete Frames." *Journal of Structural Engineering*. Vol. 121, No. 3: 580–585. ASCE.
19. Kato, B. 1986. *Seismic Design Criteria for Steel Buildings*. Proc. Pacific Structural Steel Conference. Vol. 1: 133–147. Auckland, New Zealand.
20. Kato, B., and Akiyama, H. 1982. "Seismic Design of Steel Buildings." *Journal of Structural Engineering*. Vol. 108, No. ST8, 1709–1721. ASCE.
21. Miranda, E., and Bertero, V.V. 1994. "Evaluation of Strength Reduction Factors for Earthquake-Resistant Design." *Earthquake Spectra*. Vol. 10, No. 2: 357–380. EERI.
22. NBC. 1990. *National Building Code of Canada (NBCC)*. National Research Council. Ontario, Canada.
23. Newmark, N.M., and Hall, W.J. 1982. *Earthquake Spectra and Design*. Earthquake Engineering Research Institute.
24. Osteraas, J., and Krawinkler, H. 1990. *Seismic Design Based on Strength of Structures*. Proc. 4th U.S. National Conference on Earthquake Engineering. Vol. 2: 955–964. Palm Springs, CA: Earthquake Engineering Research Institute.
25. SEAOC. 1959, 1990, 1996. *Recommended Lateral Force Requirements and Commentary*. 5th ed. San Francisco: Structural Engineers Association of California.
26. "Seismic Design Regulation of the 1976 Mexico Building Code." 1988. *Earthquake Spectra*. Vol. 4, No. 3: 427–439. Earthquake Engineering Res. Institute. (Translated by Garcia-Ranz and R. Gomez.)
27. Suyehiro, K. 1932. "Engineering Seismology—Notes on California Lectures." Vol. 58: No. 4. ASCE.
28. Uang, C.-M. 1991a. "A Comparison of Seismic Force Reduction Factors Used in U.S.A. and Japan." *Earthquake Engineering and Structural Dynamics*, Vol. 20, No. 4: 389–397.
29. Uang, C.-M. 1991b. "Establishing R (or R_w) and C_d Factors for Building Seismic Provisions." *Journal of Structural Engineering*. Vol. 117, No. 1: 19–28. ASCE.

30. Uang, C.-M. 1993. "An Evaluation of Two-Level Seismic Design Procedure." *Earthquake Spectra*. Vol. 9, No. 1, 121–135. EERI.
31. Uang, C.-M., and Bertero, V.V. 1986. *Earthquake Simulation Tests and Associated Studies of a 0.3-Scale Model of a Six-Story Concentrically Braced Steel Structure*. Report No. UCB/EERC-86/10. Berkeley. Earthquake Engineering Resource Center. University of California.
32. Uang, C.-M., and Bertero, V.V. 1991. "UBC Seismic Serviceability Regulations: Critical Review." *Journal of Structural Engineering*. Vol. 117, No. 7: 2055–2068. ASCE.
33. Uang, C.-M., and Maarouf, A. 1994. "Deflection Amplification Factor for Seismic Design Provisions." *Journal of Structural Engineering*. Vol. 120, No. 8: 2423–2436. ASCE.

Chapter 10

Stability and Rotation Capacity of Steel Beams

10.1 Introduction

A basic assumption made for the plastic design of framed structures is that flexural members shall have sufficient plastic deformation capacities while maintaining the plastic moment, M_p. Without sufficient plastic deformation capacity at the member level, the yield (or collapse) mechanism resulting from sequential plastification in the structure cannot be developed. The need for substantial deformation capacity is even greater for seismic design because it is expected not only that the yield mechanism will form, but also that the structure may displace beyond the point of incipient collapse (see point A in Figure 10.1) and cycle back and forth under severe earthquake shaking.

Although steel material of nominal yield strength not exceeding 65 ksi (448 MPa) usually exhibits high ductility, the plastic deformation capacity of a member is usually limited by instability. Instability in flexural members includes flange local buckling (FLB), web local buckling (WLB), and lateral-torsional buckling (LTB). For a given grade of structural steel, the vulnerability of each of the two local buckling modes can be measured in terms of the width-thickness ratio, b/t; the tendency for LTB is a function of the slenderness ratio, L_b/r_y, and the moment gradient. The AISC LRFD Specification (1994) defines the slenderness ratio, λ, of each buckling mode as follows:

$$\text{FLB} : \lambda = b_f/2t_f$$
$$\text{WLB} : \lambda = h_c/t_w$$
$$\text{LTB} : \lambda = L_b/r_y$$

where b_f is the flange width, t_f is the flange thickness, h_c is the clear depth between the flanges less twice the fillet radius, t_w is the web thickness, h_b is the laterally unbraced length, and r_y is the radius of gyration about the y-y axis.

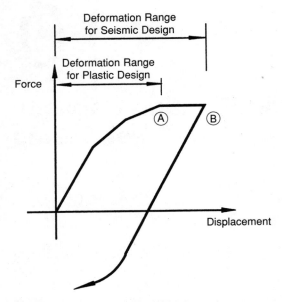

Figure 10.1 Schematic structure deformation ranges.

The AISC LRFD Specification presents three limiting slenderness ratios (λ_{pd}, λ_p, and λ_r) to classify each of the above three buckling limit states into four categories. Figure 10.2a shows this classification. If the slenderness ratio, λ, is larger than λ_r, the flexural member buckles in the elastic range (see curve 4 in Figure 10.2b). The flexural capacity is lower than the plastic moment, and the member does not exhibit ductile behavior. A wide-flange section having either a flange or a web width-thickness ratio larger than λ_r is called a *slender* section in the AISC LRFD Specification. For values of λ between λ_r and λ_p, the member buckles in the inelastic range (see curve 3 in Figure 10.2b), and the

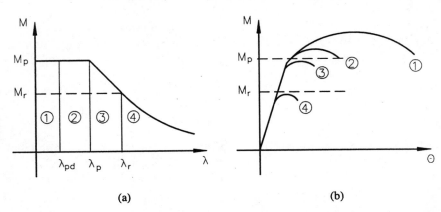

Figure 10.2 LRFD classification of flexural buckling limit states. *(Adapted from Yura et al. 1978)*

flexural capacity exceeds the elastic moment, $M_r = (F_y - F_r)S_x$, where F_y is the yield stress, F_r is the residual stress and S is the elastic section modulus. A wide flange section with flange or web width-thickness ratios falling between λ_r and λ_p is called a *noncompact* section. Flexural members with λ larger than λ_p should not be used as yielding members for either plastic or seismic design.

The plastic moment, M_p, can be reached when λ is smaller than λ_p. However, to ensure sufficient rotational capacity, λ must be further limited to λ_{pd}. The AISC LRFD Specification provides values for λ_{pd} and λ_p for the LTB limit state. If plastic design is used, the designer must limit the unbraced length (L_b) such that λ does not exceed λ_{pd}. For the FLB and WLB limit states, the AISC LRFD Specification merges λ_p and λ_{pd} into a single limiting parameter, λ_p, and λ_{pd} is not specified for these two limit states. A wide flange section whose flange and web width-thickness ratios do not exceed λ_p is called a *compact* section in the AISC LRFD Specification, and such a section is expected to be able to develop a rotation capacity of at least 3 before the onset of local buckling (Yura et al. 1978), where the rotation capacity, R, is defined as follows:

$$R = \frac{\theta_h}{\theta_p} \quad (10.1)$$

See Figure 10.3 for the definitions of θ_h and θ_p.

The inelastic deformation demand for seismic applications is typically greater than that for plastic design. Therefore, both the Uniform Building Code (ICBO 1994) and AISC Seismic Provisions (AISC1994)

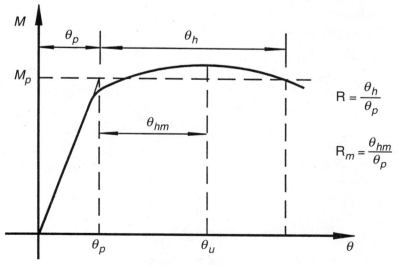

Figure 10.3 Moment-rotation relationship and definition of rotation capacity, R.

further reduce the limiting slenderness ratios, as shown in Table 10.1 such that the expected rotation capacity is about 7 to 9 (AISC 1994).

The requirement of sufficient inelastic rotation capacity is indirectly achieved when the slenderness ratio for the three instability conditions is limited. The theoretical background to these slenderness limits is presented in this chapter. It will become obvious from the presentation below that the local buckling and lateral buckling limit states are interrelated and that the treatment of this subject by code documents that assume uncoupled behavior is simplistic.

10.2 Plate elastic and post-elastic buckling behavior

A wide flange section can be viewed as an assemblage of flat plates. The flange plate is an *unstiffened* element because one edge is free (i.e., not supported). The web plate is a *stiffened* element because it is connected at two ends to the flange plates. Before the inelastic buckling behavior of wide-flange beams for plastic or seismic design is described, it is worthwhile to review briefly the elastic buckling and post-buckling behaviors of plates. Note that the postbuckling behavior of a plate is very different from that of a column because a column loses strength (with increasing deformation) once the buckling load is reached, whereas a plate can develop significant postbuckling strength beyond its theoretical critical load because of stress redistribution and transverse tensile membrane action.

The mathematical solution to the elastic buckling problem for a perfect thin plate is well known (Timoshenko and Gere 1961). The small-deflection theory assumes that the deflection is less than the thickness of the plate, the middle surface of the plate does not stretch during bending (i.e., no membrane action), and plane sections remain

TABLE 10.1 AISC/LRFD Slenderness Requirements

Limit States	Plastic Design	Seismic Design
FLB	$65/\sqrt{F_y}$ (U.S. units)	$52/\sqrt{F_y}$ (U.S. units)
	$170/\sqrt{F_y}$ (S.I. units)	$136/\sqrt{F_y}$ (S.I. units)
WLB	$640/\sqrt{F_y}$ (U.S. units)	$520/\sqrt{F_y}$ (U.S. units)
	$4413/\sqrt{F_y}$ (S.I. units)	$1365/\sqrt{F_y}$ (S.I. units)
LTB	$\dfrac{5000+3000(M_1/M_2)}{F_y}$ (U.S. units)	$\dfrac{2500}{F_y}$ (U.S. units)
	$\dfrac{34475+20685(M_1/M_2)}{F_y}$ (S.I. units)	$\dfrac{17238}{F_y}$ (S.I. units)

Note: M_1 and M_2 are smaller and larger moments at the unbraced beam ends, respectively; M_1/M_2 is positive when moments cause reverse curvature.

plane. The governing differential equation for an isotropic plate (see Figure 10.4) without initial geometrical imperfections subjected to x-direction loading per unit length, N_x ($=\sigma_x t$), is as follows:

$$\frac{\partial^4 w}{\partial x^4} + 2\frac{\partial^4 w}{\partial x^2 \partial y^2} + \frac{\partial^4 w}{\partial y^4} = \frac{1}{D} N_x \frac{\partial^2 w}{\partial x^2} \qquad (10.2)$$

where D is the flexural rigidity of the plate:

$$D = \frac{Et^3}{12(1-v^2)}$$

and where E is Young's modulus, v is Poisson's ratio (= 0.3 for steel in the elastic range), σ_x is the imposed stress, and t is the plate thickness.

The elastic critical buckling stress, obtained from solving Equation 10.2, is as follows:

$$\sigma_{cr} = k \frac{\pi^2 E}{12(1-v^2)(b/t)^2} \qquad (10.3)$$

where k is the plate buckling coefficient. For a plate simply supported on all four sides, the coefficient k is given by:

$$k = \left(\frac{1}{m}\frac{a}{b} + m\frac{b}{a}\right)^2 \qquad (10.4)$$

where m represents the number of half-waves that occur in the loaded direction of the plate, and a/b is the plate aspect ratio. The variation of k with respect to m and a/b is shown in Figure 10.5. For a long, narrow plate like the flange or web of a beam, the aspect ratio is large and the minimum value of k for given boundary conditions can be used. For example, a beam flange is similar to a plate with one simply

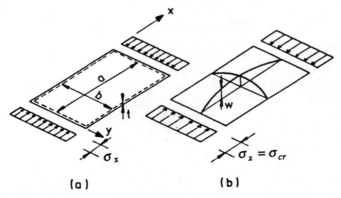

Figure 10.4 Simply supported plate. *(From Maquoi 1992 with permission)*

supported edge (at the web) and one free edge; the corresponding minimum value of k from Figure 10.5 is 0.425. Likewise, the web of a beam is somewhat restrained by the two flanges, and its unloaded edges can be treated as being semirestrained; that is, restraint somewhere between simply supported and fully fixed. The value of k can be taken as 5 in this case. This comparison of the value of k for flange (=0.425) and web (=5) buckling highlights the vulnerability of flanges to local buckling.

Prior to elastic buckling, compressive stresses are uniformly distributed along the edges of a perfectly flat plate. Once the critical buckling stress is reached, the plate still exhibits significant stiffness, and the strength can be increased further. The plate postbuckling strength is due to the redistribution of axial compressive stresses and tensile membrane action that accompanies the out-of-plane bending of the plate in both the longitudinal and transverse directions. Figure 10.6

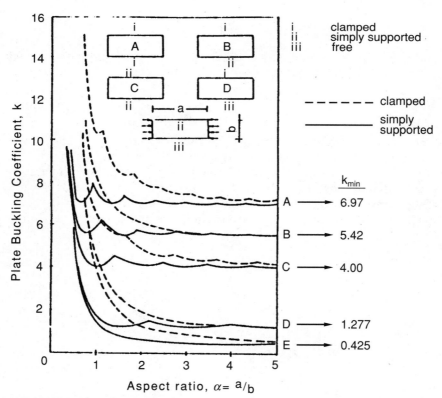

Figure 10.5 Plate elastic buckling coefficient, k. *(From Gerard and Becker 1957 with permission.)*

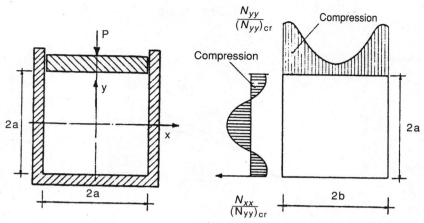

Figure 10.6 Stress redistribution of post-buckling plate. *(From Bazant and Cedolin 1991 with permission)*

shows that the distribution of compressive stresses in the loading direction is no longer uniform after plate buckling and that the compressive stresses tend to concentrate in those portions of the plate width close to the supported edges because these portions are the stiffest in a buckled plate. Such a nonuniform distribution of the stresses can be simplified through the effective width concept proposed by Von Karman et al. (1932).

Based on an energy method originally developed by Marguerre and Trefftz (1937), the longitudinal stiffness of an elastic plate after buckling has been derived by several researchers; summary data is presented in Bulson (1969). Figure 10.7 shows three cases, where ε_{cr} is the longitudinal strain corresponding to a stress level σ_{cr}, and σ_{av} is the mean value of the longitudinal membrane stresses for a given longitudinal strain ε_1, in the postbuckling range. Contrary to the case of axially loaded columns, where the postbuckling stiffness is negative, the elastic postbuckling stiffness of a plate is positive.

Out-of-plane imperfections always exist in actual plates and assemblies of plates. Figure 10.8 compares the analytically predicted response of a perfect plate and test results, both for a plate with plan dimensions of a and b. The main effects of geometric imperfections are the elimination of a well-defined buckling load and larger out-of-of-plane deflections. Nevertheless, the effect of initial geometrical imperfections on the longitudinal stiffness after plate buckling is insignificant (see Figure 10.9). Note that δ_o in Figure 10.9 is the initial deflection (i.e., imperfection) at the center of the plate.

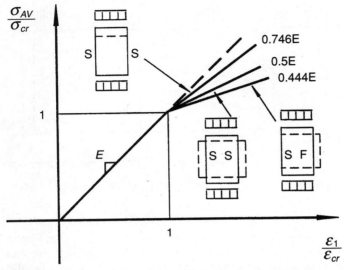

Figure 10.7 Postbuckling stiffness of loaded plate. *(Adapted from Bulson 1969)*

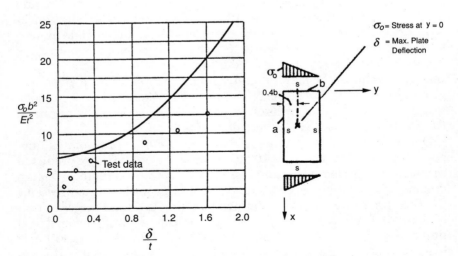

Figure 10.8 Test results and theoretical predictions for an axially loaded plate, a = 2b *(Adapted from Bulson 1969)*

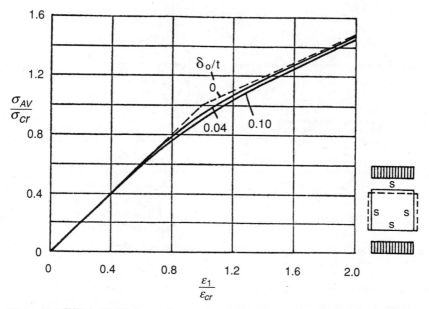

Figure 10.9 Effect of initial out-of-plane imperfection on plate post-buckling stiffness *(From Bulson 1969 with permission)*

10.3 General description of inelastic beam behavior

10.3.1 Beams with uniform bending moment

Figure 10.10a shows the typical in-plane moment versus deformation relationship of a beam under uniform moment about its strong axis. The moment-end rotation curve consists of four parts:

- The elastic range (portion OA) in which the M-θ relation is linear
- The contained plastic flow region (portion AB) in which the curve becomes nonlinear because of partial yielding
- The plastic plateau (portion BC) in which the beam is fully yielded and is not capable of resisting additional moment
- The unloading region (portion CD) in which the beam becomes unstable

Nevertheless, the in-plane behavior does not completely describe the beam behavior because of simultaneous lateral deflections. As shown in Figure 10.10b, the beam starts to deflect laterally following application of load, and the deflection becomes pronounced after the compression flange begins to yield. Lateral-torsional buckling occurs just after the

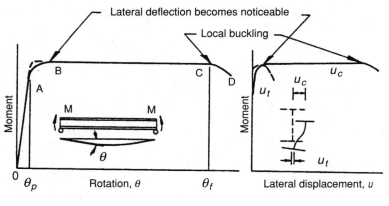

Figure 10.10 Typical beam behavior under uniform moment. *(From ASCE-WRC 1971 with permission)*

plastic moment is achieved. Despite significant lateral deflection perpendicular to the plane of loading, however, the beam is able to maintain its plastic moment capacity until local buckling occurs. During the entire loading process, the tension flange moves laterally by a only relatively small amount. Thus, the web is distorted because of the relative lateral displacement of compression and tension flanges. The lateral movement of the compressive flange also causes additional compressive strains to half of the beam flange as a result of flange bending about its strong axis. This movement may trigger flange local buckling and causes the beam's flexural strength to degrade.

When a member is under a uniform applied stress due to an axial compression force or uniform bending, the normal stress tends to magnify any lateral deflections. Even when lateral deflections are on the order of only 1 percent of the flange width, *bending yield planes* will occur at midspan, and local buckling will develop in that fully yielded region (Lay 1965). Sufficient lateral bracing must be provided for purely axial yielding to occur without any bending yield planes.

The difference in behavior between a simply supported beam and a continuous beam is important. For a continuous beam, adjacent spans may offer significant warping restraint at the ends of the buckled span, which can lengthen the plastic plateau and result in a larger plastic rotation. Tests conducted by Lee and Galambos (1962) have shown that plastic rotation capacity increases as the length of the unbraced span decreases. All the beams tested showed considerable postbuckling strength, and in each case, a plastic hinge of sufficient rotation capacity was developed. Lee and Galambos reported that if the unbraced length of a beam is greater than $45r_y$, the beam will likely fail because of lateral-torsional buckling. Otherwise, beam failure is instigated by flange local buckling. Note that the postbuckling strength observed in these tests was mainly a result of the lateral restraint provided by the adjacent elastic spans.

10.3.2 Beams with moment gradient

Moments in beams in buildings generally vary along the span, that is, beams are typically subjected to a moment gradient. Figure 10.11 shows the deformation behavior of a simply supported beam with a concentrated load at the midspan (Lukey and Adams 1969). From Figure 10.11a, it can be seen that this beam behaves elastically up to load point 4. Yielding starts to occur because of the presence of residual stresses beyond that point, which causes a slight reduction in stiffness. The rotation increases rapidly once the plastic moment, M_p, is exceeded. The moment continues to increase beyond M_p, and local buckling is observed at load point 8. Local buckling does not cause a degradation in strength. Instead, the moment increases with increasing rotation up to load point 10 before the strength starts to degrade. Figure 10.11b shows the corresponding lateral deflection of the compression flange. This lateral movement becomes pronounced only after load point 8 and increases rapidly after load point 10.

Under a moment gradient, yielding of the beam is confined to the region adjacent to the location of maximum moment, and it cannot spread along the length of the beam unless the moment is increased. However, as soon as M_p is reached, the steel strain-hardens, and the load can be further increased. This causes the midspan moment to increase and the yielded region to spread. Local buckling will start only when the compression flange has yielded over a length sufficient to accommodate a full wavelength of the buckle shape (see Section 10.5). Note that the applied moment can still be increased after local buckling and that strength degradation will not occur until lateral-torsional buckling develops.

The lateral deflection of the compression flange typically increases rapidly after M_p is exceeded; the deflected shape of the flange also changes, and curvatures producing lateral displacements are concentrated mainly in a relatively small region at midspan where the compression flange has yielded and has reduced stiffness. Post-elastic deformations will also concentrate at this location. Thus, it is not local buckling but rather lateral-torsional buckling that causes a loss in strength. After local buckling, the lateral stiffness of the compression flange is greatly reduced and lateral-torsional buckling is triggered.

The point of load application has a profound impact on the resistance to lateral-torsional buckling (Timoshenko and Gere 1961). When the load is applied close to the compression flange, the critical value of the load decreases. If it is applied to the tension flange, the critical value of the load will be much larger. For many generic solutions presented in the literature, loads are assumed to be applied at the centroid of the cross section. The effect of the load application level on the lateral-torsional buckling resistance is shown in Figure 10.12.

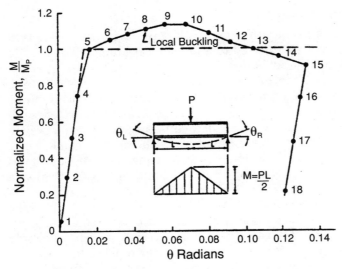

(a) Moment vs. Rotation Relationship

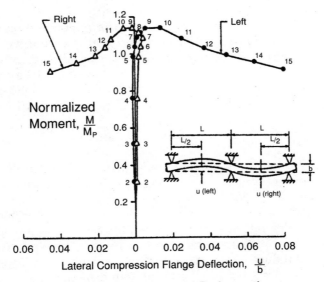

(b) Moment vs. Lateral Deformation Relationship

Figure 10.11 Typical beam behavior under moment gradient. *(From Lukey and Adams 1969 with permission.)*

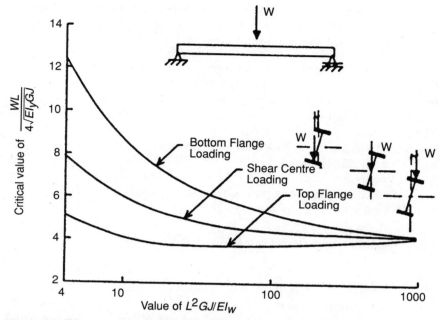

Figure 10.12 Effect of load position on lateral-torsional buckling load for parameters expressed in U.S. units only. *(From Kirby and Nethercot 1979 with permission)*

10.3.3 Comparison of beam behavior under uniform moment and moment gradient

The differences in behavior between beams with uniform bending moment and moment gradient can be summarized as follows (Galambos 1968):

- For beams subjected to uniform moment, the moment remains constant at M_p until the average strain in the compressive flange reaches the strain-hardening strain, ε_{st}, along the entire region of uniform moment. Only then can the steel strain-harden and the moment exceed M_p. In contrast, yielding in a beam under moment gradient cannot spread unless the moment is increased; therefore, strain-hardening occurs as soon as M_p is reached.

- For a beam segment under uniform moment, the unbraced compressive flange deflects laterally as soon as M_p is reached, primarily as a result of the initial out-of-straightness of the beam flange. Lateral-torsional buckling usually precedes local buckling, and moment strength is not lost until local buckling develops.

- For a beam under a moment gradient, the moment at the critical section will increase because of strain-hardening and will continue to increase until the yielded length of the compressive flange is equal to a full local buckling wavelength. The beam may still resist additional load because loss of strength will not occur until lateral-torsional buckling develops. That is, flange local buckling usually precedes lateral-torsional buckling.

Inelastic buckling occurs at an average strain of an order of magnitude larger than the yield strain. The combined phenomena of yielding, strain-hardening, in-plane and out-of-plane deformations, and local distortion all occur soon after the compression flange has yielded. They interact with each other such that individual effects can be only roughly distinguished. Residual stresses and initial out-of-straightness also impact the load-displacement relation for even the simplest beam. Because it is difficult to take each of these factors into consideration, modern steel design codes usually treat local buckling and lateral-torsional buckling separately.

10.4 Inelastic flange local buckling

10.4.1 Modelling Assumptions

For a beam under moment gradient, the yielded region will be subjected to moments of M_y or greater, where M_y $(= S\sigma_y)$ is the yield moment, S is the elastic section modulus, and σ_y is the actual yield stress (and not the nominal yield stress, F_y). The maximum (flange) strains exceed $s\varepsilon_y$ because any moment larger than M_y requires strains above $s\varepsilon_y$ (see Chapter 2). Therefore, any yielded portion of a beam flange under a moment gradient is in the fully yielded condition $(\varepsilon \geq s\varepsilon_y)$.

For a uniform moment loading condition, if all stresses in the beam are beyond the yield stress, σ_y, the material is fully yielded. However, there exists a range of partly yielded conditions at σ_y for which, according to the slip-plane theory, parts of the member are at strain ε_y and other parts have yielded and are at strain $s\varepsilon_y$. Lay (1965) has shown that even when the lateral deflections are on the order of only one-hundredth of the flange width, bending yield planes develop at midspan. This lateral movement produces a fully yielded condition (i.e., all strains at $s\varepsilon_y$) at midspan, and local buckling may occur in that fully yielded part of the member.

A physical model for flange local buckling analysis is shown in Figure 10.13. It assumes that the flange plate is subjected to a uniform stress, F_y, over the entire flange area and that the web restrains the flange at the web-to-flange junction. The flange local buckling problem is thus reduced to a classical buckling problem. The assumed cross-sectional shape before and after buckling is shown in Figure 10.14, where only the effect of local buckling on the distorted shape is shown.

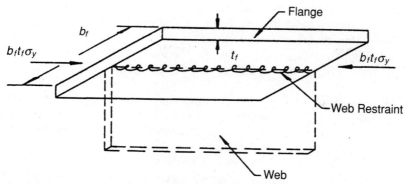

Figure 10.13 Compression flange local buckling model. *(From ASCE-WRC 1971 with permission)*

When a wide-flange steel beam is deformed well into the inelastic range, local buckling occurs and strength eventually degrades. Local buckling is either preceded or followed by lateral deflection. It is difficult to capture exactly these complex behaviors. Two approaches have been taken in the past: a solution based on buckling of an orthotropic plate and a solution based on the torsional buckling of a restrained rectangular plate. For both solutions, it is assumed that the flange is strained uniformly to its strain-hardening value, ε_{st}, and that the material has reached its strain-hardening modulus, E_{st}, that is, the plate buckles at the onset of strain-hardening. This assumption may only be reasonable for the case of uniform bending.

10.4.2 Buckling of an Orthotropic Plate

The solution based on the buckling of an orthotropic plate was presented by Haaijer and Thurlimann (1958). It assumes that the material is isotropic and homogeneous in the elastic range, that it changes during the yielding process with yielding taking place in slip bands, and that the strains in those bands jump from the value at yield ε_y to

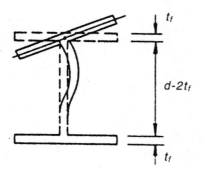

Figure 10.14 Deformed shape of cross section after local buckling. *(From Lay 1965 with permission)*

$s\varepsilon_y$ at the onset of strain-hardening. Because bands of elastic and yielded material coexist during yielding, the material is considered to be heterogeneous. After all the material in the yielded zone has been strained to ε_{st}, the material regains its homogeneous properties in the loading direction. However, slip produces changes in the composition of the material so that the material is no longer isotropic; that is, its properties are now direction-dependent. Therefore, to explain and predict the behavior of steel members under compression in the yielded and strain-hardening ranges, this condition of orthotropy must be recognized.

For a strain-hardened rectangular plate subjected to uniform compression (Figure 10.15), one can calculate the critical stress by solving the bifurcation equilibrium problem (i.e., solving the equilibrium equations formulated in the buckled configuration). When the supported edge in the direction of loading is hinged, the critical stress is calculated and expressed by the following:

$$\sigma_{cr} = \left(\frac{t}{b}\right)^2 \left[\frac{\pi^2}{12} D_x \left(\frac{b}{a}\right)^2 + G_{st}\right] \qquad (10.5)$$

where G_{st} is the strain-hardening modulus in shear, D_x is equal to $E_x/(1 - v_x v_y)$, and v_x and v_y are Poisson's ratios for stress increments in the x- and y- directions. For a long plate, the first term can be neglected.

When the supported edge in the loading direction in Figure 10.15 is changed to a fixed support, the buckling occurs when the aspect ratio is as follows:

$$\frac{a}{b} = 1.46 \sqrt[4]{\frac{D_x}{D_y}} \qquad (10.6)$$

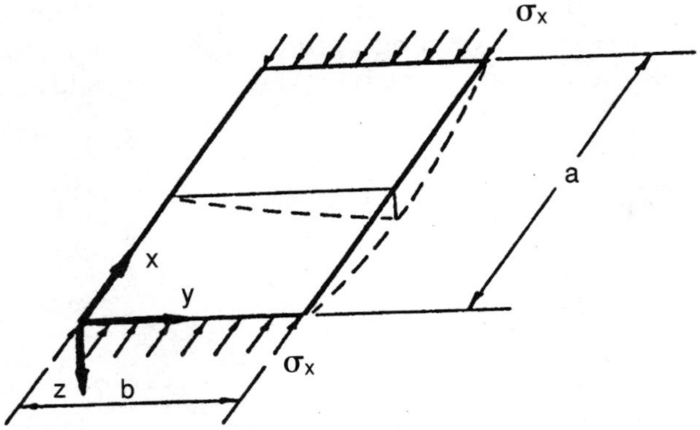

Figure 10.15 Unstiffened plate with hinged support at loaded edge. *(Haaijer and Thurlimann 1958)*

and the minimum buckling stress is:

$$\sigma_{cr} = \left(\frac{t}{b}\right)^2 \left[0.769\sqrt{D_x D_y} - 0.270(D_{xy} + D_{yx}) + 1.712 G_{st}\right] \quad (10.7)$$

where $D_{xy} = v_y D_x$ and $D_{yx} = v_x D_y$. Based on the tests of ASTM A7 steel, Haaijer (1957) derived the following numerical values for the five key coefficients: $D_x = 3{,}000$ ksi (20,685 MPa), $D_y = 32{,}800$ ksi (226,156 MPa), $D_{xy} = D_{yx} = 8{,}100$ ksi (55,850 MPa), and $G_{st} = 2400$ ksi (16,548 MPa).

In the case of a long plate with zero rotational restraint from the web, Equation 10.5 gives the following:

$$\sigma_{cr} = \left(\frac{2t_f}{b_f}\right)^2 G_{st} \quad (10.8)$$

When the flange plate is uniformly strained to the strain-hardening value, ε_{st}, and σ_{cr} is thus equal to σ_y, the plate width-thickness ratio at which local buckling of a perfect plate occurs is as follows:

$$\frac{b_f}{2t_f} = \sqrt{\frac{G_{st}}{\sigma_y}} \quad (10.9)$$

For A7 steel with a nominal yield stress of 33 ksi (228 MPa), the above equation gives $b_f/2t_f = \sqrt{2400/33} = 8.5$. If web rotational restraint is considered, a slightly different result will be obtained; based on experimental results, the limiting $b_f/2t_f$ ratio can be increased by about 9 percent, and the $b_f/2t_f$ limit becomes 9.3. Thus, Haaijer and Thurlimann recommended flange width-to-thickness limits of $b_f/2t_f \leq 8.7$ for $\sigma_y = 33$ ksi (228 MPa) and $b_f/2t_f \leq 8.3$ for $\sigma_y = 36$ ksi (248 MPa).

Note that Haaijer and Thurlimann's solution involved the determination of five material constants and a web restraint factor that had to be calculated empirically from extensive experimental work. Because their solution was applicable to only A7 steel and did not form a sufficiently complete basis for extrapolation to other steel grades, an alternative procedure was needed. The second method, proposed by Lay (1965) and presented below, requires two material constants and is applicable to different steel grades.

10.4.3 Torsional buckling of a restrained rectangular plate

It is assumed in the approach based on the torsional buckling of a restrained rectangular plate that flange local buckling occurs when the average strain in the plate reaches the strain-hardening value, ε_{st}, and that a long enough portion of the plate has yielded to permit the development of a full buckled wave. For a beam under a moment gradient, one end of the yielded region will be adjacent to a relatively stiff elastic zone, and the other end may be adjacent to a load point or connection. Both end conditions likely provide relatively stiff end restraint. In such a case, it is necessary for a longitudinal full wave length to have yielded to develop local buckling. See Figure 10.16 for the required yielded length for local buckling under uniform moment and moment gradient. Figure 10.17 shows compression flange strains measured at the onset of local buckling in tests of four beams under uniform moment. Local buckling occurs upon full yielding over half of the flange width.

The beam flange local buckling problem is solved by analogy to torsional buckling of a column. Indeed, if the flange of a wide-flange section is assumed to be unrestrained against local buckling by the web, it may be treated as a simply supported column under axial compression. Torsional buckling of such a simply supported column is described by the following equation (Bleich 1952):

$$E_{st}C_w \frac{d^4\beta}{dz^4} + (\sigma I_p - G_{st}J)\frac{d^2\beta}{dz^2} = 0 \tag{10.10}$$

where β is the torsional angle of the member, I_p ($= b_f^3 t_f/12$) is the polar moment of inertia of the flange, C_w ($= 7b_f^3 t_f^3/2304$) is the warping constant of the compression flange, J ($= b_f t_f^3/3$) is the St. Venant torsional constant for the flange, and σ is the applied compressive stress. Because warping strains occur in the loading direction, the associated strain-hardening modulus, E_{st}, must be used. Assuming $\beta = C\sin\frac{n\pi z}{L}$ and substituting this into the above equation, gives this nontrivial solution:

$$E_{st}C_w \left(\frac{n\pi}{L}\right)^4 + (G_{st}J - \sigma I_p)\left(\frac{n\pi}{L}\right)^2 = 0 \tag{10.11}$$

which can be rearranged to obtain the buckling stress:

$$\sigma_{cr} I_p = E_{st}C_w \left(\frac{n\pi}{L}\right)^2 + G_{st}J \tag{10.12}$$

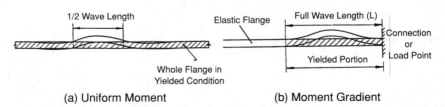

Figure 10.16 Required yielded length for local buckling. *(From Lay 1965 with permission)*

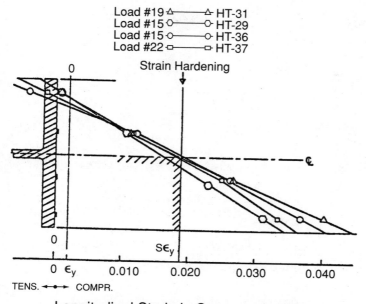

Figure 10.17 Strain distribution at local buckling for beams under uniform moment. (From Lay 1965 with permission)

where n is an integer and L/n is the half-wave length of a buckle. Ignoring the contribution from the warping resistance, which appears as the first term on the right-hand side of the equation, one can simplify Equation 10.12 as follows:

$$\sigma_{cr} = \frac{G_{st}J}{I_p} = \frac{G_{st}}{\left(\dfrac{b_f}{2t_f}\right)^2} \tag{10.13}$$

Setting $\sigma_{cr} = \sigma_y$ because the flange is fully yielded when inelastic buckling occurs, and solving for the limiting flange width-to-thickness ratio, gives the following:

$$\frac{b_f}{2t_f} = \sqrt{\frac{G_{st}}{\sigma_y}} \tag{10.14}$$

which is identical to that derived by Haaijer (see Equation 10.9).

Figure 10.14 shows the assumed buckling configuration, neglecting the relative lateral displacement between two flanges. It's also assumed that the web is fully yielded under longitudinal stresses; this roughly reflects experimental observations. In reality, however, the beam web does provide some degree of rotational restraint against local buckling to the compressive flange. The web restraint can be represented by a rotational spring of constant k, and Equation 10.10 becomes the following:

$$E_{st}C_w \frac{d^4\beta}{dz^4} + (\sigma I_p - G_{st}J)\frac{d^2\beta}{dz^2} = k\beta \qquad (10.15)$$

where k can be estimated as follows:

$$k = \left[\frac{4G_{st}}{d-2t_f}\right]\frac{t_w^3}{12} = \frac{G_{st}t_w^3}{3(d-2t_f)} \qquad (10.16)$$

The improved solution, obtained following a procedure similar to that described earlier, is:

$$\sigma_{cr}I_p = E_{st}C_w\left(\frac{n\pi}{L}\right)^2 + G_{st}J + k\left(\frac{L}{n\pi}\right)^2 \qquad (10.17)$$

The half-wave length, L/n, which produces the minimum critical stress, can be found by setting $\partial \sigma_{cr}/\partial(L/n) = 0$:

$$\frac{L}{n} = \pi\sqrt[4]{\frac{E_{st}C_w}{k}} = 0.713\frac{t_f}{t_w}\sqrt[4]{\frac{A_w}{A_f}}b \qquad (10.18)$$

where A_w ($=(d-2t_f)t_w$) and A_f ($=b_ft_f$) are web and flange areas, respectively. The half-wave length derived above is key to determining the advent of local buckling in beams.

Substituting the half-wave length into Equation 10.17 and setting σ_{cr} to σ_y makes the limiting value of $b_f/2t_f$ the following:

$$\frac{b_f}{2t_f} = \sqrt{\frac{G_{st}}{\sigma_y} + 0.381\left(\frac{E_{st}}{\sigma_y}\right)\left(\frac{t_w}{t_f}\right)^2\sqrt{\frac{A_f}{A_w}}} \qquad (10.19)$$

Two material properties in the fully yielded state are needed: the strain-hardening moduli in compression, E_{st}, and shear, G_{st}. E_{st} can be determined from a tensile or compression test, and G_{st} must be determined indirectly. The method to determine G_{st} has been controversial. Based on the discontinuous yield process described in Chapter 2, Lay (1965) derived the following expression for the strain-hardening shear modulus:

$$G_{st} = \frac{2G}{1 + \frac{E}{4E_{st}(1+v)}\tan^2\alpha} = \frac{4E}{5.2+h} \qquad (10.20)$$

where α ($= 45°$) is the slip angle between the yield plane and normal stress, h ($=E/E_{st}$) is the ratio of elastic stiffness to strain-hardening stiffness, and v ($=0.3$) is Poisson's ratio. With h equal to 33, G_{st} is 3,040 ksi (20,961 MPa). The second term on the right side of Equation 10.19

represents the beneficial contribution of web restraint against flange local buckling. For most wide-flange sections, the increase in the limiting value of $b_f/2t_f$ due to the web restraint is between 2 percent and 3.2 percent and can be ignored. The width-thickness limiting ratio is, therefore, the same as given in Equation 10.9.

The discussion presented so far is for beams under uniform moment. For beams subjected to a moment gradient, the maximum moment, M_o, can exceed M_p because of strain-hardening, and the stress in the flange will exceed σ_y in the yielded region. Based on limited experimental evidence, it was suggested by Lay (1965) that M_o could be taken as follows:

$$M_o = \frac{1}{2}\left(1 + \frac{\sigma_u}{\sigma_y}\right) M_p \qquad (10.21)$$

where σ_u is the actual ultimate stress.

The corresponding average stress over the yielded region is equal to the following:

$$\sigma_y^* = \frac{1}{4}\left(3 + \frac{\sigma_u}{\sigma_y}\right) \sigma_y \qquad (10.21b)$$

Replacing σ_y by σ_y^* and using an average web contribution of 2.6 percent in Equation 10.19 gives the following limiting width-thickness ratio for a beam under moment gradient:

$$\left(\frac{b_f}{2t_f}\right)^2 = \sqrt{\frac{4 \times 1.026 G_{st}}{(3 + \sigma_u/\sigma_y)\sigma_y}} = \sqrt{\left(\frac{3.16}{3 + \sigma_u/\sigma_y}\right)\left(\frac{1}{\varepsilon_y}\right)\left(\frac{1}{1 + h/5.2}\right)} \qquad (10.22)$$

Assuming $\sigma_u = 1.80\sigma_y$, and $h = 33$ for A36 steel, $b_f/2t_f = 8.5$. Different grades of steel will have different values of the ratio (σ_u/σ_y), yield strain (ε_y), and strain hardening ratio (h), and therefore different limiting values of $b_f/2t_f$.

In the United States, the plastic design method was first adopted in the 1961 AISC Specification. At that time, structural steels with nominal yield strength (F_y) up to 36 ksi (248 MPa) were permitted for plastic design, and the flange $b_f/2t_f$ ratio was limited to 8.5, close to Haaijer and Thurlimann's recommendations. Research conducted thereafter made it possible to extend plastic design to higher strength steels. As a result, the AISC Specification has permited steels with a yield stress of up to 65 ksi (448 MPa) to be used for plastic design since 1969. On the basis of the work of Lay and Galambos on compact beams, $b_f/2t_f$ was limited to $522/\sqrt{F_y}$ ($137/\sqrt{F_y}$ in S.I. units). For A36 steel, $522/\sqrt{F_y}$ is 8.7, which is about the same value as that computed from Equation 10.22.

It is worthwhile to mention that in 1974, the AISC Specification introduced some plastic design concepts into the allowable stress design procedures by permitting the designer to redistribute moments in a continuous compact beam. This was based on the assumption that under uniform bending, flange strains of compact beams could reach four times the yield strain, which corresponds to a rotation capacity of three. This level of plastic rotation was thought to be sufficient to justify the moment redistribution for most civil engineering structures.

The AISC limiting flange width-thickness ratios for plastic design (compact section requirements for allowable stress design) were not changed until the AISC LRFD Specification was published in 1986. The rule for flange local buckling was based on the test results of Lukey and Adams (1969). Figure 10.18 shows the relationship between the plastic rotation capacities defined in Figure 10.3 and the normalized flange width-thickness ratio. The test results presented in Figure 10.18 indicate that beams with smaller normalized width-thickness ratios not only exhibit a larger rotation capacity but also show a slower degradation in strength after the maximum flexural capacity is reached, as is evident from the larger spread between θ_{hm} and θ_h for the more compact sections in Figure 10.18. Based on Figure 10.18, the following limiting value is required for a rotation capacity, R, of 3:

$$\frac{b_f}{2t_f}\sqrt{\frac{\sigma_y E}{44 E_{st}}} \leq 78$$

To establish a conservative width-thickness ratio limit, Yura and Galambos (1978) suggested that the value of E_{st} be taken as one standard deviation (150 ksi [1,034 MPa]) below the mean value (600 ksi [4,137 MPa]). Substituting E_{st} = 450 ksi (3,103 MPa) into Equation 10.23 gives the following in U.S. units:

$$\frac{b_f}{2t_f} \leq \frac{65}{\sqrt{\sigma_y}} \quad \text{in U.S. units} \tag{10.23}$$

$$\frac{b_f}{2t_f} \leq \frac{170}{\sqrt{\sigma_y}} \quad \text{in S.I. units} \tag{10.24}$$

The AISC LRFD Specification (1986) adopted the above requirement for compact sections. Note that the difference between the actual yield stress, σ_y, and the nominal yield stress, F_y, is ignored in the above development. The above requirement also applies to plastic design in AISC LRFD Specification. This requirement is less stringent than the limit of $52/\sqrt{F_y}$ (= $137/\sqrt{F_y}$ in SI units), which was used by the AISC Specification for plastic design prior to 1986.

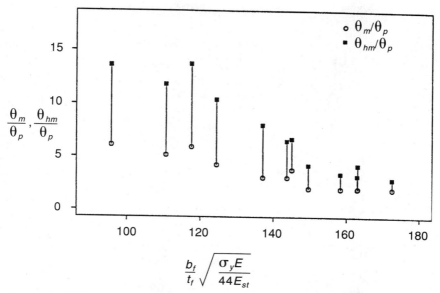

Figure 10.18 Peak and total rotation capacities. *(Adapted from Lukey and Adams 1969)*

10.5 Web local buckling

Only limited research results exist for inelastic web local buckling of wide-flange beams. Haaijer and Thurlimann (1958) studied web buckling of flexural members subjected to bending and axial load effects. Figure 10.19 shows variations in the limiting web slenderness ratio, in the form of $(d-t_f)/t_w$, with respect to the normalized axial force, $P/(\sigma_y A_w)$, and the normalized maximum compression strain in the flange, $\varepsilon_m/\varepsilon_y$. The figure shows clearly that a smaller web width-thickness ratio is required to resist higher axial loads or to achieve higher inelastic strain (or rotation capacity).

In Section 10.4, it was noted that a beam under uniform bending would be strained to $4\varepsilon_y$ for a rotation capacity, R, of 3 and that such a plastic rotation was considered in the AISC LRFD Specification to be sufficient to justify the moment redistribution for most structures. With average values of $A/A_w = 2$ and $d/(d-t_f) = 1.05$ for wide-flange shapes, the curve for $\varepsilon_m/\varepsilon_y = 4$ in Figure 10.19 is replotted as the dashed curve in Figure 10.20; this dashed curve can also be approximated by two straight lines. The results presented in Figure 10.20, however, are applicable to only A7 or A36 steel. For higher strength steel, it was suggested that the limiting width-thickness ratio be modified by $\sqrt{36/\sigma_y}$ (Adams 1966), and thus:

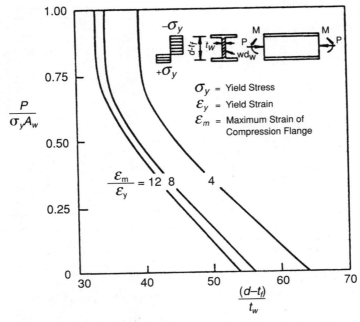

Figure 10.19 Limiting values of $(d - t_f)/t_w$ ratio of the web of a wide-flange section. *(From Haaijer and Thurlimann 1958 with permission)*

$$\frac{d}{t_w} = \frac{257}{\sqrt{\sigma_y}} \quad \text{for } \frac{P}{P_y} > 0.27 \quad \text{in U.S. units} \quad (10.25a)$$

$$\frac{d}{t_w} = \frac{675}{\sqrt{\sigma_y}} \quad \text{for } \frac{P}{P_y} > 0.27 \quad \text{in S.I. units} \quad (10.25b)$$

$$\frac{d}{t_w} = \frac{412}{\sqrt{\sigma_y}}\left(1 - 1.4\frac{P}{P_y}\right) \quad \text{for } \frac{P}{P_y} \leq 0.27 \quad \text{in U.S. units} \quad (10.25c)$$

$$\frac{d}{t_w} = \frac{1082}{\sqrt{\sigma_y}}\left(1 - 1.4\frac{P}{P_y}\right) \quad \text{for } \frac{P}{P_y} \leq 0.27 \quad \text{in S.I. units} \quad (10.25d)$$

Such a modification, which was based on satisfactory performance observed from beam and frame tests at the time, was adopted in the AISC ASD Specification for plastic design.

Based on limited test data of welded girders with unstiffened thin web, the AISC LRFD Specification specifies the following limiting value for a rotation capacity of at least 3:

$$\frac{h}{t_w} \leq \frac{640}{\sqrt{F_y}} \quad \text{in U.S. units} \quad (10.26a)$$

$$\frac{h}{t_w} \leq \frac{1680}{\sqrt{F_y}} \quad \text{in S.I. units} \quad (10.26b)$$

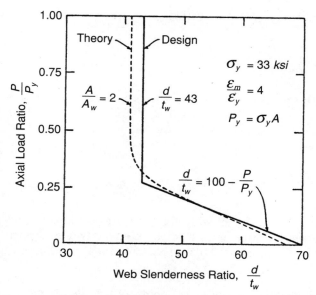

Figure 10.20 Maximum allowable web slenderness ratio for a beam subjected to combined moment and axial force. *(From Massonnet and Save 1965 with permission.)*

where h is defined as the clear distance between flanges less the fillet radius at each flange. Assuming an average value of $h/d = 0.9$, the above equation is equivalent to $d/t_w \leq 710/\sqrt{F_y}$. The AISC Seismic Provisions (1992) specify the following:

$$\frac{h}{t_w} \leq \frac{520}{\sqrt{F_y}} \quad \text{in U.S. units} \quad (10.27a)$$

$$\frac{h}{t_w} \leq = \frac{1365}{\sqrt{F_y}} \quad \text{in S.I. units} \quad (10.27b)$$

The AISC LRFD Specification beam web slenderness ratio requirements for plastic design and seismic design are not as severe as those of the AISC ASD Specification for plastic design.

Dawe and Kulak (1986) used analytical techniques to study the local buckling behavior of beam-columns. The effects of flange and web interaction, inelastic behavior, and the presence of residual stresses were considered. They proposed the following limiting ratios for plastic design:

$$\frac{h}{t_w} = \frac{430}{\sqrt{\sigma_y}}\left(1 - 0.930\frac{P}{P_y}\right) \quad \text{for } 0 \leq \frac{P}{P_y} \leq 0.15 \quad \text{in U.S. units} \quad (10.28a)$$

$$\frac{h}{t_w} = \frac{1130}{\sqrt{\sigma_y}}\left(1 - 0.930\frac{P}{P_y}\right) \quad \text{for } 0 \leq \frac{P}{P_y} \leq 0.15 \quad \text{in S.I. units} \quad (10.28b)$$

$$\frac{h}{t_w} = \frac{382}{\sqrt{\sigma_y}}\left(1 - 0.220\frac{P}{P_y}\right) \quad \text{for } 0.15 \leq \frac{P}{P_y} \leq 1.0 \quad \text{in U.S. units} \quad (10.28c)$$

$$\frac{h}{t_w} = \frac{1003}{\sqrt{\sigma_y}}\left(1 - 0.220\frac{P}{P_y}\right) \quad \text{for } 0.15 \leq \frac{P}{P_y} \leq 1.0 \quad \text{in S.I. units} \quad (10.28d)$$

A comparison of the proposed formulae with those adopted in the AISC Specifications is presented in Figure 10.21. Based on their study, Dawe and Kulak concluded that the web slenderness ratios currently specified in North American codes are conservative.

10.6 Inelastic lateral-torsional buckling

10.6.1 General

Elastic lateral-torsional buckling of wide-flange beams under uniform bending or moment gradient has been studied extensively (e.g., Bleich 1952, Timoshenko and Gere 1961). Three assumptions are common to the cited studies:

- The stiffness does not change along the length of the member.
- The shear center remains in the plane of the web.
- The cross-section of the steel member remains undistorted.

The first solution for lateral-torsional buckling of wide-flange beams in the strain-hardening range was developed by White (1956). Lee and

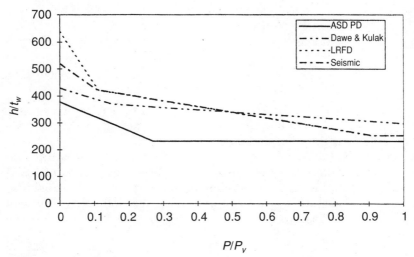

Figure 10.21 Comparison of rules for web local buckling.

Galambos (1962) extended White's work by considering unloading of the compression flange due to lateral-torsional buckling. Galambos (1963) presented a solution for determining the inelastic lateral-torsional buckling strength of rolled wide-flange beam-columns. Lay and Galambos (1965) also extended White's work and related the rotation capacity of a beam under uniform bending to the lateral support spacing. Lay and Galambos (1967) used the equivalent length concept to analyze lateral-torsional buckling of beams under moment gradient. Kemp (1984) extended the Lay and Galambos solution.

10.6.2 Beam under uniform moment

10.6.2.1 White's approach
A theoretical solution for the determination of the critical unsupported length was presented by White (1956). In that solution it was assumed that lateral-torsional buckling occurs when the material has reached the onset of strain-hardening. This assumption implies that sufficient plastic rotation has developed before lateral-torsional buckling occurs (see Section 10.3) because the beam has already undergone considerable inelastic deformation before strain-hardening is reached. In White's analysis, no elastic unloading of already-yielded fibers is assumed. Thus, this approach provides a lower bound solution analogous to that obtained by the tangent modulus concept of axially loaded columns (Shanley 1947). The critical spacing of lateral braces for all rolled wide-flange sections was found to be about $17r_y$ for a simply supported beam under a uniform bending moment equal to M_p. The theoretical basis follows.

Consider a beam with simple end restraints (i.e., the ends of the beam cannot translate or twist but are free to rotate laterally and the end sections are free to warp) as shown in Figure 10.22. When the beam is subjected to uniform moment, the elastic solution for lateral-torsional buckling is well established (Timoshenko and Gere 1961):

$$\frac{M}{EI_y} = \frac{\pi}{L_b}\sqrt{\frac{GJ}{EI_y} + \frac{\pi^2}{L_b^2}\frac{C_w}{I_y}} = \frac{\pi}{L_b}\sqrt{\frac{GJ}{EI_y} + \frac{\pi^2}{L_b^2}\left(\frac{d-t_f}{2}\right)^2} \quad (10.29)$$

in which EI_y is the flexural stiffness about the section's minor principal axis, GJ is the St. Venant torsional stiffness, and L_b is the unbraced length of the member. The above equation can also be applied to yielded members if E and G are replaced by their corresponding inelastic material properties. For relatively short beams, the equation can be further simplified if one ignores the first term, which represents the contribution due to St. Venant torsion. Substituting $M_p\ (=Z_x\sigma_y)$ for M:

$$\frac{Z_x \sigma_y}{E' I_y} = \left(\frac{\pi}{L_b}\right)^2 \left(\frac{d - t_f}{2}\right) \quad (10.30)$$

where $E'I_y$ is the effective flexural stiffness. Because White did not consider unloading of the compression flange upon buckling, $E'I_y$ can be expressed as the tangent modulus stiffness:

$$E'I_y = \frac{1}{h} E I_y \quad (10.31)$$

From Equation 10.30, the limiting slenderness for this lower bound solution can be solved as follows:

$$\frac{L_b}{r_y} = \pi \sqrt{\frac{E}{\sigma_y}} \sqrt{\frac{A(d - t_f)}{2 Z_x}} = \pi \sqrt{\frac{1}{h \varepsilon_y}} \sqrt{\frac{A(d - t_f)}{2 Z_x}} \quad (10.32)$$

If one assumes $h = 33$ for A36 steel and takes an average value of 1.2 for $A(d-t_f)/2Z_x$ for wide-flange sections, the limiting L_b/r_y ratio is about 17. This limit is optimal because closer spacing of lateral bracing would not provide greater plastic deformation capacity because of local buckling.

Lee and Galambos (1962) extended White's work by considering the unloading effect of the compression flange on lateral buckling. The method is analogous to the reduced modulus concept of axially loaded columns. Lee and Galambos showed that the critical lateral spacing, based on such an upper bound solution, is $45 r_y$. If $L_b/r_y < 45$, the Lee and Galambos experiments on continuous beams showed that failure is initiated by local buckling of the compression flange and that postbuckling strength reserve is quite large. The postbuckling strength provided by a continuous beam is mainly due to the lateral restraint provided by the adjacent elastic spans, which offer both in-plane and out-of-plane moment restraints as well as significant warping restraint at the ends of the buckled span.

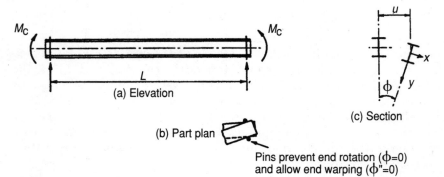

Figure 10.22 Simply supported beam under uniform bending. *(From Trahair 1993 with permission)*

10.6.2.2 Lay and Galambos' approach

Lay and Galambos (1965) extended White's work to determine the suboptimum lateral support spacing. In that work, the lateral support spacing was related to the rotation capacity. For the simply supported beam under uniform bending moment shown in Figures 10.23a and 10.23b, Figure 10.23c shows the deformed configuration of the section after buckling. It is assumed that longitudinal hinges are located at the midheight of the web and at the junction between the web and tension flange. Compared with the observed deformed shape shown in Figure 10.24b, the model simulates well the observed web deformation pattern. The major difference between the model and actual deformed shape is that, in reality, the tension flange tends to twist rather than allowing a hinge to form at its junction with the web.

The physical model shown in Figure 10.23d is therefore used to solve the bifurcation problem. This model is conservative for the lateral buckling calculation because it assumes that the beam web contributes to the loaded area without increasing the lateral flexural

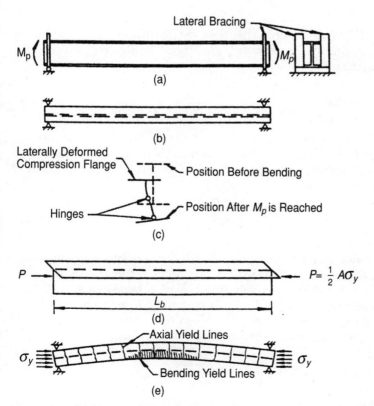

Figure 10.23 Buckling of a beam under uniform moment. *(From ASCE-WRE 1971 with permission.)*

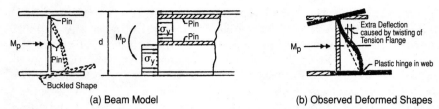

Figure 10.24 Lateral buckling model of cross section. *(From Lay and Galambos 1965 with permission)*

stiffness. Therefore, the compression portion of the beam is isolated from the beam under uniform moment, and the lateral-torsional buckling problem is reduced to that of a column under an axial force $P = A\sigma_y/2$, where A is the cross-sectional area of the beam. This compression T-column is also assumed to be pin-ended. The critical load of this T-shaped column is as follows:

$$P_{cr} = \frac{A\sigma_y}{2} = \frac{\pi^2 cEI}{L_b^2} \qquad (10.33)$$

where cEI ($=cEI_y/2$) represents the inelastic flexural stiffness of the T-column after yielding. The limiting slenderness ratio resulting from the above equation is:

$$\frac{L_b}{r_y} = \pi \sqrt{c} \sqrt{\frac{E}{\sigma_y}} \qquad (10.34a)$$

The above expression is valid only for simply supported beams. If the unbraced beam segment is continuous with adjacent spans, an effective length factor, K, may be used to account for the restraint offered by the adjacent spans:

$$\frac{KL_b}{r_y} = \pi \sqrt{c} \sqrt{\frac{E}{\sigma_y}} \qquad (10.34b)$$

Based on calculation of effective length factors and experimental observation, for the usual case of elastic adjacent spans, K may be taken as 0.54; for yielded adjacent spans, K may be taken as 0.8.

Next, consider the associated rotation capacity, R. For a beam under uniform bending, the relationship between the rotation capacity and the average strain, ε_{av}, in the beam flange is as follows:

$$R = \frac{\theta}{\theta_p} - 1 \cong \frac{\varepsilon_{av}}{\varepsilon_y} - 1 \qquad (10.35)$$

When an optimal lateral bracing is provided such that lateral-torsional buckling occurs when ε_{av} reaches ε_{st} (= $s\varepsilon_y$), the rotation capacity is:

$$R = \frac{\varepsilon_{av}}{\varepsilon_y} - 1 = \frac{s\varepsilon_y}{\varepsilon_y} - 1 = s - 1 \qquad (10.36)$$

For a suboptimum situation in which $\varepsilon_{av} < \varepsilon_{st}$, $\varepsilon_{av} = (1-\phi)\varepsilon_y + \phi s\varepsilon_y$ (see Equation 2.4), the rotation capacity is:

$$R = (s - 1)\phi \qquad (10.37)$$

It should be noted that the rotation capacity, R, in Equations 10.36 and 10.37 is valid for the case in which the inelastic action involves axial yield lines only. In reality, lateral bending is inevitable because of geometric imperfections, and bending yield lines will develop. The axial yield lines shown in Figure 10.23e are those due to the force P and spread across the whole width of the flange. The bending yield lines in the more severely compressed half of the flange are due to the lateral displacement of the compression flange upon buckling. Lay and Galambos showed that the rotation capacity considering the effects of bending and axial yield lines is as follows:

$$R = \left[\frac{2}{\pi} + \frac{1}{2}\left(1 - \frac{2}{\pi}\right)\left(1 - \frac{1}{\sqrt{h}}\right)\right](s - 1)\phi \approx 0.8(s - 1) \qquad (10.38)$$

where $h = 33$ for a fully yielded member ($\phi = 1$) is used for the simplification.

Based on a discontinuous yield concept (see Chapter 2), in which the strain is equal to ε_{st} at the yield lines and ε_y elsewhere, and a compression flange model shown in Figure 10.23e, Lay and Galambos then considered the stiffness reduction due to effects of both axial and bending yield lines to derive the coefficient c in Equation 10.34b as a function of R:

$$c = \frac{1}{1 + 0.7Rh\left(\frac{1}{s-1}\right)} \qquad (10.39)$$

In deriving this expression, it is assumed that unloading occurs once flange local buckling takes place, that is, when the average strain at the center of flange is equal to the strain-hardening strain. Substituting Equation 10.38 into Equation 10.39 gives the following:

$$c = \frac{1}{1 + 0.56h} \qquad (10.40)$$

and Equation 10.34b becomes:

$$\frac{KL_b}{r_y} \frac{\sqrt{\varepsilon_y}}{\pi} = \frac{1}{\sqrt{1 + 0.56h}} \tag{10.41}$$

If $h = 33$ and $K = 0.54$, the following limiting slenderness ratio results:

$$\frac{L_b}{r_y} = \frac{225}{\sqrt{\sigma_y}} \quad \text{in U.S. units} \tag{10.42a}$$

$$\frac{L_b}{r_y} = \frac{591}{\sqrt{\sigma_y}} \quad \text{in S.I. units} \tag{10.42b}$$

The above limiting ratio is valid only for low-yield strength steels. Some limited test results with Grade 65 steel have shown that the Equation 10.42 is not conservative for high strength steel (ASCE-WRC 1971). Instead, it was found that $L_b/r_y = 1{,}375/\sigma_y$ (=9,480/σ_y in S.I. units) not only fits the test results well for high strength steel but also approximates Equation 10.42 for lower strength steels. The limiting L_b/r_y ratio of $1{,}375/\sigma_y$ (= 9,480/σ_y in S.I. units) was adopted in the AISC Specifications for plastic design in 1969.

10.6.3 Beam under moment gradient

10.6.3.1 Equivalent length approach Lay and Galambos (1967) performed inelastic lateral buckling analysis of beams under a moment gradient using the model shown in Figure 10.25. Once again, the compression half of the beam is considered to act as an isolated column, and lateral-torsional buckling is assumed to be equivalent to buckling of an isolated column under the beam bending stresses. Mathematically, the modeling is equivalent to disregarding the contribution of St. Venant torsion to the strength of the beam. Given that the St. Venant torsion contribution is actually significant for most beams under moment gradient, the proposed solution therefore provides a lower bound estimate of a beam's lateral-torsional buckling strength.

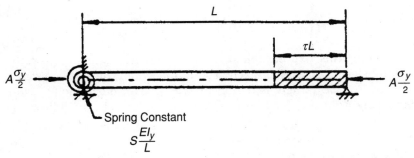

Figure 10.25 Lateral buckling model of beam under moment gradient. *(From Lay and Galambos 1967 with permission)*

In this model, the yield moment, M_y, is taken as the moment at which yielding first occurs in the flange (ignoring residual stresses). Lay and Galambos assumed an average value of $M_y = 0.94 M_p$. The variation of moment along the beam length causes not only a transition of material properties from elastic to yielded, but also a change in the compression stresses over the length of the beam. After M_p is reached, the lateral displacements usually increase rapidly. The deflected shape also changes, and the curvatures producing lateral displacement are concentrated in the yielded region. White's study (1956) showed that the extent of yielding has a much greater influence on lateral buckling behavior than does the variation in normal stress. Thus, it was further assumed that normal stresses on the compression tee remain at σ_y and do not reduce with decreasing bending moment. The conservatism introduced by this assumption is offset to some extent by the fact that normal stresses exceed σ_y in the yielded length (τL) as a result of strain-hardening. The buckling equation for the column may be expressed as follows:

$$\frac{\tan(\lambda_b \pi \tau / \sqrt{c})}{\tan(\lambda_b \pi (1-\tau))} + \frac{1}{\sqrt{c}} \left[\frac{\lambda_b \pi + S \left[\frac{1}{\lambda_b \pi} - \cot(\lambda_b \pi (1-\tau)) \right]}{\lambda_b \pi + S \left[\frac{1}{\lambda_b \pi} + \tan(\lambda_b \pi (1-\tau)) \right]} \right] = 0 \quad (10.43)$$

where c is the ratio of the lateral bending stiffness in the yielded region to the elastic value, S represents the elastic restraint of an adjacent span ($S = 0$ for a pinned end and $S \to \infty$ for a fixed end), and the limiting slenderness ratio for lateral-torsional buckling, L_b/r_y, is expressed in the nondimensional form as λ_b:

$$\lambda_b = \frac{L_b}{r_y \pi} \sqrt{\frac{\sigma_y}{E}} \quad (10.44)$$

Lay and Galambos (1965) considered the properties of bending yield planes and derived the following formula for c:

$$c = \frac{2}{h + \sqrt{h}} \quad (10.45)$$

For a given value of S, the relationship between λ_b and τ can be constructed for a particular grade of steel (see Figure 10.26). Note that Equation 10.43 was derived on the basis that the wide-flange section was compact, that is, a section that satisfies the limiting flange width-thickness ratio of Equation 10.22 and does not buckle locally until it is fully yielded. At this point, it is important to establish whether the yielded length associated with lateral-torsional buckling is greater or less than that required to cause local buckling. Given that, for a beam under moment gradient, yielding will be concentrated in a restricted region of length, τL_b, and the possibility of local buckling within this region must be considered. It has been shown in Section 10.4 that τL_b must be of a sufficient length for a full local buckling wave to form. The full wavelength of a local buckle has been given as $2l$, where, from Equation 10.18, $l/b_f = 0.71(t_f/t_w)(A_w/A_f)^{1/4}$. Therefore, the local buckling criterion to be applied to a compact section under moment gradient is that the yielded length, $\tau_{lb}L_b$, should be equal to the full wave length of the local buckle:

$$\tau_{lb}L_b = 2l = 1.42\left(\frac{t_f}{t_w}\right)\sqrt[4]{\frac{A_w}{A_f}}\, b_f \tag{10.46}$$

Based on the geometric properties of available wide-flange sections, an average value of $8.33 r_y$ can be obtained for the right-hand side of Equation 10.46. Therefore, Equation 10.46 can be expressed as the following nondimensional form (see Equation 10.44):

$$\tau_{lb}\lambda_b = 2.65\sqrt{\varepsilon_y} \tag{10.47}$$

Using Equation 10.47, the proportion (τ) of the beam length that must be yielded to trigger flange local buckling is superimposed on Figure 10.26 for a direct comparison with the required length to initiate lateral-torsional buckling. For the practical cases of inelastic beams under moment gradient, it can be seen that a smaller value of τ is

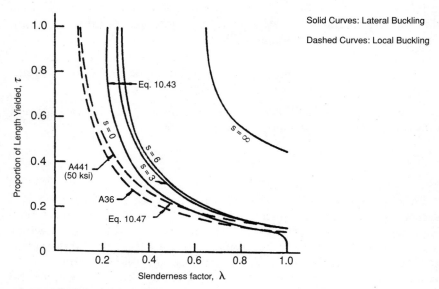

Figure 10.26 Relationship of flange local buckling and lateral-torsional buckling. *(From Lay and Galambos 1967 with permission)*

required to cause flange local buckling than to cause lateral-torsional buckling. That is, failure will generally be initiated by flange local buckling rather than by lateral-torsional buckling. Because the beam behavior under moment gradient is most likely to be governed by flange local buckling, Lay and Galambos (1967) suggested that lateral bracing be provided according to the criterion of Equation 10.41 for a beam under uniform bending moment, but with a value of R equal to 1. It was recommended that for the moment ratio $-0.5 \leq q = M_1/M_2 \leq 1.0$, where M_1 is the smaller moment at the end of the unbraced length, and M_2 is the larger moment at the end of the unbraced length, and (M_1/M_2) is positive when moments cause reverse curvature and negative for single curvature, the limiting L_b/r_y ratio be taken as 70, 55, and 45 for σ_y of 36, 50, and 65 ksi (248, 345, and 448 MPa), respectively. This recommendation can be approximated by the following:

$$\frac{L_b}{r_y} = \frac{1{,}375}{\sigma_y} + 25 \quad \text{when} - 0.5 < M_1/M_2 < 1.0 \quad \text{in U.S. units} \quad (10.48\text{a})$$

$$\frac{L_b}{r_y} = \frac{9480}{\sigma_y} + 25 \quad \text{when} - 0.5 < M_1/M_2 < 1.0 \quad \text{in S.I. units} \quad (10.48\text{b})$$

This limit on slenderness ratio has been adopted by the AISC Specifications for plastic design since 1969.

10.6.3.2 Bansal's experimental study To study the lateral instability of beams, 34 continuous steel beams having specified yield strengths of 36, 50, and 65 ksi (248, 345, and 448 MPa, respectively) were tested by Bansal (1971). Bansal reported that the deformation capacity of a beam is inversely proportional to its yield strength when the behavior is controlled by lateral instability. For given lateral slenderness ratio, type of loading, and support conditions, higher strength steel beams are more prone to lateral-torsional buckling.

Figure 10.27 shows the test results of those beams that developed a plastic mechanism. The abscissa represents the normalized lateral slenderness ratio $(L_b/r_y)(\sigma_y/36)$. Figure 10.27b shows the relationship between the slenderness ratio and the rotation capacity, R. The group of points on the left is for the beams with single curvature ($q \leq 0$); the group on the right is for the beams with reverse curvature ($q > 0$). Lines are drawn on Figure 10.27b linking data points for $q = -0.5$ (specimens 17, 18, 21) and $q = 0.85$. The modifed stenderness ratio $(L_b/r_y)(\sigma_y/36)$ necessary to achieve a rotation capacity of 3 for q equal to -0.5 and 0.5 are calculated from the intersection points of ($R = 3$, $q = -0.5$) and ($R = 3$, $q = -0.85$), respectively, to be 70 and 150, respectively. Using these data, points A and B in Figure 10.27a can be

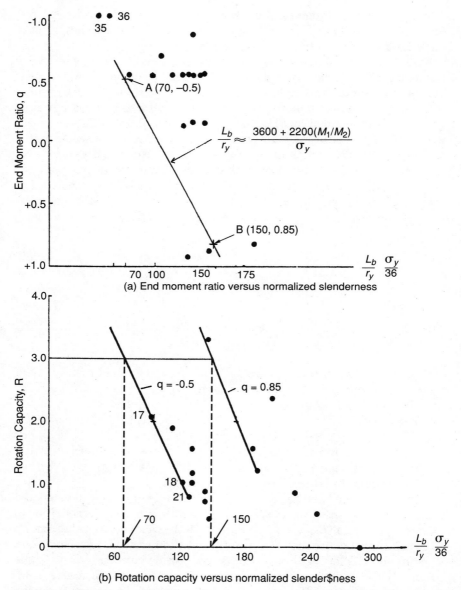

Figure 10.27 Relation of rotation capacity, moment ratio, and lateral slenderness ratio for parameters expressed in U.S. units only. *(Adapted from Bansal 1971)*

identified. The line passing through these points represents the limiting slenderness ratio for a rotation capacity of 3:

$$\frac{L_b}{r_y} \approx \frac{3600 + 2200(M_1/M_2)}{\sigma_y} \qquad \text{in U.S. units} \qquad (10.49a)$$

$$\frac{L_b}{r_y} \approx \frac{24822 + 15169(M_1/M_2)}{\sigma_y} \quad \text{in S.I. units} \quad (10.49b)$$

With the nominal yield strength, F_y, substituted for σ_y, the above limiting slenderness ratio was adopted by the AISC LRFD Specifications for plastic design.

Figure 10.28 shows a comparison of the lateral bracing requirements for two different steel strengths per the AISC LRFD Specification and AISC ASD Specification. Based on Bansal's test results, it is clear that the more relaxed lateral bracing requirements specified in the AISC LRFD Specification are justified.

10.7 Code comparisons

A worldwide survey of the classifications of beam section compactness and lateral bracing requirements in major steel design codes was made by SSRC (1991). A similar comparison of the beam section local buckling rules adopted in North America, Australia, and Eurocode 3 was also provided by Bild and Kulak (1991). Except for the design codes in the United States, the codes studied by these researchers have adopted a four-class system. Depending on the width-to-thickness ratio, λ, with respect to three limiting values $(\lambda_{pd}, \lambda_p, \text{ and } \lambda_r)$, the classification is consistent with the four curves shown in Figure 10.2. Class 1 sections, also known as the plastic sections with $\lambda \leq \lambda_{pd}$, are required for plastic design. Class 2 sections, also termed compact sections with $\lambda_{pd} < \lambda \leq \lambda_p$, do not have sufficient rotation capacity for plastic design. Note that the compact section defined in that case is not the same as that defined in the AISC LRFD Specification. As mentioned in Section 10.1, the AISC LRFD Specification values for λ_{pd} and λ_p for local buckling are merged; that is, compact sections are also plastic sections.

Expressing λ_{pd} for plastic sections in the following general format:

$$\lambda_{pd} = \delta/\sqrt{F_y} \quad \text{in U.S. units} \quad (10.50)$$

facilitates a comparison of the values of δ used in various countries to prevent flange local buckling. Results are presented in Figure 10.29a. Note that the AISC ASD and AISC LRFD Specifications provide lower and upper bounds to values of δ. Figure 10.29b presents values of δ used for seismic design. Note that the requirements for seismic design are comparable to those of the AISC ASD Specification for plastic design. A similar comparison for the web local buckling requirements is shown in Figure 10.30.

It is interesting to compare compactness requirements of seismic codes in the United States and Japan. The Japanese design code AIJ LSD (1990) considers the interaction of flange and web local buckling

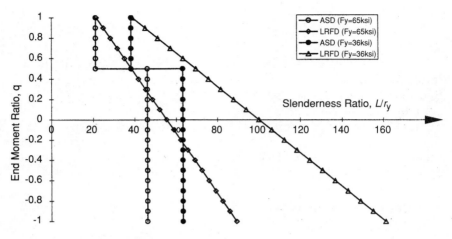

Figure 10.28 Comparison of rules for lateral-torsional buckling.

modes. For ductility Class 1, which corresponds to a rotation capacity of four (Fukumoto and Itoh, 1992), the following formula is specified:

$$\frac{\left(\dfrac{b}{t_f}\right)^2}{\left(\dfrac{200}{\sqrt{\sigma_{yf}}}\right)^2} + \frac{\left(\dfrac{d}{t_w}\right)^2}{\left(\dfrac{1270}{\sqrt{\sigma_{yw}}}\right)^2} \leq 1 \quad \text{in S.I. units} \quad (10.51)$$

where σ_{yf} and σ_{yw} are the actual yield stresses of the flange and web materials, respectively. A comparison of the above equation with the AISC LRFD requirements for seismic design is shown in Figure 10.31.

10.8 Interaction of beam buckling modes

The AISC LRFD Specification sets limits on the width-thickness ratio of beam flanges and webs to avoid premature local buckling and on the maximum laterally unsupported length (L_b) to delay lateral-torsional buckling. The AISC LRFD Specification considers each buckling mode as an independent limit state. The advantage of this design approach lies in its simplicity. However, uncertainties arise because this simple approach ignores the interaction between local buckling and lateral-torsional buckling.

Much test evidence shows that flange local buckling, web local buckling, and lateral-torsional buckling interact. The flange restrains the web and vice versa. This can be easily seen from the physical model of flange local buckling shown in Figure 10.13. In Lay's derivation of the slenderness ratio limit for flange local buckling, web restraint was treated as a spring with positive stiffness. However, if the web buckles prior to the flange, the beneficial restraint from the beam web is

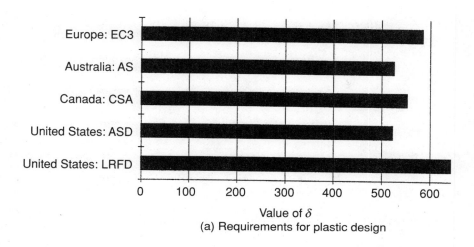
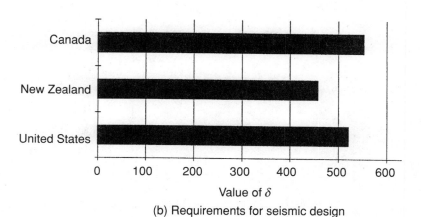

Figure 10.29 Comparison of flange local buckling requirements.

lost, and the buckled web may be regarded as a spring with negative stiffness. As a result, the amplitude and wavelength of flange local buckling will be affected by web local buckling.

In Section 10.3, the relation between local buckling and lateral-torsional buckling under different moment patterns was discussed. Recall that under uniform moment, lateral-torsional buckling triggers flange local buckling, whereas under moment gradient, if adequate lateral bracing is provided, flange buckling occurs first, provided that the yielded zone is sufficiently long to accommodate a full wavelength of the buckle. In the latter case, even though flange buckling will not cause an

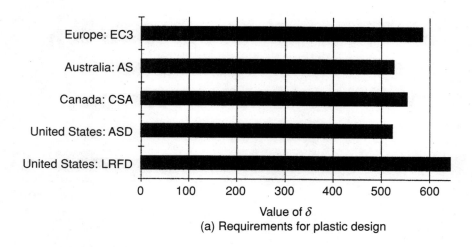

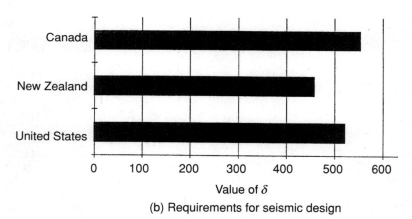

Figure 10.30 Comparison of web local buckling requirements.

immediate loss of strength, it triggers lateral-torsional buckling after which strength is lost. To evaluate the beam rotation capacity, one should consider the interactions of different modes.

To describe this interaction, consider a beam under moment gradient in which flange local buckling has occurred in the yielded compression region. The beam has not buckled laterally, so both halves of the compression flange participate in the flange local buckling. When lateral-torsional buckling occurs, the half-flange that undergoes additional compression in the process of lateral deflection will lose stiffness rapidly because of the local buckle, whereas the other half-

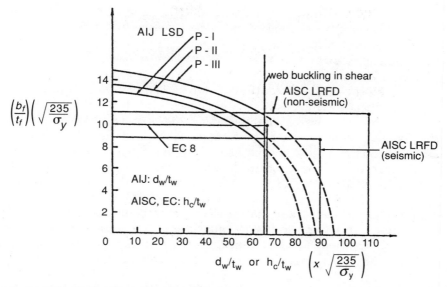

Figure 10.31 Comparison of local buckling requirements for Japanese and U.S. seismic provisions for σ_y in MPa *(Fukumoto and Itoh 1992)*

flange will behave in a manner similar to that for an unbuckled flange. If the bending stiffness of such a flange is taken to be $c_b EI_y$, where c_b is a factor with a value less than 1.00, the bending stiffness of the unbuckled half-flange can be approximated as $c_b EI_y/8$. When this value is used, the solid curves shown in Figure 10.26 are obtained for lateral-torsional buckling after a local buckle has occurred. Clearly, the beam is considerably weakened as a result. It may be concluded that local buckling in a beam under moment gradient will lead to lateral buckling and that these two effects in combination are likely to cause a loss in strength.

The onset of local buckling of the compression flange is not by itself the cause of strength deterioration, and additional ductility can be realized once local buckling commences. The compatibility requirement for longitudinal strains across the compression flange in the region of local buckling induces membrane effects that resist the out-of-plane deformations of the flange. However, the onset of lateral buckling in a locally buckled beam results in immediate attainment of the maximum moment and subsequent unloading. This phenomenon is caused by two contributing factors. First, the lateral displacement permits strain compatibility to be maintained as one half of the flange continues to buckle locally in compression. Second, local buckling on the same half of the flange significantly reduces the resistance of the cross section to lateral buckling.

That phenomenon was demonstrated by Dekker (1989). In order to investigate the effect of flange local buckling alone under moment gradient, Dekker fabricated several T-shaped beams in such way that their plastic neutral axis was sufficiently close to the compression flange to eliminate web local buckling. The compression flange of each beam was provided with continuous lateral support to avoid lateral-torsional buckling. Interestingly, flange local buckling was observed to develop in a symmetrical manner about the web (i.e., both halves of the flange buckled toward the neutral axis), and the conventional antisymmetric flange buckling mode with each half of the compression flange moving in opposite directions was not observed. Figure 10.32 summarizes the width-thickness ratios and the rotation capacities of all the specimens tested. Note that $b_f/2t_f$ ratios were as high as 14.5. Dekker reported that flange local buckling did not cause a loss in strength until large rotation capacities (>20) were reached. The typical failure mode involved tensile fracture of the T-beam.

Kemp (1986, 1991, 1996) proposed that beam rotation capacity be related to the slenderness ratio for each of three buckling modes. Kemp normalized the slenderness ratio by the material yield stress of each buckling mode as follows:

$$\text{FLB:} \qquad \lambda_f = \frac{b_f/2t_f}{\sqrt{\sigma_{yf}/36}} \qquad \text{in U.S. units} \qquad (10.52a)$$

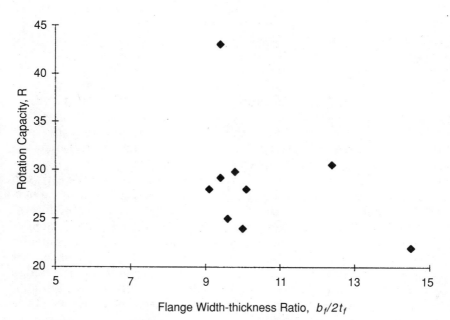

Figure 10.32 Flange width-thickness ratio versus rotation capacity based on Dekker's (1989) isolated flange local buckling test results.

$$\lambda_f = \frac{b_f/2t_f}{\sqrt{\sigma_{yf}/250}} \quad \text{in S.I. units} \quad (10.52\text{b})$$

WLB:
$$\lambda_w = \frac{d/t_w}{\sqrt{\sigma_{yw}/36}} \quad \text{in U.S. units} \quad (10.52\text{c})$$

$$\lambda_w = \frac{d/t_w}{\sqrt{\sigma_{yw}/250}} \quad \text{in S.I. units} \quad (10.52\text{d})$$

LTB:
$$\lambda_L = \frac{L_i/r_{yc}}{\sqrt{\sigma_{yf}/36}} \quad \text{in U.S. units} \quad (10.52\text{e})$$

$$\lambda_L = \frac{L_i/r_{yc}}{\sqrt{\sigma_{yf}/250}} \quad \text{in S.I. units} \quad (10.52\text{f})$$

where σ_{yf} and σ_{yw} are the actual yield stresses of the flange and web materials, respectively; L_i is the length from the section of maximum moment to the adjacent point of inflection; r_{yc} is the radius of gyration of the portion of the elastic section in compression; and other terms have been defined previously.

Based on 44 beam test results reported by Adams et al. (1965), Lukey and Adams (1969), Kemp (1985, 1986), and Kuhlmann (1989), Kemp showed weak relationships between the rotation capacity and the normalized slenderness ratios for each local buckling mode (see Figures

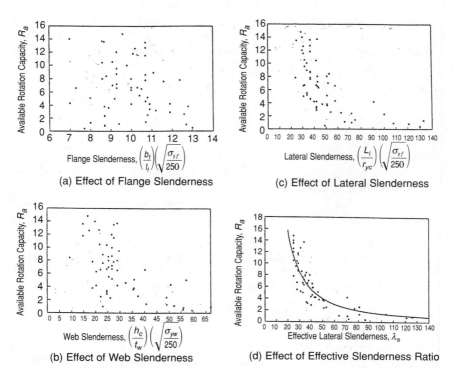

(a) Effect of Flange Slenderness

(b) Effect of Web Slenderness

(c) Effect of Lateral Slenderness

(d) Effect of Effective Slenderness Ratio

Figure 10.33 Relationship between rotation capacity and slenderness ratios; units of σ_y are MPa. *(From Kemp 1996 with permission)*

10.33a and 10.33b). Figure 10.33c shows that the correlation is improved if the rotation capacity is plotted against $(L_i/r_{yc})(\sqrt{\sigma_{yf}/250})$ for lateral-torsional buckling. To consider all three buckling modes simultaneously, Kemp defined the following parameter:

$$\lambda_e = \lambda_L \left(\frac{\lambda_f}{9}\right)\left(\frac{\lambda_w}{70}\right) \tag{10.53}$$

Figure 10.33d demonstrates a clear relationship between the rotation capacity, R, and λ_e. An empirical fit to the data yields the following relationship:

$$R = 3.01\left(\frac{60}{\lambda_e}\right)^{1.5} \tag{10.54}$$

The database used by Kemp is based on monotonic testing of small-scale beam specimens. The results shown in Figure 10.33 imply that the buckling rules adopted in modern design codes may need to be improved.

Based on the discussion presented above, several observations can be made (Kemp 1986):

- The antisymmetric flange local buckling may be initiated under the conditions defined in Equation 10.22. The buckling amplitude will not grow appreciably because of the membrane force associated with compatibility constraints to warping across the flange, unless these constraints are released with the onset of web local buckling or lateral-torsional buckling.
- Web local buckling may occur independently of flange buckling. However, this does not cause significant loss of strength. Web local buckling may release the compatibility constraints and hasten the development of flange buckling.
- Lateral-torsional buckling may be an independent mode of failure or may develop rapidly after flange buckling because of the reduced stiffness of the flange in the plastic region. Lateral-torsional buckling may release the constraints to flange buckling. Once lateral-torsional buckling occurs, the strength of a beam will degrade.
- The current practice of separating beam flexural buckling behavior into three independent limit states may not be appropriate. A design criterion that considers the interaction of three buckling modes is necessary to ensure adequate plastic rotation capacity.

10.9 Cyclic beam buckling behavior

A moment-resisting frame is expected to form a yield mechanism and to displace laterally for a number of cycles during a design earthquake. Because the range of deformation (see Figure 10.1) exceeds the incipient collapse threshold for plastic design, beam plastic rotation demand is generally higher for seismic design.

Unfortunately, the cyclic nature of earthquake-induced excitations also reduces beam plastic rotation capacity. At the material level, the low-cyclic fatigue endurance limit is usually much higher than that required for seismic design, but the Bauschinger effect will reduce the buckling strength of steel elements. For a given displacement amplitude, a beam that would not buckle under monotonic loading may buckle under cyclic loading. Furthermore, a beam that buckles inelastically exhibits permanent deformation after the load is removed. Such residual deformations at both the section and member levels serve as geometric imperfections. Thus, the amplitude of the buckles increase as the number of cycles increases.

The effect of large alternating strains on the cyclic behavior of steel beams has been investigated by Bertero and Popov (1965). Small-scale cantilever beams, M4 × 13 (U.S. shapes) of A7 steel with a mean yield stress of 41 ksi (283 MPa) and a span of 35 inches (89 cm), were loaded cyclically to failure. The value of $b_f/2t_f$ (= 5.3) of those beams was within the limit for plastic design, the web was very compact (h/t_w = 10.4), and L_b/r_y (= 37) was small. The number of cycles required to cause complete fracture and flange local buckling versus the maximum strain at the fixed end of the beam, is shown in Figure 10.34a. Of the 11 beams tested, none exhibited local buckling during the first half of first loading cycle. This was true even in the experiment with imposed strains of 2.5 percent, which was greater than the strain at the onset of strain-hardening of the material, ε_{st}. Figure 10.34a shows that the number of cycles required to trigger flange local buckling is significantly reduced for cycles at larger strains. However, once flange local buckling occurs, Figure 10.34b also shows that more cycles are needed to fracture the steel beam. The large number of cycles resisted prior to fracture indicates that the low-cycle fatigue endurance of steel itself is generally not a governing factor in seismic applications.

Factors that influence inelastic beam buckling under monotonic loading also apply to cyclic loading. For example, a beam under moment gradient that satisfies the compactness requirements of modern seismic codes is less likely to show significant strength degradation under cyclic loading. Further, it is commonly observed in cyclic testing that once flange local buckling commences, it interacts with web local buckling and, more importantly, lateral-torsional buckling.

Takanashi (1974) conducted monotonic and cyclic testing of simply supported beams of two different grades of steel. The 8-inch (20.3 cm) deep wide-flange steel beams tested had $b_f/2t_f$ = 6.3 and d/t_w = 36. Lateral bracing was provided at both ends of the beam and at midspan to prevent lateral and torsional movements; the value of L_b/r_y was 65 and close to the limit presented in the 1969 AISC Specification.

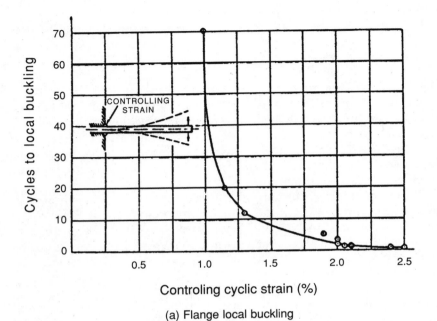

(a) Flange local buckling

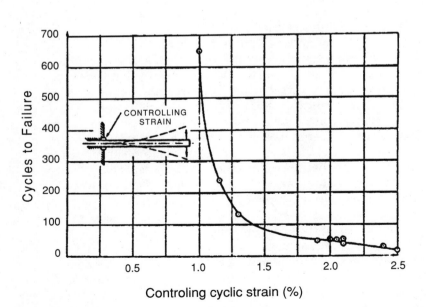

Figure 10.34 Number of cycles to cause flange local buckling and fracture. *(Bertero and Popov 1965)*

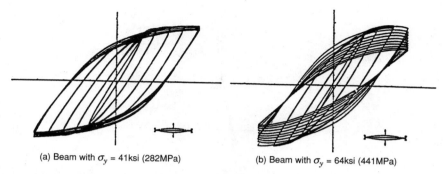

(a) Beam with σ_y = 41ksi (282MPa)   (b) Beam with σ_y = 64ksi (441MPa)

Figure 10.35 Cyclic behavior of beams with different yield strengths. *(Takanashi 1973.)*

Figure 10.35 shows the cyclic response of two beams with different yield stresses. For the beam with a yield stress of 41 ksi (282 MPa), Figure 10.35a shows that the beam was able to achieve a cyclic rotation ductility of 3 (or a rotation capacity of 2) after 40 cycles of increasing amplitudes. For a beam with a yield stress of 64 ksi (441 MPa), inferior behavior was observed (Figure 10.35b). Stable hysteresis loops could be maintained only for the first 30 cycles up to a rotation ductility of 2. The beam strength degraded significantly because of lateral buckling at a rotation ductility of 2.5. Based on those results, Takanashi concluded that cyclic rotation capacity is significantly lower than monotonic rotation capacity. It should be noted, however, that his conclusion was based on the loading sequence shown in Figure 10.35 for a large number of inelastic cycles. Most steel beams are unlikely to experience that many inelastic cycles in a design earthquake. The cyclic rotation capacities reported by Takanashi may be conservative for seismic design.

Vann et al. (1975) also conducted cyclic and monotonic tests of beams and beam-columns of mild steel. Each specimen was clamped at one end, and a load was cyclically applied at the free end, which was braced laterally. Beam sections chosen were W8×13 (W200×19 in S.I. units) ($b_f/2t_f$ = 7.8, h/t_w = 29.9) and W6×16 (W150×24 in S.I. units) ($b_f/2t_f$ = 5.0, h/t_w = 19.1). Specimens of each cross section were tested with unsupported lengths of $30r_y$ and $60r_y$.

Figure 10.36 shows the hysteresis behavior of a W8×13 beam with an unbraced length of $30r_y$. The specimen was cycled at a ductility ratio of 7.2. For such a properly braced beam, flange local buckling was first observed in the second half-cycle. Although flange local buckling did not cause an immediate degradation of strength, it did induce web local buckling. The loss of load capacity may be attributed primarily to web local buckling, which began in the fifth half-cycle. Figure 10.36 also demonstrates that the beam's monotonic behavior at large deflections is a reliable indicator of the cyclic behavior; that is, the drop in load following web local buckling in the monotonic test

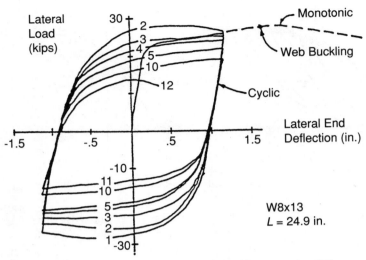

Figure 10.36 Beam behavior under cyclic loading. *(Vann et al. 1973)*

(shown in dashed line) was also observed in the corresponding cyclic test. Note that, in such cyclic tests, beam fracture usually will occur in the crest of a flange buckle and only after many cycles. Occasionally, the fracture occurs between the web and one flange, where the curvature may be high because of web local buckling.

When a beam (W6×16) of a more compact section but with twice the slenderness ratio, $L_b/r_y = 60$, was tested, it was observed that the specimen lost strength because of lateral-torsional buckling. Vann et al. observed that the loss of stiffness is much more significant when lateral-torsional buckling and not local buckling, is the dominant mode of failure. They also concluded that the deterioration of strength is severe only when flange local buckling is combined with either web local buckling or lateral-torsional buckling.

The effect of beam slenderness ratio on cyclic response was demonstrated by full-scale testing of an interior beam-column assembly (Noel and Uang 1996). Figure 10.37 shows the dimensions and member sizes of the test specimen. Table 10.2 lists the key parameters for both beams. The plastic moduli of both beams were similar, but because Beam 1 was shallower, its flanges were more compact. Equal and opposite displacements of increasing amplitude were imposed to the beam ends during testing. Each beam-column connection was strengthened with a cover plate and a haunch on the top and bottom flanges, respectively.

The load versus beam tip deflection relationships of the beams are shown in Figure 10.38. Figure 10.39 shows the buckled beams. As expected, Beam 2 experienced more severe buckling. Flange local

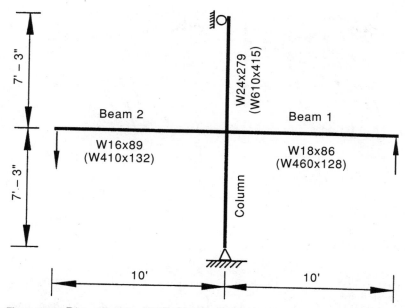

Figure 10.37 Dimensions and U.S. member shapes of test specimen (S.I. shapes in parentheses) (*Noel and Uang 1996*)

buckling of the Beam 2 bottom flange was first observed during the first half-cycle at 3.5 inch (= 8.9 cm) displacement amplitude at the cantilever tip. During the nine cycles of testing at this amplitude, both the strength and stiffness of Beam 2 degraded continuously. The corresponding hysteresis loops of Beam 1, however, were very stable and highly reproducible. Figure 10.40 demonstrates the variations of the beam 2 strength in these nine cycles and two subsequent cycles of a displacement amplitude of 5 inches. The amplitudes of local buckling and lateral-torsional buckling were measured and are included in Figure 10.40. Although flange local buckling occurred in the first cycle, the beam strength did not degrade until after the third cycle. From then on, web local buckling and lateral-torsional buckling accompanied flange local buckling, and the strength degraded with each cycle.

TABLE 10.2 Beam Properties

Beam	U.S. shapes (S.I. shapes)	F_{yf} (ksi) (MPa)	F_{yw} (ksi) (MPa)	$b_f/2t_f$	h/t_w	L/r_y	Z_x (in3) (mm3)
Beam 1	W16x89 (W410x132)	47.7 (329)	47.5 (328)	5.9	27	39	175 (2,868,000)
Beam 2	W18x86 (W460x128)	50.0 (344.8)	51.5 (355)	7.2	33.4	36.5	186 (3,048,000)

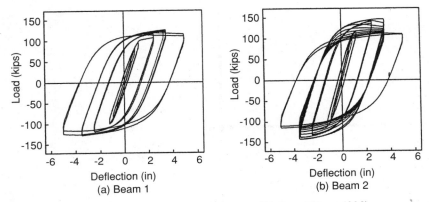

Figure 10.38 Load-displacement hysteresis curves. *(Noel and Uang 1996)*

10.10 References

1. Adams, P.F.; Lay, M.G.; and Galambos, T.V. 1965. "Experiments on High Strength Steel Members." Bulletin No. 110, Welding Research Council, 1–16.
2. AISC. 1989. *Manual of Steel Construction.* Allowable Stress Design, 9th ed. Chicago: American Institute of Steel Construction.
3. AISC. 1994. *Load and Resistance Factor Design Specification for Structural Steel Building.* 2nd ed. Chicago: American Institute of Steel Construction.
4. ASCE-WRC. 1971. *Plastic Design in Steel: A Guide and Commentary.* New York: ASCE.
5. Bansal, J.P. 1971. "The Lateral Instability of Continuous Steel Beams." Thesis submitted in partial fullfilment of the requirement for the Degree of Doctor of Philosophy in Engineering. Austin: University of Texas.

Figure 10.39 Beam buckling patterns. *(Courtesy of Forrell/Elsesser Engineers, Inc.)*

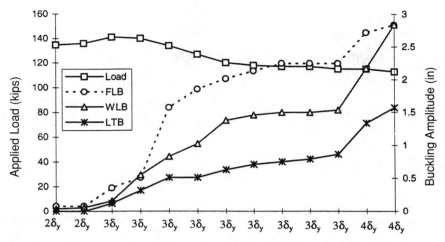

Figure 10.40 Variations of buckling amplitudes. *(Noel and Uang 1996)*

6. Bazant, Z.P., and Cedolin, L. 1991. *Stability of Structures.* New York: Oxford University Press.
7. Bertero, V.V., and Popov, E.P. 1965. "Effect of Large Alternating Strains of Steel Beams." *Journal of Structural Division.* Vol. 91, No. ST1, 1–12. ASCE.
8. Bild, S., and Kulak, G.L. 1991. "Local Buckling Rules for Structural Steel Members." *Journal of Construction Steel Research,* 20, 1–52.
9. Bleich, F. 1952. *Buckling Strength of Metal Structures.* New York: McGraw-Hill.
10. Bulson, P.S. 1969. *The Stability of Flat Plates.* New York: American Elsevier.
11. Cherry, S. 1960. "The Stability of Beams with Buckled Compression Flange." *The Structural Engineer.* September, 1960 277–285.
12. Dawe, J.L., and Kulak, G.L. 1986. "Local Buckling Behavior of Beam-Columns." *Journal of Structural Engineering.* Vol. 112, No. 11, 2447–2461. ASCE.
13. Dekker, N.R. 1989. "The Effect of Noninteractive Local and Torsional Buckling on the Ductility of Flanged Beams." *Civil Engineer in South Africa.* April. 121–124.
14. Fukumoto, Y., and Itoh, Y. 1992. "Width-to-Thickness Ratios for Plate Elements in Earthquake Engineering Design of Steel Structures," in *Stability and Ductility of Steel Structures Under Cyclic Loading* (ed. Y. Fukumoto and G. Lee). Florida: CRC Press.
15. Galambos, T.V. 1963. "Inelastic Lateral Buckling of Beams." *Journal of Structural Division.* Vol. 89, No. ST5: 217–241. ASCE.
16. Galambos, T.V., and Lay, M.G. 1965. "Studies of the Ductility of Steel Structures." *Journal of Structural Division.* Vol. 91, No. ST4: 125–151. ASCE.
17. Galambos, T.V. 1968. "Deformation and Energy Absorption Capacity of Steel Structures in the Inelastic Range." AISI. Bulletin No. 8.
18. Galambos, T.V. 1988. *Guide to Stability Design Criteria for Metal Structures.* New York: John Wiley & Sons.
19. Gerard, G., and Becker, H. *Handbook of Structural Stability: Part I—Buckling of Flat Planes,* Technical Note 3871 (Section 7.3), National Advisory Committee for Aeronautics, Washington, DC, 1957.
20. Haaijer, G. 1957. "Plate Buckling in the Strain-Hardening Range." *Journal of Engineering Mechanics Division.* Vol. 83, No. EM2, Paper 1212, 1–45. ASCE.
21. Haaijer, G., and Thurlimann, B. 1958. "On Inelastic Buckling in Steel." *Journal of Engineering Mechanics Division.* Vol. 84, No. EM2, Paper 1581, 1–47. ASCE.
22. Kato, B. 1989. "Rotation Capacity of H-Section Members as Determined by Local Buckling." *Journal of Construction Steel Research.* Vol. 13: 95–109.

23. Kármán Theodore von, Sechler E.E., and Donnell L.H. "The strength of Thin Plates in Compression," *Transactions*. ASME, 54, APM–54–5 (1932), 53.
24. Kato, B. 1990. "Deformation Capacity of Steel Structures." *Journal of Constructional Steel Research*. Vol. 17: 33–94.
25. Kemp, A.R. 1984. "Slenderness Limits Normal to the Plane of Bending for Beam-Columns in Plastic Design." *Journal of Constructional Steel Research*. Vol. 4: 135–150.
26. Kemp, A.R. 1985. "Interaction of Plastic Local and Lateral Buckling." *Journal of Structural Engineering*. Vol. 111, No.10, 2181–2196. ASCE.
27. Kemp, A.R. 1986. "Factors Affecting the Rotation Capacity of Plastically Designed Members." *The Structural Engineer*. London, England. Vol. 69, No. 5, 28–35.
28. Kemp, A.R. 1991. "Available Rotation Capacity in Steel and Composite Beam." *The Structural Engineer*. London, England. Vol. 69, No. 5, 88–97.
29. Kemp, A.R. 1996. "Inelastic Local and Lateral Buckling in Design Codes." *Journal of Structural Engineering*. Vol. 122, No. 4, 374–382. ASCE.
30. Kirby, P.A., and Nethercot, D.A. 1979. *Design for Structural Stability*. New York: John Wiley & Sons.
31. Korol, R.M. and Daali, M.L. "Local Buckling rules for rotation capacity," *Engineering Journal*. Vol. 31, No 2, 41–47. AISC, 1994.
32. Kuhlmann, U. 1989. "Definition of Flange Slenderness Limits on the Basis of Rotation Capacity Values." *Journal of Constructional Steel Research*. Vol. 14: 21–40.
33. Lay, M.G., and Galambos, T.V. 1965. "Inelastic Steel Beams under Uniform Moment." *Journal of Structural Division*. Vol. 91, No. ST6: 67–93. ASCE.
34. Lay, M.G. 1965a. "Yielding of Uniformly Loaded Steel Members." *Journal of Structural Division*. Vol. 91, No. ST6: 49–65. ASCE.
35. Lay, M.G. 1965b. "Flange Local Buckling in Wide-Flange Shapes." *Journal of Structural Division*. Vol. 91, No. ST6: 95–116. ASCE.
36. Lay, M.G., and Galambos, T.V. 1966. "Bracing Requirements for Inelastic Steel Beams." *Journal of Structural Division*. Vol. 92, No. ST2: 207–228. ASCE.
37. Lay, M.G., and Galambos, T.V. 1967. "Inelastic Beams under Moment Gradient." *Journal of Structural Division*. Vol. 93, No. ST1: 381–399. ASCE.
38. Lee, G.C., and Galambos, T.V. 1962. "Post-Buckling Strength of Wide-Flange Beams." *Journal of Engineering Mechanics Division*. Vol. 88, No. EM 1: 59–75. ASCE.
39. Lukey, A.F., and Adams, P.F. 1969. "Rotation Capacity of Beams under Moment Gradient." *Journal of Structural Division*. Vol. 95, No. ST6: 1173–1188. ASCE.
40. Maquoi, R. 1992. "Behavior of Plate Components," in *Stability Problems of Steel Structures* (ed. M. Ivanyi and M. Skaloud). New York: Springer-Verlag.
41. Massonet, C. E., and Save, M.A. 1965. *Plastic Analysis and Design, Volume 1: Beams and Frames*. New York: Blaisdell Publishing Company.
42. Noel, S., and Uang, C.M. 1996."Cyclic Testing of Steel Moment Connections for the San Francisco Civic Center Complex (Test Report to HSH Design/Build)." *TR-96/07*. University of California, San Diego
43. *Uniform Building Code*. 1994. International Conference of Building Officials.
44. Shanley, F.R. "Inelastic column theory", *Journal of Aeronautical Sciences*. Vol. 14, No 5: 261–264, 1947.
45. SSRC 1991. *A World Review: Stability of Metal Structures*. edited by Beedle, L.S. Pennsylvania: Structural Stability Research Council.
46. Takanashi, K. 1973. *Inelastic Lateral Buckling of Steel Beams Subjected to Repeated and Reversed Loadings*. Proc. 5th World Conf. Earthquake Engineering. Vol. 1, 795–798, International Association for Earthquake Engineering. Rome, Italy.
47. Timoshenko, S.P., and Gere, J.M. 1961. *Theory of Elastic Stability*. New York: McGraw-Hill.
48. Trahair, N.S. 1993. *Flexural-Torsional Buckling of Structures*. Florida: CRC Press.
49. Vann, W.P.; Thompson, L.E.; Whalley, L.E.; and Ozier, L.D. 1973. *Cyclic Behavior of Rolled Steel Members*. Proc. 5th World Conf. Earthquake Engineering. Vol. 1: 1187–1193, International Association for Earthquake Engineering. Rome, Italy.
50. White, M.W. 1956. "The Lateral-Torsional Buckling of Yielded Structural Steel Members." *Fritz Engineering. Lab. Rep. No. 205 E8*. Bethlehem, PA: Lehigh University.
51. Yura, J.A.; Galambos, T.V.; and Ravindra, M.K. 1978. "The Bending Resistance of Steel Beams." *Journal of Structural Engineering*. ASCE. Vol. 104, No. 9, 1355–1370.

Chapter 11

Passive Energy Dissipation Systems

11.1 Introduction

Conventional seismic design practice in the United States utilizes design forces substantially smaller than those calculated for elastic response. Such practice is based on the premise that inelastic or nonlinear actions in a well-detailed structure will provide the building with sufficient energy dissipation capacity to enable it to survive the design earthquake without catastrophic damage. In a steel moment-resisting frame (Chapter 8), energy dissipation resulting from flexural hinging in the beams adjacent to the beam-to-column connection is sought (and assumed). In a concentrically braced steel frame (see Section 7.2), brace buckling is assumed to provide the requisite energy dissipation; in an eccentrically braced steel frame, the design objective is to focus the energy dissipation in the shear-yielding links (see Section 7.3). Inelastic behavior in beams, braces, and shear links, although able to dissipate large amounts of earthquake-induced energy, corresponds to structural damage. Generally, the level of damage increases with repeated inelastic cycling.

One key drawback to conventional design practice is that energy dissipation (and structural damage) is focused in framing elements that form part of the gravity-load-resisting system. This drawback can be mitigated and perhaps eliminated if the earthquake-induced energy is dissipated in supplemental energy dissipators (also termed dampers) placed in parallel with the gravity-load-resisting system. This strategy is attractive for two primary reasons:

- Damage to the gravity-load-resisting system is substantially reduced, likely leading to major reductions in post-earthquake repair costs
- Earthquake-damaged dampers can be easily replaced without the need to shore the gravity framing

Recognizing that damage to structural and nonstructural framing is primarily due to excessive interstory drift, one can effectively use seismic dampers to reduce displacement response in a building. For low to moderate levels of damping, this reduction in displacement response will generally result in lower inertial forces, smaller deformation demands on the seismic and gravity framing members, and lower acceleration, velocity, and displacement demands on nonstructural framing members and building contents. Such reductions in response will generally result in reduced earthquake-repair costs following minor to moderate earthquake shaking. The effects of viscous damping on the seismic response of single degree-of-freedom systems is introduced in Section 11.2 to demonstrate the effects of adding damping to a building.

Much seismic damping hardware was developed for controlling the response of structures to wind-induced vibration, and man-made shock loads. Examples of such hardware include tuned mass damping systems (passive, semi-active, and active) and viscoelastic dampers (Mahmoodi et al. 1987) to reduce wind-induced response in slender structures, vibration isolation hardware for mechanical engineering systems, and fluid viscous shock isolation dampers to mitigate blast and shock effects. Early seismic damper applications can be found in nuclear power plants wherein mechanical and hydraulic snubbers were used to reduce displacements in piping systems.

Specific hardware for the passive seismic control of building response was first developed in New Zealand in the late 1960s and early 1970s. Dampers based on the yielding of steel plates and bars (Kelly et al. 1972) and on the extrusion of lead through an orifice (Robinson et al. 1976) are products of these early development efforts.

The analysis and design of seismic framing systems incorporating passive energy dissipation systems is beyond the scope of this book. Linear and nonlinear methods of analysis that are consistent with the new analysis procedures set forth in FEMA 273, *Guidelines for the Seismic Rehabilitation of Buildings*, have been developed. The reader is referred to FEMA (1997) for detailed information.

The following sections describe the types of damper hardware that could be implemented in a steel-framed building. The nomenclature developed for FEMA 273 is used throughout this chapter. Emphasis is placed on damping hardware based on the yielding of steel (Section 11.4). Strategies for installing dampers in building frames are introduced in Section 11.5.

11.2 Response modification due to added damping

Some level of equivalent viscous damping in seismic framing systems has long been assumed for seismic design in the United States.

Damping in building frames ranging between 2 percent and 7 percent of critical have been reported (DoD, 1986) at low levels of excitation. For seismic design of most conventional construction, damping equal to 5 percent of critical is assumed—design response spectra based on 5 percent of critical viscous damping are generally reported in seismic codes and guidelines.

The effect of increased damping on elastic building response can be seen in Figure 11.1. In this figure, the free vibration decay of a linearly elastic single-degree-of-freedom oscillator is presented for 2 percent, 5 percent, 10 percent, and 20 percent equivalent viscous damping. The fundamental period of the oscillator is 0.5 second, and the initial displacement is 1.00 inch. The seismic response of a building frame to an earthquake ground motion can be considered to be equal to its response to the individual earthquake impulses that form the ground motion record, so it is clear that displacement response is reduced as the level of damping increases.

The effects of earthquake shaking on building response are typically represented by either acceleration or displacement response spectra. Increased damping results in smaller response spectrum ordinates for both pseudo-acceleration and displacement spectra. However, increased damping may increase spectral accelerations in yielding buildings if the level of damping is sufficiently high because of contributions of the damping force to the inertial force.

With the 5 percent spectral ordinates used as benchmark values, reductions in displacement response for levels of viscous damping

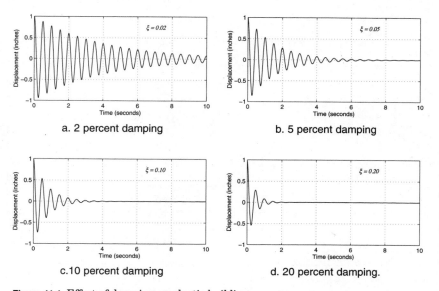

Figure 11.1 Effect of damping on elastic building response.

exceeding 5 percent of critical are presented in Table 11.1. It is evident from the data presented in this table that moderate increases in viscous damping, for example, from 5 percent of critical to 20 percent of critical, can produce substantial reductions in displacement response, irrespective of the fundamental period of the building.

11.3 Types of damper hardware

Passive seismic damping hardware is classified in FEMA 273 as displacement dependent, velocity dependent, or other. This classification system is used throughout this chapter. Dampers whose response is dependent on both displacement and velocity are classed as velocity-dependent dampers. Only those types of dampers that can be characterized as either displacement dependent or velocity dependent are described below. The reader is referred to FEMA (1997), Hanson et al. (1993), and Soong and Constantinou (1994) for additional information.

11.3.1 Displacement-dependent dampers

Examples of displacement-dependent dampers include flexural-yielding steel plates, torsional-yielding steel bars, lead-extrusion dampers, and friction-slip dampers. Dampers composed of shape-memory alloys can be designed to respond in a manner similar to that demonstrated by yielding metals. The response of displacement-dependent dampers is substantially independent of the rate of loading. The force-displacement relations for a metallic-yielding damper and a friction-slip damper are shown in Figure 11.2. Displacement-dependent dampers generally

TABLE 11.1 Displacement Response Reductions Due to Increased Damping

Damping (% of critical)	Acceleration domain	Velocity domain	Displacement domain
5	1.00	1.00	1.00
10	0.77	0.83	0.86
15	0.65	0.72	0.78
20	0.55	0.65	0.73
30	0.42	0.56	0.65
40	0.33	0.48	0.59
50	0.26	0.43	0.55

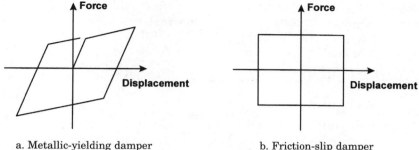

a. Metallic-yielding damper b. Friction-slip damper

Figure 11.2 Hysteresis curves for displacement-dependent dampers.

exhibit either bilinear or trilinear hysteresis, elastic-plastic response, or rigid plastic response in the case of the friction-slip damper.

Steel-yielding dampers are described in detail in Section 11.4. Such dampers have been used for new and retrofit construction in Japan, Mexico, New Zealand, and the United States. Friction dampers dissipate energy by sliding friction and exhibit hysteresis similar to that shown in Figure 11.2b. The friction damper originally proposed by Pall and Marsh (1982) is composed of a friction pad inserted at the junction of the braces in an X-braced frame and steel-plate links arranged in a rectangle surrounding the brace intersection point that enforce the required displacement field on the damper. A schematic diagram of the friction damper developed by Pall and Marsh is shown in Figure 11.3; this damper has been widely used in Canada. Correct selection of the sliding interface materials is key to the dependable performance of a friction damper because otherwise friction dampers may exhibit substantial changes in properties with time due to corrosion, bi-metallic contact, load-dwell effects, and relaxation. The reader is referred to Soong and Constantinou (1994) for more information on these subjects.

11.3.2 Velocity-dependent dampers

The class of velocity-dependent dampers includes dampers that utilize viscoelastic solid materials, viscoelastic fluid materials (e.g., viscous-damping walls), and dampers that operate by forcing a fluid through an orifice (e.g., fluid viscous dampers). The force-displacement relations for a solid or fluid viscoelastic damper and a fluid viscous damper are shown in Figure 11.4.

Solid viscoelastic dampers are generally composed of constrained layers of viscoelastic copolymers. The hysteresis for this type of damper is dependent on frequency, ambient temperature, tempera-

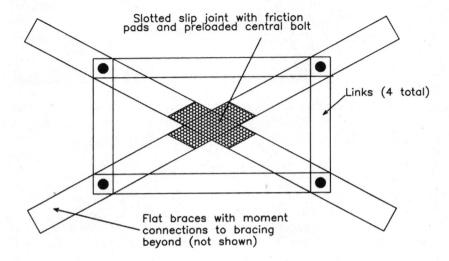

Figure 11.3 Schematic diagram of the Pall Friction Damper.

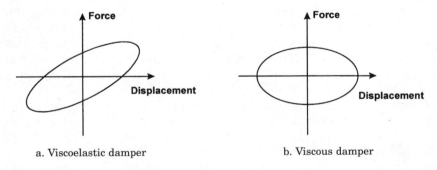

Figure 11.4 Hysteresis curves for velocity-dependent dampers.

ture rise, and amplitude of motion. The force (F) in a solid viscoelastic damper may be expressed as:

$$F = k_{eff}d + cv \tag{11.1}$$

where k_{eff} is the effective stiffness of the damper, d is the relative displacement between the ends of the damper, c is the damping coefficient of the damper, and v is the relative velocity between the ends of the damper. Fluid viscoelastic dampers operate by shearing viscoelastic fluids. This type of damper exhibits hysteresis similar to that of a solid viscoelastic damper except that fluid viscoelastic dampers have no effective stiffness under static loading conditions.

Pure viscous behavior can be realized by forcing fluid through an orifice. Fluid viscous dampers may exhibit stiffness. However, in the absence of substantial stiffness, the force (F) in a fluid viscous damper may be calculated as:

$$F = C_0 |v^a| sgn(v) \qquad (11.2)$$

where C_0 is the damper constant, v is the relative velocity between the ends of the damper, α is the velocity exponent, and sgn is the signum function. The simplest form of the fluid damper is the linear fluid damper for which α is equal to 1.0. Typical values for α range between 0.5 and 1.0.

11.4 Steel-yielding dampers

The key objective of introducing passive dampers into a building frame is to reduce or eliminate nonlinear response in structural framing. One realizes such improvements in response by reducing interstory drifts. This is generally best achieved through the addition of damping and stiffness to the structural framing. Passive dampers based on plastically deforming solid steel components in flexure, shear, torsion, or combinations thereof are ideally suited for this task because they generally exhibit high elastic stiffness and can sustain many cycles of stable postyielding deformation, resulting in high levels of energy dissipation or damping.

Four types of steel-yielding seismic dampers were developed in New Zealand in the early 1970s (Kelly et al. 1972, Skinner et al. 1975). Shown schematically in Figure 11.5, these dampers were (a) a flexural-yielding damper utilizing a U-shaped steel plate rolled between two surfaces in parallel relative motion, (b) a torsional- and flexural-yielding damper utilizing square or rectangular steel billet, (c) an omni-directional flexural-beam damper utilizing square steel billet or circular rod, and (d) a single-axis, rectangular flexural-beam damper deforming in single or double curvature. Multiple-plate, single-axis flexural-beam dampers, conceptually similar to the single-axis damper of Figure 11.5d, were described by Skinner et al. (1975). Early applications of steel-yielding dampers are reported by Skinner et al. (1980).

More recent developments in steel-yielding damper hardware are closely linked to the first generation of dampers developed in New Zealand. Most of these dampers are multiple-plate single-axis, triangular- or hourglass-shaped flexural-beam devices. The profiles of these dampers are chosen to promote yielding over the full-depth of the steel plate, thereby avoiding high plastic strains at low levels of relative deformation in the damper (Tyler 1978, Steimer et al. 1981).

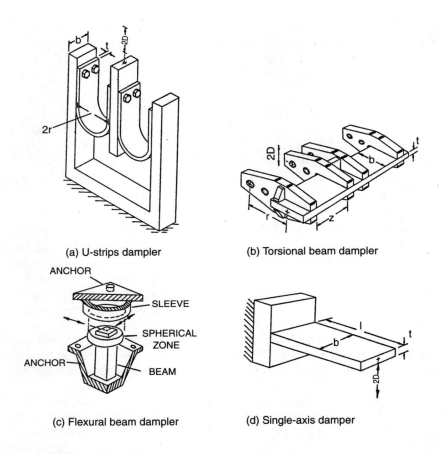

Figure 11.5 Steel-yielding dampers. *(Skinner et al. 1975)*

Triangular-shaped flexural-beam dampers deforming in single curvature were first used as energy dissipators in the late 1970s in the Dunedin Motorway Overbridge (Skinner et al. 1980). In the early 1980s, triangular-shaped dampers were introduced into a sleeved-pile isolation system supporting a 12-story building (Boardman et al. 1983).

The multiple-plate, hourglass-shaped *Added Damping and Stiffness* (ADAS) flexural-beam damper (Figure 11.6) developed for building applications is a direct descendent of similar shaped single-plate dampers used as energy-dissipating supports for piping systems (Steimer et al., 1981). ADAS dampers are typically installed immediately below stiff floor framing and must be supported by rotationally stiff support framing in order to produce double curvature in the steel plates.

Figure 11.6 Steel-yielding ADAS elements. *(Courtesy of CounterQuake Corporation)*

Figure 11.7 presents a plastic mechanism for a single-bay frame incorporating an ADAS element mounted atop chevron bracing; rigid plastic behavior is assumed for the components of the frame. The plastic rotational demand (γ_p) on the ADAS element can be estimated as:

$$\gamma_p = \frac{H}{h}\theta_p \qquad (11.3)$$

where H is the story height, h is the height of the ADAS element, and θ_p is the plastic story drift angle. For a plastic story drift angle of ±0.015 radian and H/h equal to 10, the required plastic rotational demand in the ADAS element is 0.15 radian.

A schematic elevation of a one-plate idealized ADAS element, the idealized displaced shape, the idealized moment diagram, and the idealized curvature distribution are shown in Figures 11.8a, 11.8b, 11.8c, and 11.8d, respectively. The hourglass shape of the ADAS element, idealized as an X-plate for this discussion, is chosen such that the

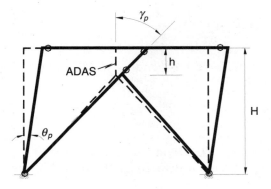

Figure 11.7 Kinematics of a single-story frame with an ADAS element.

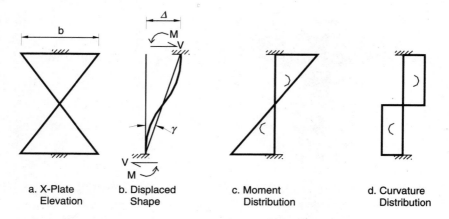

a. X-Plate Elevation b. Displaced Shape c. Moment Distribution d. Curvature Distribution

Figure 11.8 Deformation and force distribution in ADAS elements.

curvature is constant over the height of the damper, thus ensuring that yielding occurs simultaneously and uniformly over the full height of the damper.

The mechanical characteristics of an X-plate ADAS element can be simply estimated. Ignoring shear deformations and assuming that the ADAS element is restrained against rotation top and bottom, the elastic stiffness (k_e) of the ADAS element is given by:

$$k_e = \frac{2Ebt^3N}{3h^3} = \frac{2NEb}{3}\left(\frac{t}{h}\right)^3 \quad (11.4)$$

where E is Young's modulus, b is the base width of the X-plate, t is the thickness of the X-plate, N is the number of plates in the ADAS element, and h is the height of the X-plate. The yield strength (V_y) of the ADAS element is equal to:

$$V_y = \frac{F_y bt^2 N}{3h} \quad (11.5)$$

where F_y is the steel yield stress. The corresponding yield lateral displacement (Δ_y) is given by:

$$\Delta_y = \frac{F_y h^2}{2Et} \quad (11.6)$$

where all terms are defined above. The yield rotational angle (γ_y) is defined as the yield lateral displacement divided by the height of the X-plate:

$$\gamma_y = \frac{F_y h}{2Et} \qquad (11.7)$$

The plastic strength (V_p) of an N-plate ADAS element is equal to:

$$V_p = \frac{F_y b t^2 N}{2h} \qquad (11.8)$$

The height-to-thickness ratio (h/t) of the X-plate has a profound impact on the stiffness of the ADAS element. For example, a twofold increase in the value of t for a constant value of h will increase the elastic stiffness of the ADAS element by a factor of eight. Careful selection of values for h and t enable the engineer to achieve both a target yield displacement (or rotational angle) and the requisite elastic stiffness. Such flexibility provides the engineer with opportunities to tune the design of a seismic framing system.

Detailed testing of individual ADAS dampers has proven that steel-yielding dampers can exhibit high stiffness and stable hysteresis (Bergman and Goel 1987, Whittaker et al. 1989, Whittaker et al. 1991). Sample hysteresis for a scale model of a seven-plate ADAS element is presented Figure 11.9 (Whittaker et al., 1989). The yield lateral displacement (Δ_y) for this damper was approximately 0.16 inch, which corresponds to a yield rotational angle (γ_y) of 0.033 radian. The

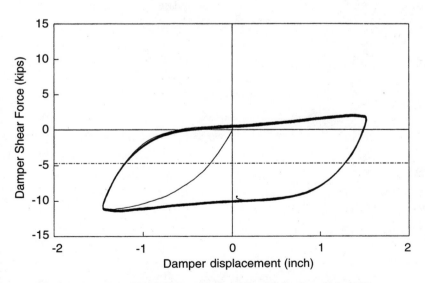

Figure 11.9 Seven-plate ADAS element hysteresis. *(Whittaker et al. 1989)*

maximum lateral displacement for this damper was 2.15 inches, which corresponds to a total rotational angle of 0.43 radian. The plastic rotational angle was equal to 0.40 radian—more than twice the target plastic rotational demand of 0.15 radian identified earlier.

Earthquake simulator testing of a three-story steel moment frame incorporating six ADAS dampers (Whittaker et al., 1989, 1991) provided proof-of-concept data that led to the use of ADAS dampers for the retrofit of a nonductile, two-story reinforced concrete building in downtown San Francisco (Fiero et al. 1993). The earthquake simulator testing program demonstrated that a properly designed steel-yielding energy dissipation system can substantially protect gravity framing systems from damage in severe earthquake shaking. The efficacy of the ADAS elements can be seen in Figure 11.10 (Whittaker et al. 1989): an energy response history for the three-story frame subjected to the 1940 El Centro earthquake record scaled to a peak acceleration of 0.33g. The vertically hatched zone corresponds to the energy dissipated by the steel moment-resisting frame; the remaining energy was dissipated by the ADAS elements.

A triangular-plate version of the ADAS damper, labeled T-ADAS, has been developed for building applications. Key differences between the ADAS and T-ADAS dampers are the use of vertically slotted holes in selected T-ADAS damper-to-framing connections to eliminate axial loads in the steel plates and the relatively small rotational stiffness required of the framing immediately below the T-ADAS damper. Tests of individual T-ADAS dampers (Figure 11.11) and T-ADAS elements

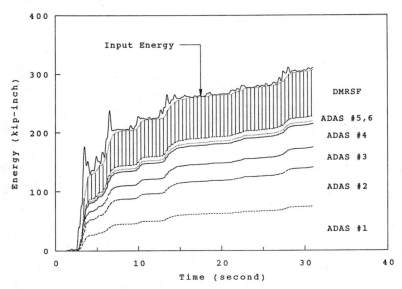

Figure 11.10 Energy response history *(Whittaker et al. 1989)*

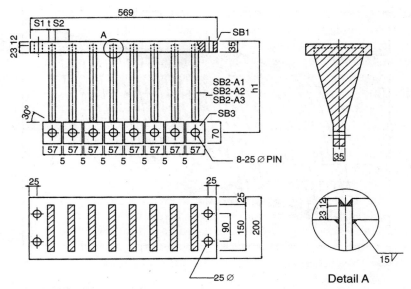

Figure 11.11 T-ADAS element *(Tsai et al. 1993)*

installed in a two-story steel frame produced excellent results (Tsai et al. 1993). The force-displacement relation for an eight-plate T-ADAS element tested by Tsai is presented in Figure 11.12.

The plastic mechanism presented in Figure 11.7 is also applicable to a single-bay frame incorporating an T-ADAS element mounted atop chevron bracing. The relationship between plastic rotational demand (γ_p) in the T-ADAS element and the plastic story drift (θ_p) is given by Equation 11.3.

A schematic elevation of one-plate idealized T-ADAS element, the idealized displaced shape, the idealized moment diagram, and the idealized curvature distribution are shown in Figures 11.13a, 11.13b, 11.13c, and 11.13d, respectively. The shape of the T-ADAS element is chosen such that the curvature is constant over the height of the damper, thus ensuring that yielding occurs simultaneously and uniformly over the full height of the damper.

Mechanical characteristics of an T-ADAS element have been estimated by Tsai et al. (1993). Ignoring shear deformations and assuming that the T-ADAS element is restrained against rotation at its base, the elastic stiffness (k_e) of the T-ADAS element is given by:

$$k_e = \frac{Ebt^3N}{6h^3} = \frac{NEb}{6}\left(\frac{t}{h}\right)^3 \tag{11.9}$$

where E is Young's modulus, b is the base width of the plate, t is the thickness of the plate, N is the number of plates in the T-ADAS

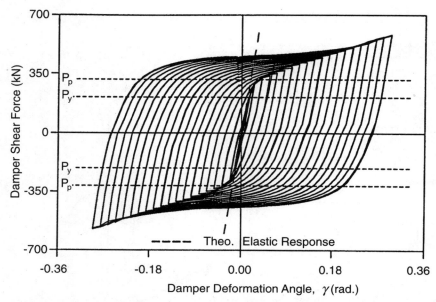

Figure 11.12 8-plate T-ADAS element hysteresis *(Tsai et al. 1993)*

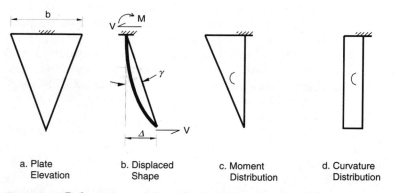

a. Plate Elevation b. Displaced Shape c. Moment Distribution d. Curvature Distribution

Figure 11.13 Deformation and force distributions in T-ADAS elements

element, and h is the height of the plate. The yield strength (V_y) of the T-ADAS element is equal to:

$$V_y = \frac{F_y b t^2 N}{6h} \tag{11.10}$$

where F_y is the steel yield stress. The corresponding yield lateral displacement (Δ_y) is given by:

$$\Delta_y = \frac{F_y h^2}{Et} \qquad (11.11)$$

where all terms are defined above. The yield rotational angle (γ_y) is defined as the yield lateral displacement divided by the height of the plate:

$$\gamma_y = \frac{F_y h}{Et} \qquad (11.12)$$

The plastic strength (V_p) of an N-plate ADAS element is equal to:

$$V_p = \frac{F_y b t^2 N}{4h} \qquad (11.13)$$

As with the ADAS element, the height-to-thickness ratio ($=h/t$) of the T-ADAS plates has a profound impact on the stiffness of the damper. Careful selection of values for h and t will enable the engineer to achieve both the required stiffness and a target yield displacement (or rotational angle).

Other steel-yielding dampers have been developed and implemented by Japanese construction companies. One such system is the so-called honeycomb damper, which is composed of multiple, in-line, hourglass-shaped steel plates (Kajima 1991). The honeycomb damper is conceptually identical to the ADAS damper but is deformed in double curvature in the plane of the steel plate. Omnidirectional single-taper (Bell) and double-taper (Tsudumi) dampers, similar to the omnidirectional damper reported by Skinner et al. (1975), have also been developed and implemented (Kajima 1991).

11.5 Implementation of dampers in building frames

To optimize the benefits afforded by a seismic damper, the relative displacement and velocity between the ends of the damper must be maximized. Assuming that seismic dampers are installed between floor levels in a building, it is desirable to eliminate (to the degree possible) deformations and relative velocities in the framing supporting the dampers, that is, to ensure that the relative displacement and velocity between the ends of the damper are approximately equal to the interstory displacement and velocity.

Typical seismic damper installation details are shown in Figure 11.14. The stiffness of the support framing shown in this figure must be accounted for in the analysis, and the minimum strength of the

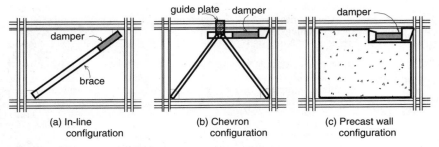

Figure 11.14 Schematic damper installation details.

support framing must be greater than the maximum force output of the damper.

The uniaxial, diagonal brace installation of Figure 11.14a has been used for viscoelastic (Aiken and Kelly 1990) and viscous dampers. Steel-yielding, friction, lead-extrusion, and viscous dampers have been implemented through use of the inverted chevron brace installation of Figure 11.14b. The precast wall panel mounting frame of Figure 11.14c has been used to implement friction dampers, lead extrusion dampers, and fluid viscoelastic dampers.

Seismic dampers have been implemented in new and retrofit steel construction in the United States. Most applications have involved moment-resisting frames: a framing system that is typically characterized by large flexibility and large interstory drifts in severe earthquake shaking. In such framing systems, seismic dampers have been used to reduce interstory drifts and floor accelerations thereby reducing plastic hinge rotation demands for severe earthquake shaking and reducing nonstructural damage. Seismic dampers will likely be more effective for flexible framing systems (e.g., moment frames) than for stiff framing systems (e.g., braced frames).

References

1. Aiken, I.D., and J.M. Kelly. 1990. *Earthquake Simulator Testing and Analytical Studies of Two Energy-Absorbing Systems for Multi-Story Structures.* Report No. UCB/EERC-90/03. Berkeley: Earthquake Engineering Research Center, University of California.
2. Bergman, D.M., and Goel, S.C. 1987. *Evaluation of Cyclic Testing of Steel-Plate Devices for Added Damping and Stiffness.* Report UMCE 87-10. MI: Civil Engineering Department, University of Michigan.
3. Boardman, P. R., Wood, B. J., and Carr, A. J. 1983. "Union House—A Cross-Braced Structure with Energy Dissipators." *Bulletin of the N.Z. National Society for Earthquake Engineering.* Vol. 16, No. 2. June.
4. DoD. 1986. *Seismic Design for Essential Buildings.* TM-5-809-10-1. Washington, DC: Departments of the Army, Navy, and Air Force.
5. Fiero, E.; Perry, C.; Sedarat, H.; and Scholl, R. 1993. "Seismic Retrofit in San Francisco Using Energy Dissipation Devices." *Earthquake Spectra*, Vol. 9, No. 3. Oakland, CA: Earthquake Engineering Research Institute. August.

6. FEMA. 1997. *Guidelines for the Seismic Rehabilitation of Buildings.* Volumes 1 (Guidelines), 2 (Commentary), and 3 (Example Applications). Washington, DC: Federal Emergency Management Agency.
7. Hanson, R.D.; Aiken, I.; Nims, D.K.; Richter, P.J.; and Bachman, R. 1993. *Passive Energy Dissipation, Active Control, and Hybrid Control Systems.* Report ATC-17-1, Volume 2. Redwood City: Applied Technology Council.
8. Kajima. 1991. *Honeycomb Damper Systems* Japan: Kajima Corporation.
9. Kelly, J.M.; Skinner, R.I.; and Heine, A.J. 1972. "Mechanisms of Energy Absorption in Special Devices for Use in Earthquake-Resistant Structures." *Bulletin of the N.Z. National Society for Earthquake Engineering.* Vol. 5, No. 3. September.
10. Mahmoodi, P.; Robertson, L.E.; Yontar, M.; Moy, C.; and Feld, L. 1987. "Performance of Viscoelastic Dampers in World Trade Center Towers." *Proc., 5th ASCE Structures Congress.* Orlando, Florida. April.
11. Pall, A.S., and Marsh, C. 1982. "Response of Friction Damped Braced Frames." *Journal of the Structural Division.* Vol. 108, No. 6: 1313—1323. June. ASCE.
12. Robinson, W.H., and Greenbank, L.R. 1976. "An Extrusion Energy Absorber Suitable for the Protection of Structures During an Earthquake." *Earthquake Engineering and Structural Dynamics.* Vol. 4.
13. Skinner, R.I.; Kelly, J.M.; and Heine, A.J. 1975. "Hysteretic Dampers for Earthquake-Resistant Structures." *Earthquake Engineering and Structural Dynamics.* Vol. 3.
14. Skinner, R.I.; Tyler, R.G.; Heine, A.J.; and Robinson, W.H. 1980. "Hysteretic Dampers for the Protection of Structures from Earthquakes.' *Bulletin of the N.Z. National Society for Earthquake Engineering.* Vol. 13, No. 1. March.
15. Soong, T.T., and Constantinou, M. C. 1994. *Passive and Active Structural Control in Civil Engineering.* Vienna and New York: Springer-Verlag.
16. Steimer, S.F.; Godden, W.G.; and Kelly, J.M. 1981. *Experimental Behavior of a Spatial Piping System with Steel Energy Absorbers Subjected to a Simulated Differential Seismic Input.* Report No. UCB/EERC-81/09. Berkeley: Earthquake Engineering Research Center, University of California.
17. Tsai, K.-C.; Chen, H.-W.; Hong, C.-P. P.; and Su, Y.-F. 1993. "Design of Steel Triangular Plate Energy Absorbers for Seismic-Resistant Construction." *Earthquake Spectra.* Vol. 9, No. 3: 505—528. August. Oakland, CA: Earthquake Engineering Research Institute,.
18. Tyler, R.G. 1978. "Tapered Steel Energy Dissipators for Earthquake-Resistant Structures." *Bulletin of the N.Z. National Society for Earthquake Engineering.* Vol. 11, No. 4. December.
19. Whittaker, A.S.; Bertero, V.V.; Thompson, C. L.; and Alonso, L. J. 1989. *Earthquake Simulator Testing of Steel Plate Added Damping and Stiffness Elements.* Report No. UCB/EERC-89/02. Berkeley: Earthquake Engineering Research Center, University of California.
20. Whittaker, A.S.; Bertero, V.V.; Thompson, C.L.; and Alonso, L.J. 1991. "Seismic Testing of Steel Plate Energy Dissipation Devices." *Earthquake Spectra.* Vol. 7, No. 4: 563—604. November. Oakland, CA: Earthquake Engineering Research Institute.

Index

AASHTO 178
actual plastic moment 330, 373
ADAS 471-475, 478
AISC Group 4 and 5 73
AISC LRFD Specification 204, 448
algorithm 72, 89, 168, 173
allowable stress factor 391
allowable stress method 3
alternating plasticity 42, 81
anisotropy 34
Autostress design method 178

backup bar 319, 327, 334, 338, 348, 356
base shear coefficient 384
basic independent mechanism 122, 172
Bauschinger 2
Bauschinger effect 22, 25, 211, 456
beam mechanism 122, 124
beam-sway mechanism 188, 280
bending yield planes 421, 425, 444
bifurcation equilibrium problem 427
bolted connection 351, 353
brace slenderness 192, 197, 198, 208, 233
brittle failure 27, 31, 35, 36, 274, 334
brittleness 1, 27
BSL 382
buckling wavelength 424, 445, 450

capacity design 182, 232, 242, 251, 284, 364
captive columns 187
carbon equivalent 29, 31
Charpy V-notch test 12, 13, 16
clear span 209, 275, 289, 373
collapse 73, 315, 326, 357, 359, 381, 399, 403, 411, 456

column buckling 279
column hinging 188, 266
column splice 187, 245, 247, 279, 402
column-sway mechanism 188
combined mechanism 124
compact section 178, 263, 284, 414, 433, 445, 448
composite action 138, 168, 259, 261, 326, 334
concentrically braced frames 203
concrete infill 220
connection detail 265, 302, 310, 337, 363, 407
contained plastic flow 57, 67, 420
continuity plates 263, 285, 287, 331, 335, 336, 345, 357
crack propagation 39, 42, 328, 335, 345
cyclic push-over analysis 198

DBTT 12, 13
deflection stability 145
degree of indeterminacy 153
design base shear ratio 397
design interaction curve 88, 89
design spectrum 383, 384, 386
direct combination of mechanism 153, 154
discontinuous yield concept 442
displacement amplification factor 385, 389, 390, 392, 393, 401
distortion 32, 37, 285, 290
doubler plate 263, 287, 300, 301, 307, 328, 335
drift 194, 198, 201, 231, 249, 253, 273, 277, 370, 382, 386-389, 465
drift limit 273, 278, 388, 395, 397, 398, 402
dual-certification 21
dual-certified steel 259, 328, 367

482 Index

ductile Braced Frame 201
ductile moment frame 275, 302, 307, 364
ductility 1, 16, 29, 34, 40, 263, 302, 310, 312, 332, 384, 390, 391
ductility reduction factor 391, 393

eccentrically braced frames 247
elastic core 67, 90
elastic design spectrum 383, 384, 386
elasto-plastic models 45, 46, 81
elongation 8, 21, 57
end-condition 214
end-restraint 216
energy dissipation 12, 23, 43, 197, 198, 207, 208, 247, 277, 290, 383, 464, 470, 475
energy dissipation 12, 23, 208, 209, 247, 338, 351, 383, 463, 475
energy dissipation devices 470
equilibrium method 110
equivalent load-set 137, 142
Euler buckling 209
event-to-event calculation 106
external work 115, 124

fillet welds 261, 307, 356, 359
flange local buckling 350, 411, 421, 422, 425, 428, 433, 442, 445, 448, 449, 453, 455
flexural link 257
flux-cored arc welding 27, 307
fracture appearance 12, 14
fracture Appearance Transition Temperature 14
fracture mechanics 37, 327
fracture toughness 13, 40, 41, 355
frame mechanism 122
friction damper 468, 479
full penetration weld 357

gable mechanism 124
geometric imperfections 418, 442, 456
gravity load 143, 150, 187, 191, 233, 234, 242, 243, 326, 382, 386, 401
gusset plates 203, 232

heat-affected zone 27, 29, 74, 321, 358
historical description 2
Hooke 98
hysteretic energy 1, 23, 43, 197, 207, 208, 280, 345, 391
hysteretic model 45

imperfections 43, 209, 416, 418, 442, 456
inclusions 34, 35
incomplete plastic collapse mechanism 113
indeterminacy 109, 111, 153
infill 184, 187, 201, 220
inflection point 104, 216, 280, 370
inspection 33, 307, 314-316, 318, 327
instantaneous center of rotation 132, 134
internal work 115, 124
interpass temperature 29
interstory drift 231, 249, 278, 382, 465, 470

joint mechanism 124

kinematic method 4, 106, 114, 120, 121, 124
kink 147, 149, 182, 206, 207, 294, 368
knee-braced 201

lamellar tearing 34, 35, 321
lateral bracing 145, 278, 285, 350, 421, 439, 442, 446, 448, 456
lateral torsional buckling 77, 216, 258, 261, 278, 300, 350, 411, 420, 422, 424, 437, 449
Limit States Design 3, 73
linear programming 153, 173
link plastic rotation 253
link rotation angle 260, 261, 263
Load and Resistance Factor Design 3, 73
local buckling 25, 43, 77, 182, 213, 219, 279, 300, 345, 357, 376, 411, 415, 421, 422, 425, 434, 448, 449

long link 256-258, 260
lower bound theorem 121
Lüder lines 24

martensite 27, 31
member mechanism 122
Menegotto-Pinto 46, 53, 71
method of inequalities 153, 165
microcracks 29, 34, 38, 42
mill test certificate 73
minimum reliable strength 21
modulus of elasticity 8, 74, 297
Mohr's circle 92, 332, 334
moment curvature 69, 89, 181
moment redistribution 177, 181, 433, 434
monotonic loading 197, 351, 456
monotonic push-over analysis 192

NBCC 388
necking 8
NEHRP 204, 382, 384
neutral axis 64, 65, 67, 70, 83, 88, 90, 334, 453
Nil-Ductility-Transition Temperature 14
nominal plastic moment 280, 286, 307
nominal strength 183, 220, 231, 241, 283, 286
noncompact section 181, 414
nonstructural damage 273, 313, 381, 383, 385, 397, 479
Northridge 2, 200, 274, 277, 278, 300, 302, 312, 362, 401
notch toughness 16, 307, 334, 356
number of possible basic mechanism 153
number of potential plastic hinge location 153, 154

optimization 57, 127, 132, 173, 178
out-of-straightness 209, 213, 424, 425
overlaid welds 344
overly complete plastic mechanism 130
overstrength 263, 384, 390-392, 399, 400, 403

panel mechanism 122
panel zone 275, 277, 285, 300
partial penetration weld 40, 280
passive energy dissipation 465
period of vibration 267, 334, 384, 387, 393, 394, 405, 407, 466
plastic collapse load 46, 110, 111, 115, 120, 129
plastic collapse mechanism 109-111, 113, 120, 183
plastic drift 253, 255
plastic hinge 104, 106
plastic hinge location 136-138, 140, 144, 153, 154, 159, 168, 178
plastic moment 43, 63, 65, 68, 81, 102
plastic plateau 8, 24, 421
plastic rotation 79, 107, 116, 125, 132, 149, 162, 178, 181, 188, 253, 259, 277, 278, 300, 312, 336, 337, 353, 421, 433, 456, 472, 475
plastic rotation demand 188, 277, 456
plastic section modulus 69, 93, 97, 100, 283
plastic story drift 472, 476
plastic-design tools 175
Poisson's ratio 332, 416, 427, 431
post-buckling 223, 229
postbuckling stiffness 223, 418
postbuckling strength 415, 417, 421, 422, 439
postyield 42, 46, 277, 295, 297, 470
postyield stiffness 295
power function 46, 48, 49, 53
Prandtl 98
preheat 22, 29, 32
probable plastic moment 366
probable strength 21, 284, 366
push-over 191, 192, 198
radius of gyration 209, 278, 412, 454
Ramberg-Osgood 46, 50, 71

redistribution factor 109
redundancy 307, 349, 364, 370, 401
reloading 22, 180, 207
replacement load 137
residual stresses 12, 26, 32, 37, 59, 74, 77, 209, 280, 332, 340, 422, 425

Index

response modification factor 205, 232, 260, 392, 399
response spectrum 383, 384, 393, 394, 466
restrained weld 32, 35, 36
rigid frame action 273
rigid plastic model 43
root pass 37, 319
rotation capacity 414, 415, 433-435, 438, 440-442, 446, 447, 449, 451, 453-455, 458
runoff tabs 328, 336, 338, 355

Saint-Venant 2
sand-heap analogy 97, 99
section shape 159, 208, 216
seismic coefficient 384, 402, 404-406
seismic design 21, 193, 199, 203, 208, 238, 259, 287, 301, 381, 382, 411, 436, 448, 456, 464
seismic force reduction 278, 384, 386, 389, 392
semirigid connections 4, 353
shakedown theorem 145
shape factor 69, 70, 104, 373
shear distortion 290, 291, 293, 294
shear failure 183, 187, 226
shear link 249, 256, 259, 464
shear strain hardening 300
shielded metal arc welding 27
short link 256, 260, 262
short-columns 187
sidesway mechanism 122
sign convention 111, 116, 159, 162, 166
simplex method 173
size effects 307
slender section 413
slenderness ratio 181, 192, 198, 207, 208, 411, 413, 415, 444, 446, 453
slip band 426
slip planes 24, 27
slippage 263, 304, 335
St. Venant torsion 429, 438, 443
statical method 106, 110, 111, 113, 121, 177
steel-yielding dampers 468, 470, 474, 478

step-by-step method 106
stiffeners 257, 258, 260, 261, 285, 349, 351, 360
stocky section 208, 213, 226, 231
story drift 194, 382, 383, 387, 391, 465
story plastic collapse mechanism 197
strain hardening 8, 46, 68, 79, 181, 257, 284, 330, 367, 432
strain rate 13, 14, 16, 17, 73, 334, 336, 337, 345, 391
strain-hardening shear modulus 431
strain-hardening stiffness 46, 49, 431
stress concentration 40, 42, 287, 331, 349
stress reversal 22
stress-intensity-factor 40
strong-column/weak-beam 188, 226, 280, 284, 328, 339, 368, 374
structural damage 381, 383, 395, 398, 464
structural ductility factor 390, 392
structural overstrength 384, 391-393, 399-401, 403
structural overstrength factor 391, 392, 399, 401

T-ADAS 475-478
tangent modulus 8, 24, 438, 439
tangent modulus of elasticity 24
tangent stiffness 53, 211, 225, 236
tensile strength 21
tension-only braced frames 204
tension-only braces 201
tension-only design 192, 198
testing protocol 211, 366
torsional plastic section modulus 100
Tresca 2
Tresca yield condition 90
triaxial stress condition 332, 338, 368

ultrasonic inspection 327, 340
Uniform Building Code 204, 274, 385
uniqueness theorem 121
unloading 22, 58, 180, 206, 207, 438, 439, 442
upper bound theorem 120

virtual-work method 106, 114, 115
viscous damping 465-467
Von Mises yield criterion 3, 90, 95, 101

warping 421, 429, 455
warping constant 429
warping restraint 421, 439
wavelength 424, 445, 450
web local buckling 350, 411, 434, 448, 453, 455

web-flange core 16, 41, 74
weld access hole 31, 358
weld runoff tabs 328, 336, 338, 356
weld shrinkage 32, 37, 340
width-to-thickness 43, 181, 210, 213, 214, 219, 238, 376, 428
working stress method 3
workmanship 40, 307, 327

yield strength 8, 12, 16, 17, 21, 73
Young modulus 209, 416

Coláiste na hOllscoile Gaillimh